Cellular and PCS

The Big Picture

Other McGraw-Hill Communications Books of Interest

To order or receive additional information on these or any other McGraw-Hill Titles, in the United States please call 1-800-722-4726. In other countries, contact your local McGraw-Hill representative.

Cellular and PCS

The Big Picture

Lawrence Harte
Richard Levine
Steve Prokup

McGraw-Hill

New York San Francisco Washington, D.C. Auckland
Bogotá London Madrid Mexico City
Milan Montreal New Delhi
San Juan Singapore
Sydney Tokyo Toronto

McGraw-Hill

1 2 3 4 5 6 7 8 9 0   DOC/DOC   9 0 2 1 0 9 8 7

ISBN 0-07-026944-0

Printed and bound by R. R. Donnelley & Sons Company.

McGraw-Hill books are available at special quantity discounts to use as premiums and sales promotions, or for use in corporate training programs. For more information, please write to the Director of Special Sales, McGraw-Hill, 11 West 19th Street, New York, NY 10011. Or contact your local bookstore.

 This book is printed on recycled, acid-fee paper containing a minimum of 50% recycled, de-inked fiber.

ACKNOWLEDGEMENTS

We thank all the gifted people who gave their technical and emotional support for the creation of this book. In many cases, published sources were not available on this subject area. Experts from manufacturers, service providers, trade associations, and other telecommunications related companies gave their personal precious time to help us and for this we sincerely thank and respect them.

Of the numerous manufacturer representatives, we thank Lee Horsman at Allen Telecom, Eric Unruh of AMD, David DeVaney with Astronet, Lucian Dang at Casio Manufacturing Corporation, Carlton Peyton, Dick Hayter, Eric Stasik, Ron Bohaychuk, Sandeep Chennakeshu, Ted Ericsson, Tony Gorse, Tony Sammarco, Lars Wilhelmsson, from Ericsson, Jim Mullen and Stan Kay of Hughes Network Systems, Mat Kirimura with Japan Radio Company, Imamura-San of Mitsubishi, Steve Jones with NEC, Brian Murphy, Cindy Goluboff, and Richard Caplan from Nortel, Al Bartle and Mike Wise at Oki Telecom, John Avery with Panasonic, Tom Crawford of Qualcomm, Tom Ohlsson and Vibeke Arenz with Spectralink, Raul Carr, of Tellabs, Marybeth Flanders with Telular, and Paul McLellan at VLSI.

Thanks to the service provider representatives: Chris Lawrence and Jim Lipsit at AT&T Wireless Services, John Boduch with BellSouth Cellular Corp, P.F. Ng of Cantel, David Danaee' at Claircom, Jim Durcan with Comcast, Angie McKimmon and Bob Zicker at GTE Mobilnet, Hilbert Chan with Mobility Canada, Ernesto Ramos of MovilNet, Richard Dreher with Pocket Communications and Ken Corcoran from Southwestern Bell.

Thanks also to the following research and consulting experts: Michael Vadon of BRC Consultancy Ltd, Dave Crowe with Cellular Networking Perspectives, Scott Goldman at Communications Now, Elliott Hamilton at EMCI, Herschel Shosteck of Herschel Shosteck Associates, Michael H. Sommer with Information Technologies Inc., David Balston from Intercai Mondiale, Linda Gossack with the US Dept of Commerce, Sasan Ardalan from XCAD corporation, Konstantin Zsigo with Zsigo Wireless Data Consultants, Marty Nelson of Communications Services Inc., and Deborah Jones.

Specific mention must go to our trade press, industry associations, and education experts: Bradley Eubank from PCIA, Eric Schimmel with the TIA, Sang-Lin Han of Chung Nam National University, Paul Wilkinson at Audiovox, Gregg Aurelius of TCOM Test Associates, Jill Roumeliotis at ArrayComm, Inc, Gary Schober and Scott Schober at Berkley Varitronics, Al Baughman with LCC Inc., George Peponides from PCSI.

We say thanks to our financial experts and industry analyst contributors who include Alex Cena of Bear Sterns, Jeff Hines with NatWest Securities, Walt Piecyk from Paine Webber, Tony Robertson at Robertson Stevens, Jeffrey Schlesinger of UBS.

The editors and illustrators who made this book read well and look good include Eric Byrd (illustrator), Jacquelyn Gottlieb (editor/illustrator), and Nancy Knoernschild (editor/layout).

About the Authors

Lawrence Harte is the president of APDG, a provider of expert information to the telecommunications market. Mr. Harte has over 17 years of experience in the electronics industry including company leadership, product management, development, marketing, design, and testing of telecommunications (cellular), radar, and microwave systems. He has been issued patents relating to cellular technology and authored over 75 articles on related subjects. Mr. Harte earned his Bachelors degree from University of the State of New York and an MBA at Wake Forest University. During the IS-54 TDMA cellular standard development, Mr. Harte served as an editor for the Telecommunications Industries Association (TIA) TR45.3, the digital cellular standards committee.

Richard Levine is the founder and principal engineer of Beta Scientific Laboratory and is also Adjunct Professor of Electrical Engineering at Southern Methodist University. He is active as a technology consultant to many firms developing new cellular and PCS systems and products used in Brazil, Canada, England, France, Germany, Israel, Korea, Mexico, and the United States. He is also a well-known teacher of cellular and PCS technology to people in the industry. He was formerly the chairman of several working groups in the North American digital cellular standards development. Levine earned the Bachelor, Master and Doctor of Science degrees from M.I.T., is licensed as a Professional Engineer, and has earned both amateur and professional radio operator licenses. He has been issued several patents on telecommunications, computer systems, and related technologies.

Steve Prokup is vice president of engineering for ReadyCom, a provider of cellular voice paging service. Mr. Prokup has spent the last several years developing and managing software for cellular systems in the North American market. He has worked with several different cellular standards including analog and digital cellular systems. He has been involved with the development of an IS-136 TDMA cellular phone, including support of the industry specification validation efforts. Mr. Prokup earned his Bachelors degree in Computer Science from Ohio State University. He previously worked on software development for Department of Defense applications. Mr. Prokup is an inventor of several technologies relating to cellular telephony and has authored numerous papers describing digital cellular technology.

The creation of this book would not have been possible without the support and understanding of our families:

I dedicate this book to Jacquelyn Gottlieb—the love of my life, my parents Virginia and Lawrence M. Harte, Lawrence William and Danielle Elizabeth—my children, and all the rest of my loving family. - Lawrence

I dedicate this book to my wife Sara and the next generations: Naomi, Yossi, Kyle, Susan, Earl, and David. I am proud of all of you, and I hope my part of this book will make you proud as well. Love and thanks - Richard

I dedicate this book to Nancy, Christopher, Sara, and Alexander for their assistance, patience, and understanding. Without their help this would not have been possible. I love you all. - Steve

TABLE OF CONTENTS

Preface

In 1996, there were over 75 million cellular customers and over 90 million cordless telephones in use worldwide. Predictions show that over 300 million customers and will be using cellular systems by the year 2,000. Wireless phones and systems that are being introduced to the market have advanced features, services, and cost advantages over the older cellular and cordless technologies. This book is a guide which provides the big picture of these new technologies, features, costs, and services for handsets and systems.

This book offers a balance between marketing and technical issues. It covers what's new in cellular and cordless technology, explains how it works using over 300 illustrations, and describes the marketing aspects of the new technologies and services. Over 100 industry experts have reviewed the technical content of this book. Many of the industry buzzwords are defined and explained.

To help meet the growing demand for cost effective cellular service and advanced features, several digital cellular technologies are available in over 60% of the world markets. Theoretically, analog cellular systems might have provided for cost-effective expansion indefinitely, but for a variety of reasons, they have not. Furthermore, each digital cellular technology's unique advantages and limitations offer important economic and technical choices for managers, salespeople, technicians, and others involved with wireless telephones and systems. *Cellular and PCS, The Big Picture* provides the background for a good understanding of the technologies, issues, and options available.

The chapters in this book are organized to help technical and non-technical readers alike to find the information they need. These chapters are divided to cover specific technologies, economics, and services and may be read either consecutively or individually.

Chapter 1. Provides an introduction to wireless technologies including cellular, wireless office, cordless, and PCS. Advanced wireless messaging services such as advertising, imaging, and monitoring are identified and explained. This chapter is an excellent introduction for newcomers to wireless technology.

Chapter 2. Explains analog cellular technology and its evolution. There are several different analog cellular systems used throughout the world. A description of these include Advanced Mobile Phones Service (AMPS), Total Access Communications Systems (TACS), Nordic Mobile Telephone (NMT) and other cellular systems is included. This chapter discusses the basic operation and attributes of these systems.

Chapter 3. Describes digital technology and services and presents a comparative analysis of analog and digital wireless systems. An overview of Frequency Division Multiple Access (FDMA), Time Division Multiple Access (TDMA), and Code Division Multiple Access (CDMA) is included. A non-technical description of advanced digital services common to all digital technologies is discussed followed by a summary of the potential new services and how they may be implemented.

Chapter 4. Explains the North American TDMA digital cellular system, including its radio frequency (RF) channel structure, signaling, and key attributes. This chapter covers the IS-54 standard and the new IS-136 TDMA standard with its new Digital Control Channel (DCC).

Chapter 5. This chapter provides an explanation of how CDMA systems operate and the major differences between CDMA and other cellular technologies. The CDMA IS-95 technology, its channel structure, signaling, and numerous significant features will be explained.

Chapter 6. An explanation the Global System for Mobile Communication (GSM) digital standard is provided in this chapter, as well as the closely related DCS-1800 (UK) and PCS 1900 (North America) systems. Important characteristics of GSM, such as the RF channel structure and signaling, will also be covered in the discussion.

Chapter 7. Provides an overview of Wireless Office Telephone Systems (WOTS) including Digital European Cordless Telephone (DECT),

Business Link (IS-94), unlicensed Spread Spectrum systems, and other technologies.

Chapter 8. Describes cordless telephone technology including the first generation of Cordless Technology (CT0) through CT2, Telego, and unlicensed spread spectrum cordless telephones.

Chapter 9. Explains the fundamental building blocks of wireless telephones (mobile telephone equipment), encompassing radio, digital signal processing, and audio sections. Optional accessories are listed and described and. include battery power supply considerations and the differences between analog and digital mobile telephone designs.

Chapter 10. Includes wireless network requirements, equipment, implementation methods, a high-level overview of the public switched telephone network (PSTN), cellular network interconnections (such as IS-41), cellular system equipment, and system planning.

Chapter 11. A review of cellular system economics, including costs of digital subscriber units and system equipment, and an analysis of cellular network capital and operational costs is discussed. Revenue producing services, distribution channels, churn, and activation subsidy are further explained.

Chapter 12. Describes future cellular technologies. Some of the systems that will be discussed include Integrated Dispatch Enhanced Network (iDEN), cellular digital packet data (CDPD), enhanced time division multiple access (E-TDMA), cellular voice paging. spatial division multiple access (SDMA), and satellite cellular systems.

Appendixes are provided which include acronyms and industry definitions.

This book is based on current standards, but the TIA, Electronics Industry Association (EIA) and European Telecommunications Standard Institute (ETSI) are continually improving and changing specifications. This book provides an understanding of the systems. For more specific information on each technology, be sure to consult the latest revision of each applicable standard.

Chapter 1
Wireless Basics

Wireless Basics

During 1996, one out of ten Americans (over 30 million customers) had a cellular phone [1], over 40 million wired office telephones were candidates for wireless office systems [2], over 60 million cordless phones were in use [3], and commercial PCS service was introduced in Washington, DC.

Over the past 10 years, the wireless telephony end-user equipment size, weight, and costs have dropped over 20% per year [4]. This progress has allowed for wearable microportable communicators such as the 3.3 oz wireless telephone that was introduced in 1996 shown in figure 1.1.

The ability to transmit information through the air is considered a natural resource and is, therefore, regulated. The Federal Communications Commission (FCC) regulates the use of radio spectrum in the US, while the Department of Communications (DOC) regulates it in Canada. In other countries, government regulatory agencies have other names such as Secretary or Minister of Telecommunications.

Figure 1.1, Microportable Cellular Telephone
Source: Motorola

The use of the radio spectrum can be divided into licensed and unlicensed frequency bands. Licensed frequency bands require that the user (or service provider) apply for the right to transmit radio energy. Unlicensed frequency bands allow users (or service providers) to communicate without applying for a license. Unlicensed radio transmission must conform to pre-established regulations which specify the frequency bands and amount of radio energy which can be used. Unlicensed users typically have very limited, if any, rights when they experience radio interference.

All wireless technologies have the ability to offer advanced information services such as messaging and interactive services. These services can be offered using audio or digital message technology. While advanced services can be offered through analog technology, new digital wireless technologies can provide these services more efficiently and offer some features analog wireless cannot.

Wireless Technology

Wireless systems link customers and information services via a wireless communication path or channel. A typical wireless communications system uses mobile or fixed radios which communicate with a fixed radio tower (called a base station), and (which in turn) connects to the public switched telephone network (PSTN).

Cellular systems provide radio coverage to a wide area, such as a city, through the use of many radio towers (25 to 500 per city). Wireless office systems typically use 5 to 20 small radio base stations to offer radio coverage in small areas such as a school campus or hospital building. Cordless telephones typically allow one handset to communicate with a single radio base station within a home. Personal Communications Systems (PCS) are considered an integration of cellular, wireless office, and cordless with the addition of advanced information services. Regardless of the type of system, wireless systems all use radio channels to communicate with wireless telephones.

Cellular

The cellular concept originated at Bell Laboratories in 1947 [5], the first automatic cellular system started operation in Japan in 1979 [6], and the first cellular system in the United States started in October 1983 in Chicago. A single cellular system uses many small radio coverage areas (cells) to serve hundreds of square miles. Radio frequencies in cellular systems are re-used in distant cells, and telephone calls are automatically switched between neighboring cell sites when the mobile telephone moves out of range of the serving cell. Neighboring cellular systems often allow customers from other cellular systems to use their service which is called ROAMing. The basic goals for a cellular system include affordability, nationwide (and possibly global) compatibility, the ability to provide efficient service to many customers, and the ability to serve many types of telephones (fixed, mobile and portable) at the same time.

The cellular concept employs a central switching office called a Mobile-service Switching Center (MSC) to interconnect small radio coverage areas into a larger system. To maintain a call when the cellular telephone moves to another coverage area, the cellular system switches the phone's radio channel frequency to a frequency in use at an adjacent cell site. The cellular concept also allows a frequency to be used by more than one customer (called a subscriber) at a time so that subscribers using the same channel will not interfere if they are far apart. Cellular systems take advantage of this by breaking the coverage area into many small cells. Each cell site base station can simultaneously transmit on several different radio channel frequencies. Adjacent cells use different frequencies to avoid interference, but widely separated cells can reuse the same frequencies. This allows the system to repeatedly reuse radio channels and increase the number of subscribers they can serve with a limited number of channel frequencies. Each cell site radio coverage area is determined by the base station's transmitter power. Consequently, lowering power decreases the coverage area.

Figure 1.2 shows the basic parts of a cellular system. The cellular telephone (sometimes called a mobile station) has the ability to tune in to many different radio channels or frequencies. The base station commands the mobile telephone on which frequency to use in order to communicate with another base station that may be from two to fifteen miles away. The base station routes the radio signal to the MSC either by wire (for example, a leased telephone line), microwave radio link, or fiberoptic line. The MSC connects the call to the PSTN which then connects the called to its destination (for example, office telephone).

Since each base station typically has several different radio frequencies available for use, a single base station is able to communicate with several mobile stations at the same time. When a base station has reached its capacity or maximum number of radio channels, additional customers cannot access the cellular system through that base station. The cellular system can expand by adding more radio channels to the base station or by adding more cell cites with smaller coverage areas.

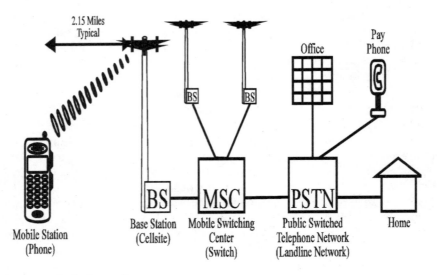

Figure 1.2, Cellular System

To maintain a call while a subscriber is moving throughout several cell site areas, the call is transferred between adjacent cell sites while in progress. This call transfer is termed "switching" or "handoff." The trends in cellular systems include changing from analog (FM radio transmission) to digital technology and in providing advanced information services such as short message delivery.

Wireless Office Telephone System (WOTS)

A Wireless Office Telephone System (WOTS) is similar to a miniature (mini) cellular system. Because there are usually a limited number of users associated with WOTS systems and small radio coverage areas, they often offer advanced features such as four digit dialing, internal paging and messaging, smaller handset size, and long battery life. Over the past three to four years, the trend has been to use the same type of phone for both office and cellular systems.

Figure 1.3 shows the basic parts of a wireless office telephone system. Micro base stations are located throughout a building or area to provide radio coverage. These micro base stations are connected to a switch which is similar to a miniature cellular switch. While this switch is connected to the PSTN, inter-office calls can be directly routed to each other without connecting to the PSTN.

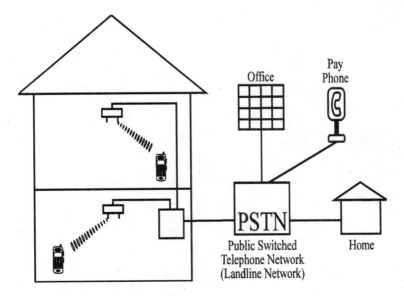

Figure 1.3, Wireless Office Telephone System

Cordless

Cordless telephones transmit at very low radio power, and they transmit on unlicensed radio frequencies. A very limited number of frequencies used by early cordless telephones created an interference problem (known as cross talk) when cordless telephones were used near each other. As a result, most cordless telephones currently sold can automatically tune to several different radio channel frequencies. When the cordless telephone senses interference from another phone, it will automatically change to a new frequency. Because cordless telephones typically do not have handoff capability, the call will be disconnected when the cordless phone moves far enough away from its base station.

Multiple radio channel capability and digital transmission has led to some cordless phones being produced which allow customers to use the same phone in the home as in the cellular system

Figure 1.4 shows the basic parts of a cordless telephone system. Typically, a single cordless base station is located in a building or home which provides a limited radio coverage up to approximately 1000 feet. These cordless base stations are connected to a standard phone line to allow for connecting to the PSTN.

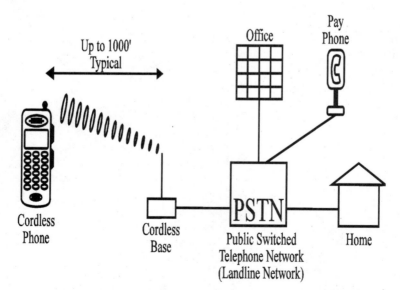

Figure 1.4, Cordless Telephone System

Personal Communication Systems (PCS)

It is sometimes perceived that Personal Communications Systems (PCS) are the combination of cellular, wireless office telephone systems, cordless telephone systems, and advanced intelligent features in one portable device. In 1995, the United States began licensing bands of frequencies for new PCS systems. While each of the digital PCS technologies may offer some unique services, PCS companies typically provide similar services as cellular systems.

Personal Communications Systems (PCS) are new wireless networks that use cellular technology (frequency reuse) to provide telecommunication services in North America to customers through the use of new radio frequency bands in the 1900 MHz range.

To use the PCS licensed radio spectrum as a resource, a PCS company must obtain a license from the Federal Communications Commission (FCC) and change frequencies (called relocate) for existing users of the radio spectrum. A licensee is permitted to use a radio frequency band as a legal "common carrier" which implies certain legal obligations such as serving public users compared to creating private wireless networks.

To use the unlicensed PCS frequency band, a PCS company must use equipment that will conform to the FCC unlicensed requirements and pay a fee to relocate existing users (such as pre-existing point-to-point microwave systems). Unlicensed requirements include low power transmission that does not typically interfere with other users in the same frequency band. The fee for relocating existing users is paid by the manufacturers of the unlicensed radio equipment and will involve an arbitrator who will assist in the relocation process.

Licenses for new PCS radio channel frequency bands were auctioned by the federal government. As of December 1995, the US government has raised $7.7 billion from the sale of radio spectrum [7]. Because there were existing licensed users in the radio spectrum chosen for PCS service operators, the government requires the existing users to relocate over time. The existing users of the PCS radio spectrum are primarily microwave point-to-point communications. These users will be re-assigned to other frequencies, and the equipment will be changed at no charge to them. The new users of the PCS radio channels must pay to convert these existing users to new frequencies.

Figure 1.5 shows a PCS telephone. This PCS telephone allows for several types of services such as voice, paging, and messaging.

NOKIA 2191

Figure 1.5, PCS Telephone
source: Nokia

Messaging Services

There are many new messaging services that add value to basic wire-less telephone service. These services render more value for the sub-scriber and add new revenue sources for wireless service providers. Some message services will be paid for by subscribers, while others will be paid by merchants who sell information services. Messaging services can be delivered by digital messages which appear on the display or by audio messages. Audio messages can be delivered to any of the existing wireless customers.

In order to determine the potential market for wireless service, it may be helpful to separate messaging services designated for "human" and "non-human" users. Human uses for messaging services include pag-ing, news and traffic reporting, direction routing, etc. Non-human uses include monitoring and remote control of machines such as elec-tric and water meters, gas valves, and thousands of other devices.

The potential for non-human applications of wireless messaging is much greater than human applications. As the number of telephones per person has become higher in rural areas because of the greater need to communicate, the need for wireless monitoring and remote control should be just as great in rural areas because of the need. Some farmers have even used pagers, which hang on the necks of the farm animals, in order to round up the cattle. As you can see, there are many innovative uses for wireless messaging services.

Advertising

The money paid by companies for advertising helps offset the cost of products and services delivered on radio, television, and the internet. However, unlike advertising on radio and television, two-way wireless advertising can confirm reception of the message to an individual. Because each wireless customer has a unique identifier, specific advertising messages can be directed to targeted customers, according to profile or geographic location. For example, an advertising message for a hamburger lunch should not be sent to a vegetarian.

Subscribers could elect to receive advertising by selecting a reduced rate program (for example, $5 less per month). The service provider could charge the business for advertising per confirmed receipt. The subscriber could fill out a profile sheet when selecting the rate plan which would focus the types of advertising that would be sent to the subscriber.

By queuing advertising messages based on available radio channel capacity, service providers could send messages during periods of low demand, and charge variable rates based on peak traffic load. Such a system would reduce advertisers' costs, as well as use the cellular system more efficiently.

Figure 1.6, Wireless Advertising

Figure 1.6 shows a sample advertising message. During the lunch rush (for example, 11:30 a.m. to 1:30 p.m). , advertising messages could be sent only to wireless customers traveling down the interstate towards an exit where a restaurant is located. After the subscriber unit receives and displays the message, it confirms receipt. A menu or other option can then be selected. The subscriber could also have some interactive response that allows ordering of the product and service.

Weather and Traffic Reports

Service providers could send current weather and traffic reports to individuals or groups of subscribers. For individual customers, the wireless system would send a message with a unique identifier. Messages to groups of subscribers can be sent using the multicast or broadcast message service. When a multicast or broadcast message service is used, a message is transmitted with a generic address code. If the subscribers' mobile phone can decode the address, the message will be displayed.

In communicating with appropriately situated mobile telephones, the wireless system can send traffic report information specific to the unit's location. In addition, subscribers could select reports on a "Pay per Jam" basis. Figure 1.7 shows a sample traffic reporting message. The cellular or PCS system broadcasts a message to all subscribers within 10 miles of the traffic jam. The message can be sent as a text message or audio message. Alternatively, the customer could call in for localized traffic reports.

Figure 1.7, Traffic Reporting

Direction Routing/Maps

Directions or routing maps can be sent to a mobile station via audio, text, or graphic form. Directions can be requested by a subscriber or provided automatically via a dispatch center. Unlike computer based mapping systems, directions by wireless can be adjusted for traffic, weather, and construction changes.

Point to point message services can send directions directly to the requesting subscriber. The subscriber could use a map to find a reference marker (for example, A-37) and enter the reference mark via the keypad. Dispatch centers would then send directions to the individual or multiple vehicles. The dispatcher could also send delivery directions to a truck or pickup directions for a taxi.

Figure 1.8 shows a sample direction routing system. The wireless system tracks the position of the mobile station and sends a message whenever a new direction is required. In this example, the mobile station is connected to a vehicle adapter kit which has a routing display device. The routing display indicates how much distance remains before the next turn. Either a right or left turn indicator will be flashed to indicate the direction of the next turn. If a routing display is not available, the message could be sent in text or audio form.

Figure 1.8, Direction Routing

Telemetry/Monitoring

Wireless service is a low cost way to transfer telemetry and monitoring information without wire line connections. Telemetry applications include monitoring utility meters, gas lines, vending machines, critical equipment, environmental sensors (water level, earthquake, fire), and many others. For applications that only transmit small amounts of information, a packet transmission system may be cost effective. When monitoring devices require the transfer of large amounts of data, circuit switched data transfer is better suited. Figure 1.9 shows an electric meter which can be monitored remotely via wireless service.

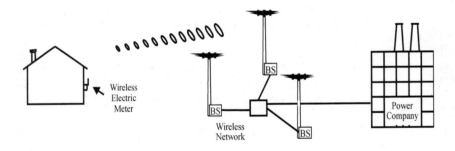

Figure 1.9, Telemetry Monitoring

Remote Control

In distant and rural areas, it is often impractical to install telephone or remote control lines. Radio receivers can be installed to control devices such as gates, vents, water valves, gas valves, audio alert devices (for example, sirens), and visual displays (for example, road signs). If power is not available to control these devices, power could be obtained from batteries, solar panels, or even from a generator which runs from the flow of gas or water. Because low cost radio coverage was not available in rural areas until a few years ago, wireless control service was not typically used.

Fax Delivery

Fax delivery is the transfer of scanned or pre-stored information such as purchase orders, invoices, brochures, or any other supporting documentation. Remote FAX machines are now beginning to appear in airports, hotels, conference centers, and business centers making remote fax delivery more convenient.

Figure 1.10 shows a fax display device which can be connected to a cellular phone. When the faxed information is received, the cellular phone routes the audio signal to the fax receiver in the display device. The information is then stored in the memory of the fax display. To view the fax, the subscriber looks into the display unit and the information appears. Simple menu control and zoom features are provided to allow the subscriber to navigate through multiple pages. To send a fax, the menu can be used to forward a fax and create a text message.

In cases where fax delivery is attempted and the mobile station is either not available or appears in a poor radio coverage area, some wireless systems can store faxed messages and forward them when delivery becomes possible. Advanced fax delivery systems also make it possible to correct radio transmission errors not commonly corrected by fax machines. This is especially important during handoffs and poor signal reception.

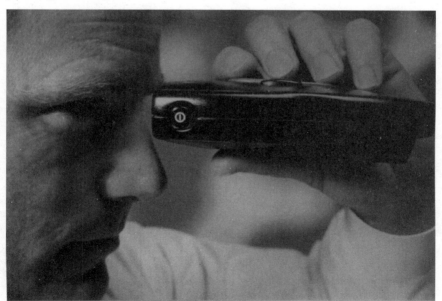

Figure 1.10, Portable Fax Display Device
Source: Reflection Technologies

Image Transfer

Still image capture provides the ability to send pictures from a digital camera to any place in the world. Digital cameras allow visual information to be captured, transferred and stored without film almost instantly and efficiently. Still image capture is used by insurance agents, security services, reporters, and police, among others. The benefits for a wireless digital image transfer system includes rapid information capture (for example, on site insurance adjustment), instant distribution and the fact that no film, film developing, postage or physical storage of the picture is required.

Radio channels sometime distort signals and cause file transfer errors which then could distort the images. Image data files that must be transferred will benefit from a reliable direct digital channel. Unfortunately, direct digital channels connect only the mobile telephone and the wireless system. For data to reach a remote computer, it must be converted between the wireless system and the PSTN.

Figure 1.11 shows a wireless image capture system. The digital camera is used to capture a digital image. The image is digitally compressed and then connected to a cellular or PCS phone. After compression, the wireless phone sets up a call to a remote computer. The

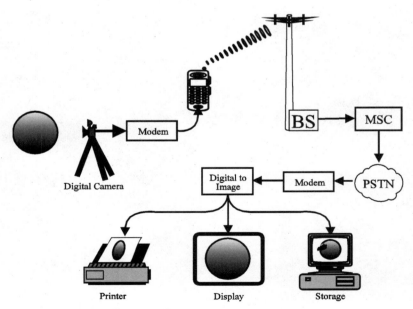

Figure 1.11, Image Capture System

remote computer receives the data file and requests information to be retransmitted if it is received in error. The digital image file can be printed, viewed on a display, or stored for subsequent retrieval.

Video Transfer

Wireless systems can send and receive non-real time (stored or delayed) and real time (instantaneous) video. Newscasters sometimes use non-real time video transfer to send high resolution video over a single cellular radio channel. Newscasters already use wireless phones to transfer digitized video.

Because high resolution video clips are comprised of large amounts of information, they cannot be sent over a single cellular radio channel at the same time they are recorded. When sending a high resolution video clip via a single wireless channel, it is sent in non-real time. The video signal is first digitized, then compressed and sent as a large data file. Even with a significant amount of data compression, the data file will take longer to send than the recording time. When the data file is received, it is uncompressed and it can be played back at the same rate that it was recorded.

Low resolution real time video, such as that used for security monitoring, can be sent over a single radio channel. To send high quality digitized video in real time, the digital signal must be compressed and

Figure 1.12, Cellular Video Transmission System
Source: Barron Technologies

divided so that it can be simultaneously sent over multiple cellular channels. When the multiple digitized signals are received, they are re-combined and uncompressed to create the original high quality digitized video signal. In the future, high speed direct digital connections will make it possible to transfer digitized video files faster and more effectively.

Figure 1.12 illustrates a real time high quality video transmission system which is used by newscasters. This system uses multiple cellular phones to increase the information carrying capacity of the channel to more than 50. This video transmission system holds up to four cellular radios. Each of the four cellular phones is connected to a multiple channel receiver which can combine the digitized video signals into the original high resolution video signal.

Location Monitoring

Location monitoring means locating and tracking mobile telephones. It can be used for safety (911), vehicle tracking (such as trucks), survey site locations, security (i.e. personal location), and for inventory and asset (for example, equipment) location.

New digital wireless systems have built-in potential for location monitoring because nearly all digital systems must compensate for the subscriber unit's distance from the cell site to adjust for transmission time delays. This built-in distance monitoring makes it likely that future digital systems will offer some mobile station location services.

Another way to monitor location is to attach a global positioning satellite (GPS) receiver to a mobile station. The GPS device determines its position from satellites orbiting the earth. After the GPS unit has determined its location, the mobile station can transfer the position data to a computer in the wireless system via the radio channel.

Figure 1.13 shows a sample position location system. The very first step in the position location system is to determine what cell site area the mobile telephone is operating in. The cell site identification can locate a mobile station within a few miles. All cellular and PCS systems already have this capability. By using a directional antenna system or by monitoring the same signal on nearby antennas (called triangulation), the relative angle (and possibly position) from the cell

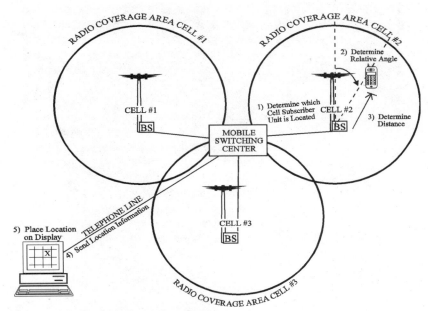

Figure 1.13, Position Location System

site can be determined. If the wireless system has the capability to track the transmission delay, an approximate distance from the cell site antenna can also be used to help pinpoint the location of the mobile telephone.

Point of Sale Credit Authorization

Wireless radio is sometimes used to authorize credit cards without a wired telephone connection. Wireless point of sale credit authorization increases mobility, eliminates the cost of telephone line installation and can bypass local telephone message unit tariffs in some cities. With the introduction of low cost wireless data, per transaction costs for wireless credit card transactions can be below 2 cents each. Wireless packet systems such as CDPD can offer transaction times of less than 6 seconds. Figure 1.14 shows a picture of a wireless credit card machine.

A wireless credit card machine can have a radio transmitter built in (see figure 1.15), while a standard credit card machine can connect through a modem to a wireless phone. Figure 1.15 shows a wireless credit card system. When a credit card is either swiped through the machine or manually keyed in, the radio transmitter accesses the

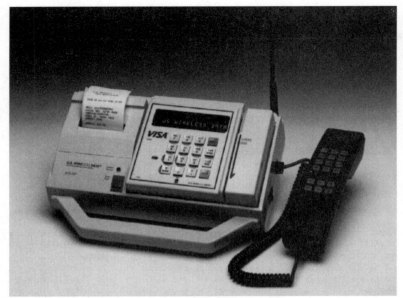

Figure 1.14, Wireless Credit Card Machine
Source: U.S. Wireless

wireless system to send the data to the bank processing center. If the wireless credit card system is capable of transferring packets of information, short messages are sent (less than 1/10th of a second) which contain their destination address. If a cellular phone or PCS phone is used, the phone will dial the bank processing center's telephone number. In either access method, the data is routed through either a public (for example, the phone system or internet) or private (for example, direct bank network connection) system so that the transaction can be authorized.

Two Way Paging

Pagers are simple communication devices which alert users when they receive simple messages. Two way pagers transmit a confirmation message after the receipt of a page message. This two-way communication capability also allows users to make simple responses to messages they have received (for example, Dinner Tonight? Yes/No). While some of the new PCS radio channels are dedicated to two-way paging service, cellular and PCS wireless telephones can be used in a similar way.

Paging systems have a significant economic advantage over wireless voice systems. The average duration of a cellular telephone call is approximately 2 minutes while the average duration of a numeric

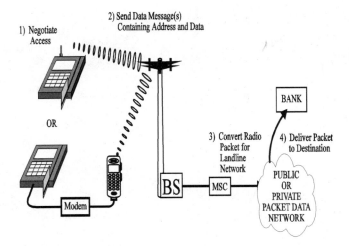

Figure 1.15, Wireless Credit Card System

paging message is less than 1/10th of a second. For each cellular call, approximately 1200 paging messages can be sent. Because paging messages can be delayed for several minutes without significant disadvantages to the receiver of the message, it allows the short paging messages to be placed in a queuing sequence. Wireless voice services, such as cellular, cannot queue (delay) an incoming call for more than a few seconds or the caller will probably hang up. To ensure calls get through on a cellular system, extra communication channels (radio channels) are required. With the advantages of short numeric messages and the ability to queue messages, numeric paging systems can serve over 100,000 customers on a single radio channel compared to a cellular system which can serve approximately 20-32 customers per radio channel.

Two-way pagers have an added advantage that messages can be sent only to the region where the two-way pager has last responded. In a one-way paging system, the location of a pager is usually unknown. When a message is sent to the pager in a particular region, every tower in that region must transmit the page message to ensure the page is received. When a page message is sent by a tower which cannot be received by the pager, it is a waste of capacity and cost. Two-way pagers, on the other hand, can be located and paging messages can be directed into the region where the pager has been located last.

Voice Paging

Unlike dedicated paging systems, wireless voice systems can transfer voice and data information in both directions. Voice messages sent on a wireless voice paging system, such as cellular or PCS, can be stored (queued), compressed, and sent when the cellular or PCS system is not busy. A voice pager stores several minutes of voice messages in its memory so the customer can listen to the messages after then have been transferred.

News Services

Wireless news services can deliver specific news topics as selected by subscribers. Subscribers can receive any news category identifiable by name, such as, information about a certain city, business stories, stock quotes, or articles on hobbies.

Some internet on-line news services provide brief abstracts before sending whole articles. The short overview lets the subscriber decide whether to get the article, and whether they need it immediately or not. The extensive transfer time for long or illustrated articles can be expensive. Delaying delivery can let the subscriber find a more economical way to receive the information.

Sports Betting

Paging already offers sports information via alpha paging. With two-way communication capability, this type of service can also include sports betting. Wireless customers could receive sports scores and bet on teams.

Wireless short message services can provide instant score updates along with league standing, player lineup, point spreads (based on a betting agent), and sports statistics. Using short messages, customers will be able to track favorite teams and place bets using the existing mobile telephone keypad. Figure 1.16 shows a sample sports betting process. The subscriber could select the team (press 1 for Dallas, press 2 for Buffalo), key in the amount, and press Send to confirm. Such a message might be sent to a betting agency via the internet. When the betting agency accepts the bet, the acceptance notification would be returned and appear on the mobile phone (for example, Bet Confirmed). Likewise, the betting agency would be able to return the final results of the bet once the event was over.

Figure 1.16, Sports Betting

Customers may be able to obtain these services either directly from wireless providers or from other firms with the capability to send short messages (for example, Las Vegas Casinos). Cellular or PCS carriers could broadcast short messages to multiple users, or send them to individual "Sports Net" subscribers. If responses are limited to menu items, transactions would use little air time - as little as 1/10th of a second. Even if a service provider charged only 1/2 a cent per message, average revenue would exceed $3.00 per minute, not including a premium fee for the service.

Remote Vending

Even in urban locations, high installation and service costs prohibit connecting vending machines to a dedicated wire line. Wireless remote vending combines the benefits of point-of-sale credit card authorization, inventory control, and advertising. Vending machines that can authorize and accept charge cards or other types of money cards make it possible to sell more merchandise without requiring the customer to have enough or the exact change. Wireless remote vending also informs distributors of inventory levels in the machines, removing guesswork from ordering and dispatching. Finally, vending machines can be a good point for advertising.

References:

1. Heads Up, Daily News Wire Service, 27 February, 1995.
2. SuperPhone conference, Institute for International Research, New York NY, January 22-24, 1997.
3. Ibid.
4. Ibid.
5. The Bell System Technical Journal, Vol. 58, No. 1, American Telephone and Telegraph Company, Murray Hill, New Jersey, January, 1979.

Cellular and PCS, The Big Picture

6. Balston, D.M., "Cellular Radio Systems," Artech House, MA, 1993, p. 169.
7. Hines, Jeffrey, "Wireless Update," Paine Webber, December 15, 1995, p.1.

Chapter 2

Analog Cellular Systems

Analog Cellular Systems

Today's cellular systems consist of three basic elements: a mobile telephone (mobile radio), cell sites, and Mobile Switching Center (MSC). Figure 2.1 shows a basic cellular system in which a geographic service area such as a city which is divided into smaller radio coverage area cells. A mobile telephone communicates by radio signals to the cell site within a radio coverage area. The cell site's base station (BS) converts these radio signals for transfer to the MSC via wired (land line) or wireless (microwave) communications links. The MSC routes the call to another mobile telephone in the system or the appropriate land line facility. These three elements are integrated to form a ubiquitous coverage radio system that can connect to the public switched telephone network (PSTN).

History

There several types of analog cellular systems throughout the world. In the Americas, Advanced Mobile Phone Service (AMPS) is the primary analog cellular system. In Europe, there are several different types of cellular systems including Total Access Communications System (TACS) and Nordic Mobile Telephone (NMT) system. In Asia, TACS and NMT are the primary analog cellular systems. While these analog systems are not directly compatible, they share many of the same system features associated with cellular technology.

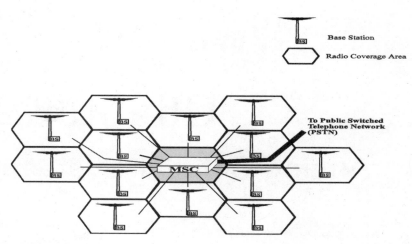

Figure 2.1, Cellular System

The first commercial analog cellular system was started by NTT in Tokyo Japan on December 3rd, 1979 [1]. In 1981, a commercial NMT system was started in the Nordic countries [2]. Although there was a AMPS test system operating in 1979, the first commercial AMPS system was introduced in the United States in 1983. By 1985, a commercial TACS system began in the United Kingdom [3]. Since their introduction, these analog cellular technologies continue to evolve to provide new capacity and features.

Mobile Telephone

A mobile telephone (typically called a mobile station) contains a radio transceiver, user interface, and antenna assembly (see figure 2.2) in one physical package. The radio transceiver converts audio to a radio frequency (RF) signal and RF signals into audio. A user interface provides the display and keypad which allow the subscriber to communicate commands to the transceiver. The antenna assembly couples RF energy between the electronics within the mobile telephone and the outside "air" for transmission and reception.

Analog mobile cellular telephones have many industry names. These names sometimes vary by the type of cellular radio. Handheld cellular radios are often referred to as "portables." Cellular radios which are installed in cars are typically called "mobiles." Cellular radios mounted in bags are often called "bag phones." In most cases, these

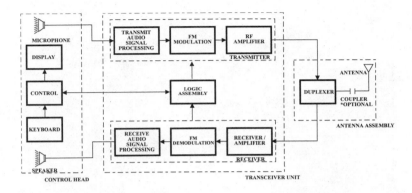

Figure 2.2, Mobile Telephone Block Diagram

three types and sizes also correspond to three distinct maximum power levels: 600 milliwatts, 1.8 watts, and 3 watts, or classes I, II and III. We will refer to any type of mobile cellular radio as a mobile telephone.

The evolution of portable cellular telephones has resulted in approximately 20% reduction in weight [4], and 24% reduction in cost [5] each year over the past 5 years. Figure 2.3 shows the progression of portable telephones over the last few years.

Each mobile telephone must have distinct signaling identification when operating in a cellular system and contain feature options specific to the customer. To make each mobile telephone unique, several types of information is stored in its internal memory. In the United States, this internal memory is called a Number Assignment Module (NAM). The NAM contains the Mobile Identification Number (MIN) which is the telephone number, home system identifier, access classification, and other customer features. The internal memory which stores the telephone number and system features can be modified either by changing an integrated circuit (chip) stored inside the mobile telephone or by programming the phone number into memory through special keypad instructions.

Figure 2.3 Portable Mobile Telephone Evolution
source: Ericsson

Mobile telephones also contain a unique Electronic Serial Number (ESN) which is not supposed to be changed. If the ESN could be easily changed, it would be possible to duplicate (called cloning) another mobile telephone's identification to make fraudulent calls. Because duplications of mobile telephone numbers and ESN's is technically possible, advanced authentication programs which validate pre-stored information have been created to provide a more reliable unique identification system.

Initially, information stored in a NAM was programmed into a standard Programmable Read Only Memory (PROM) chip. Because of the cost of the chips and that special programming devices were required, manufacturers now make the NAM information programmable via the handset keypad. The information is stored internally in an Electrically Alterable PROM (EPROM). This is also referred to as a non-volatile memory, since the information contents stay intact even if power is not available, such as when a battery is replaced.

Cell Site

A cell site is the link between the mobile telephone and the cellular system switching center. A cell site consists of an base station (BS), transmission tower, and antenna assembly. The base station is the radio portion of the cell site which converts radio signals to electrical signals for transfer to and from a switching center.

A base station contains amplifiers, radio transceivers, RF combiners, control sections, communications links, a scanning receiver, backup power supplies, and an antenna assembly (see figure 2.4). The transceiver sections are similar to the mobile telephone transceiver as they convert audio to RF signals and RF signals to audio signals. The transmitter output side of these radio transceivers is supplied to a high power RF amplifier (typically 10 to 50 Watts). The RF combiner allows separate radio channels to be combined onto one or several antenna assemblies without interfering with each other. This combined RF signal is routed to the transmitter antenna on top of the radio tower via low energy loss coaxial cable.

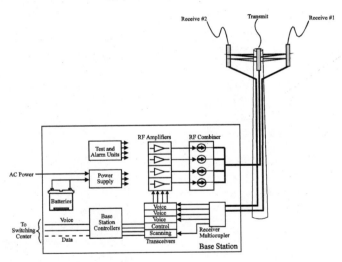

Figure 2.4, Cell Site and Base Station Block Diagram

Cell sites typically have two receiver antennas to allow for selection of the strongest radio signal (minimizing radio signal fading). Receiver antennas are connected to the RF multicoupler via low loss coaxial cable which splits the received signals to multiple transceivers. The receiver portion of the transceiver converts the RF signal to an audio signal which is routed to the communication links. Communication links route audio and control information between the base station and a telephone switching center. A scanning receiver measures the signal strength on any of the cellular channels which is used to determine if a call transfer to another cell site is required. The backup power supply maintains radio equipment and cooling system operation when primary power is interrupted. Many sections of the base station are duplicated to maintain functioning if equipment fails.

Mobile Switching Center

The Mobile Switching Center (MSC) is the control center of the cellular system. It monitors the location and call quality of mobile telephones and switches the mobile telephone call between cell sites and the public switched telephone network (PSTN). The MSC is sometimes called different names such as Motile Telephone Switching Office (MTSO), Mobile-service Switching Center (MSC) or Mobile Telephone Exchange (MTX).

The MSC consists of controllers, switching assembly, communications links, operator terminal, subscriber database, and backup energy sources (see figure 2.5). The controllers, which are each powerful computers, are the brains of the entire cellular system, guiding the MTSO through the creation and interpretation of commands to and from the base stations. In addition to the main controller, secondary controllers devoted specifically to control of the cell sites (base stations) and to handling of the signaling messages between the MSC and the PSTN are also provided. A switching assembly routes voice connections from the cell sites to each other or to the public telephone network. Communications links between cell sites and the MSC may be copper wire, microwave, or fiber optic. An operator terminal allows operations, administration and maintenance of the system. A subscriber database contains customer specified features and billing records. Backup energy sources provide power when primary power is interrupted. As with the base station, the MTSO has many standby duplicate circuits and backup power sources to allow system operation to be maintained when a failure occurs.

System Overview

The cellular system provides telephone service to many customers through duplex radio channels, frequency reuse, cost effective capacity expansion, and coordinated system control. To conserve the limited

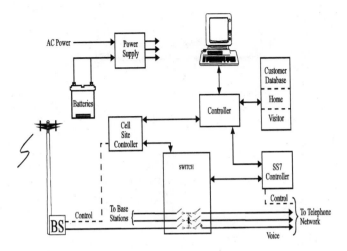

Figure 2.5, Cellular Mobile Telephone Switching Office

amount of radio spectrum, cellular systems reuse the same channels many times within a geographic coverage area. The technique, called frequency reuse, makes it possible to expand system capacity by increasing the number of channels that are effectively available for subscribers. As the subscriber moves through the system, the Mobile Telephone Switching Office (MTSO) centrally transfers calls from one cell to another and maintains call continuity. In fact, without frequency reuse, it would not be economically feasible to provide cellular or PCS service, unless all other radio frequency bands (broadcasting, emergency radio systems, ship to shore, military, etc.) were shut off and their spectrum capacity were also used for cellular/PCS.

Frequency Duplex

To allow simultaneous transmission and reception (no need for push to talk), the base stations transmit on one set of radio channels, called forward channels and they receive on another set of channels, called the reverse channels. The transmit and receive channels assigned for a particular cell are separated by a fixed amount of frequency. Figure 2.6 displays a base station transmitting to the mobile telephone at 875 MHz on the forward channel. The mobile telephone then transmits to the base station at 830 MHz on the reverse channel. Figure 2.6 (b) shows the base station transmitting at 890 MHz resulting in the mobile telephone transmitting at 845 MHz.

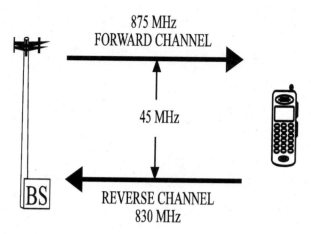

875 MHz
FORWARD CHANNEL

45 MHz

REVERSE CHANNEL
830 MHz

Figure 2.6, Duplex Radio Channel Spacing

Frequency Reuse

In early mobile radio systems, one high-power transmitter with a modest allocation of frequency spectrum served a large geographic area. Because each cellular radio requires a certain bandwidth, the resulting limited number of radio channels kept the serving capacity of such systems low. The customer demand for the few available channels was very high. For example, in 1976, New York City had only 12 radio channels to support 545 subscribers and a two-year long waiting list of typically 3,700 [6].

To increase the number of radio channels in where the frequency spectrum allocation is limited, cellular providers must reuse frequencies. One strategy for reusing frequencies relies on the fact that signal strength decreases exponentially with distance, so subscribers who are far enough apart can use the same radio channel without interference (see Figure 2.7).

To minimize interference in this way, cellular system planners position the cell sites that use the same radio channel far away from each other. The distances between sites are initially planned by general RF propagation rules, but it is difficult to account for enough propagation factors to precisely position the towers, so the cell site position and power levels are usually adjusted later.

The acceptable distance between cells that use the same channels is determined by the distance to radius (D/R) ratio. The D/R ratio is the

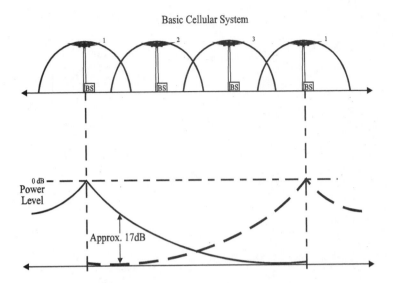

Figure 2.7, Frequency Reuse

ratio of the distance (D) between cells using the same radio frequency to the radius (R) of the cells. In today's system, a typical D/R ratio is 4.6: a channel used in a cell with a 1 mile radius would not interfere with the same channel being reused at a cell 4.6 miles away.

Capacity Expansion

As cellular systems mature, they must serve more subscribers, either by adding more radio channels in a cell, or by adding new cells. To add radio channels, cellular systems use several techniques in addition to strategically locating cell sites that use the same frequencies. Directional antennas and underlay/overlay transmit patterns improve signal quality by focusing radio signals into one area and reducing the interference to other areas. The reduced interference allows more frequency reuse. Directional antennas can be used to sector a cell in to wedges so that only a portion of the cell area (for example, 1/3 or 120 degrees) is used for a single radio channel. Such sectoring reduces interference with the other cells in the area. Figure 2.8 shows cells that are sectored into three 120 degree sectors.

Another technique, called cell splitting, helps to expand capacity gradually. Cells are split by adjusting the power level and/or using reduced antenna height to cover a reduced area (see figure 2.9). Reducing a coverage area by changing the RF boundaries of a cell site has the

31

TOP VIEW

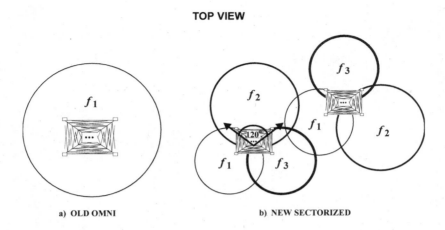

a) OLD OMNI   b) NEW SECTORIZED

Figure 2.8, Cell Site Sectorization

same effect as placing cells farther apart, and allows new cell sites to be added. However, the boundaries of a cell site vary with the terrain and land conditions, especially with seasonal variations in foliage. Coverage areas actually increase in fall and winter as the leaves fall from the trees.

Current analog systems serve only one subscriber at a time on a radio channel, so system capacity is influenced by the number of radio channels available. However, a typical subscriber uses the system for only a few minutes a day, so on a daily basis, many subscribers share a single channel. Typically, 20 - 32 subscribers share each radio channel [7], depending upon the average talk time per hour per subscriber. Generally, a cell with 50 channels can support 1000 - 1600 subscribers.

When a cellular system is first established, it can effectively serve only a limited number of callers. When that limit is exceeded, callers experience too many system busy signals (known as blocking) and their calls cannot be completed. More callers can be served by adding more cells with smaller coverage areas - that is, by cell splitting. The increased number of smaller cells provides more available radio channels in a given area because it allows radio channels to be reused at closer geographical distances.

System planning must also account for present and future coverage requirements. After the cellular service provider is granted a license, the cellular service providers typically have only a few years to pro-

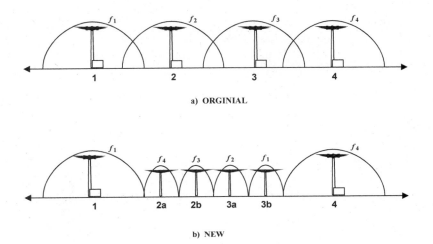

a) ORGINIAL

b) NEW

Figure 2.9, Cell Splitting

vide coverage to almost all of their licensed territory [8]. To accomplish this, and to ensure that the system will be efficient and competitive, cellular carriers must plan and design the system in advance.

Radio Interference

Radio interference limits the number of radio channels that can be used in a single cell site and how close nearby cell sites that use the same frequency can be located together. The main types of interference are co-channel, adjacent channel, and alternate channel interference.

Co-channel Interference

Co-channel interference occurs when two nearby cellular radios operating on the same radio channel interfere with each other. Co-channel interference at a particular location can be measured by comparing the received radio signal power (signal strength) from the desired signal, compared to the signal strength of the interfering signal. Today's analog systems are designed to assure that interfering signal strength remains approximately less than 2 percent of desired signal strength. At this level, the desired signal is nearly undistorted.

To minimize interference, cellular carriers frequently monitor the received signal strength by regularly driving test equipment throughout the system. This testing determines whether the combined interference from cells using the same channel exceeds the 2 percent level of the desired channel (which is 17 dB below the desired signal). This information is used to determine whether radio channels and/or power levels at each cell need to be changed. Figure 2.10 shows how co-channel interference occurs.

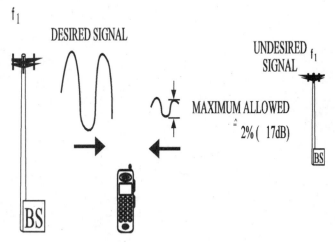

Figure 2.10, Co-Channel Interference

Radio technology that provides a higher tolerance to co-channel interference (i.e. exceeding 2 percent with no distortion) would allow system operators to reuse the same frequencies more often, thus increasing the system capacity. This higher tolerance is an advantage of next-generation digital cellular technologies.

Adjacent Channel Interference

Adjacent channel interference occurs when one radio channel interferes with a channel next to it (for example, channel 412 interferes with 413). Each radio channel has a limited amount of bandwidth (10 kHz to 30 kHz wide), but some radio energy is transmitted at low levels outside this band. A cellular radio operating at full power can produce enough low-level radio energy outside the channel bandwidth to interfere with cellular radios operating on adjacent channels. Because of alternate channel interference, radio channels cannot be spaced adjacent to each other in a single cell site (for example, channel 115 and 116). A channel separation of 3 channels is typically sufficient to protect most radio channels from adjacent channel interference. However, for frequency planning reasons (discussed in chapter 10), the radio channel frequencies at each cell site are selected so they are typically separated by 21 channels from other radio channels in that base station or sector. Figure 2.11 displays adjacent channel interference.

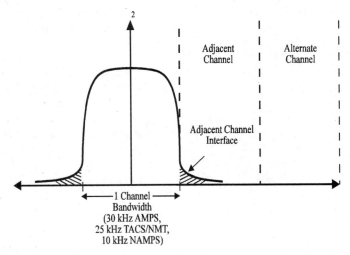

Figure 2.11, Adjacent Channel Interference

Alternate Channel Interference

Alternate channel interference occurs when radio energy from a transmitter which is located two radio channels away interferes with the desired signal.

Interleaved Radio Channels

Because a majority of the radio energy is in the center of the band, some cellular systems allow radio channels to be interleaved (offset) at fi channel bandwidth (for example, 12.5 kHz offset in a 25 kHz channel) to allow radio channel spacing to be as close as possible. Since a goal of the cellular system is to reuse as many radio channel frequencies as possible, placing more radio channels in each cell site increases the capacity of the cellular system. By careful frequency planning, the use of interleaved radio channels can increase system capacity by more than 100% [9].

Basic Cellular Operation

In early mobile radio systems, a mobile telephone scanned the limited number of available channels until it found an unused one which allowed it to initiate a call. Because the analog cellular systems in use today have hundreds of radio channels, a mobile telephone cannot scan them all in a reasonable amount of time. To quickly direct a mobile telephone to an available channel, some of the available radio channels are dedicated as control channels. Most cellular systems use two types of radio channels, control channels and voice channels. Control channels carry only digital messages and signals which allow the mobile telephone to retrieve system control information and compete for access. Control channels never carry voice (in the AMPS system). Voice channels are primarily used to transfer voice information, but also send and receive some digital control messages, and certain super-audible tones used for call processing control. Figure 2.12 displays that some channels are dedicated as control channels which coordinate access to the voice channels. After the access to a voice channel has been authorized, the control channel sends out a channel assignment message which commands the mobile telephone to tune to a voice channel.

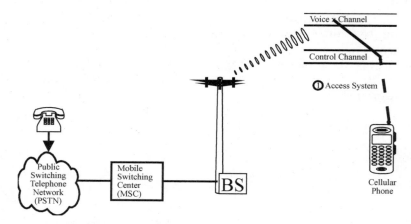

Figure 2.12, Control Channels and Voice Channels

When a mobile telephone is first powered on, it initializes itself by scanning the predetermined set of control channels and then tuning to the strongest one. Figure 2.13 shows that during this initialization mode, it retrieves system identification and setup information.

After initialization, the mobile telephone enters the idle mode and waits to be paged for an incoming call and senses if the user has ini-

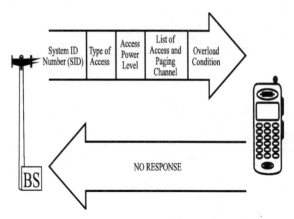

Figure 2.13, Cellular System Broadcast Information

tiated (dialed) a call (access). When a call begins to be received or initiated, the mobile telephone enters system access mode to try to access the system via a control channel. When it gains access, the control channel sends an initial voice channel designation message indicating an open voice channel. The mobile telephone then tunes to the designated voice channel and enters the conversation mode. As the mobile telephone operates on a voice channel, the system uses Frequency Modulation (FM) similar to commercial broadcast FM radio. To send control messages on the voice channel, the voice information is either replaced by a short burst (blank and burst) message or in some systems, control messages can be sent along with the audio signal.

Access

A mobile telephone's attempt to obtain service from a cellular system is referred to as "access." Mobile telephones compete on the control channel to obtain access from a cellular system. Access is attempted when a command is received by the mobile telephone indicating the system needs to service that mobile telephone (such as a paging message indicating a call to be received) or as a result of a request from the user to place a call. The mobile telephone gains access by monitoring the busy/idle status of the control channel both before and during transmission of the access attempt message. If the channel is available, the mobile station begins to transmit and the base station simultaneously monitors the channel's busy status. Transmissions must begin within a prescribed time limit after the mobile station finds that the control channel access is free, or the access attempt is stopped on the assumption that another mobile telephone has possibly gained attention of the base station control channel receiver. Figure 2.14 shows a sample access process.

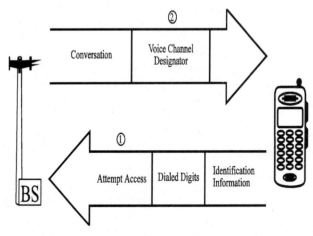

Figure 2.14, Cellular System Call Origination Radio Channel Access

If the access attempt succeeds, the system sends out a channel assignment message commanding the mobile telephone to tune to a cellular voice channel. Figure 2.15 displays the access process when a call is placed from the mobile telephone to the cellular system (called "origination"). The access attempt message is called a Call Setup message and it contains the dialed digits and other information. The system will assign a voice channel by sending a voice channel designator message, if a voice channel is available. If the access attempt fails, the mobile telephone waits a random amount of time before trying again. The mobile station uses a random number generating algorithm internally to determine the random time to wait. The design of the system minimizes the chance of repeated collisions between different mobile stations which are both trying to access the control channel, since each one waits a different random time interval before trying again if they have already collided on their first, simultaneous attempt.

An access overload class (ACCOLC) code is stored in the mobile telephone's memory which can inhibit it from transmitting when the system gets to busy. When an access overload class category is sent on the control channel which matches its stored access overload class, the mobile telephone is inhibited from attempting to access the cellular system. This process allows the cellular system to selectively reduce the number of access attempts and only allow particular groups of mobile telephones to access the system. The higher level access classes are reserved for usage by emergency personnel.

Paging

To receive calls, a mobile telephone is notified of an incoming call by a process called paging. A page is a control channel message which contains the telephone's Mobile Identification Number (MIN or telephone number of the desired mobile phone) and it responds automatically with a system access message of a type called a Page Response. This indicates that an incoming call is to be received. After the mobile

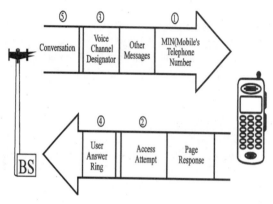

Figure 2.15, Cellular System Paging

telephone receives its' own telephone number, the mobile telephone begins to ring. When the customer answers the call (user presses SEND), the mobile telephone transmits a service request to the system to answer the call. It does this by sending the telephone number and an electronic serial number to provide the users identity. Figure 2.15 shows that if the mobile telephone is paged in the system and wishes to receive the call (user presses SEND), it responds to the page by attempting access to the system.

Discontinuous Reception

The use of discontinuous reception allows the mobile telephone's receiver to turn itself off ("sleep") for brief periods of time (typically less than 2 seconds) to save battery energy. Because the cellular system coordinates discontinuous reception, it knows when the mobile telephone has turned the receiver off and will hold pages until it knows the mobile telephone's receiver is turned back on ("awake").

Conversation

After a mobile telephone has been commanded to tune to a radio voice channel, it sends mostly voice or other customer information. Periodically, control messages may be sent between the base station and the mobile telephone. Control messages may command the mobile telephone to adjust it's power level, change frequencies, or request a special service (such as three way calling).

Discontinuous Transmission

To conserve battery life, a mobile phone may be permitted by the base station to only transmit when it senses the mobile telephone's user is talking. When there is silence, the mobile telephone may stop transmitting for brief periods of time (several seconds). When the mobile telephone user begins to talk again, the transmitter is turned on again.

Handoff

Handoff is a process where the cellular system automatically switches channels to maintain voice transmission when a mobile telephone moves from one cell radio coverage area to another. The MSC's switching equipment transfers calls from cell to cell and connects the call to other mobile telephones or the land line telephone network. The MSC creates and interprets the necessary command signals to control

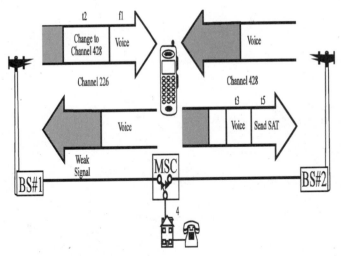

Figure 2.16, Cellular System Handoff

mobile telephones via base stations. This allows the switching from channel to channel as the mobile telephone moves from one coverage area to another.

Figure 2.16 shows the cellular handoff process. Initially, base station #1 is communicating with the mobile telephone (t1). Because the signal strength of the mobile telephone has decreased, it has become necessary to transfer the call to a neighboring cell, base station #2. This is accomplished by base station #1 sending a handoff command to the mobile telephone (t2). The mobile telephone tunes to the new radio channel (428) and begins to transmit a control tone which indicates it is operating on the channel (t3). The system senses that the mobile telephone is ready to communicate on channel 428 and the MSC switches the call to base station #2 (t4). The conversation can then continue (t5). This entire process is usually accomplished in less than 1/4 of a second.

When a cellular radio moves far away from the cell that is serving it, the cellular system must transfer service to a closer cell. Figure 2.17 illustrates the process. To determine when handoff is necessary, the serving base station continuously monitors the signal strength of the cellular radio. When the cellular radio's signal strength falls below a minimum level of signal strength, the serving base station requests

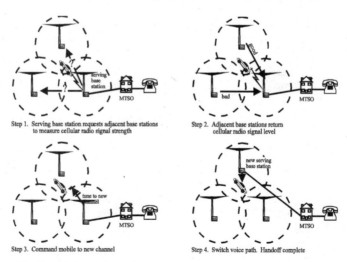

Figure 2.17, Handoff Messaging

adjacent base stations to measure that radio's signal strength (step 1). The adjacent base stations tune to the cellular radio's current operating channel and measure the signal strength. When a closer adjacent base station measures sufficient signal strength (step 2), the serving base station commands the cellular radio to switch to the new base station (step 3). After the cellular radio starts communicating with the new base station, the communication link carrying the landline voice path is switched to the new serving base station to complete the hand-off (step 4).

Mobile Reported Interference (MRI)

Some mobile telephones can transmit their radio channel quality information back to the base station to assist with handoff decisions. The process, called Mobile Reported Interference (MRI), sends channel quality information via the sub-band digital audio channel. Radio channel quality is measured using message parity bits to count the bits on the sub-band digital audio channel that are received in error. The base station sets a channel quality threshold level so that mobile telephone can determine when the signal strength and interference levels become unacceptable. When signal levels fall below the threshold, the mobile telephone informs the cellular system of poor radio channel conditions, and a handoff request is processed.

Figure 2.18 illustrates the MRI process. The process begins as the serving base station sends the minimum acceptable signal strength level to the mobile telephone. This information sets a minimum signal strength threshold in the mobile telephone (step 1). The mobile

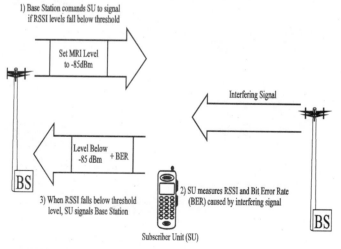

Figure 2.18, Mobile Reported Interference

telephone continues to monitor the Received Signal Strength Indicator (RSSI) and the Bit Error Rate (BER) of the sub-band digital signaling channel until the threshold is reached (step 2). The BER is an indicator of co-channel interference. When the threshold is crossed, the mobile telephone sends a single message to the base station indicating that received signal strength and BER rates are beyond tolerance. The base station then uses the information to assist the hand-off decision. If the base station wants another measurement from the mobile telephone, it sends a new message indicating a new threshold level.

RF Power Control

Mobile telephones are typically classified by their maximum amount of power output, called the "power class." Mobile telephone power output is adjusted by commands received from the base station to reduce the transmitted power from the mobile telephone in smaller cells. This reduces interference to nearby cell sites. As the mobile telephone moves closer to the cell site, less power is required from the mobile telephone and it is commanded to reduce its transmitter output power level. The base station transmitter power level can also be reduced although the base station RF output power is not typically reduced. While the maximum output power varies for different classes of mobile telephones, typically they have the same minimum power level. Figure 2.19 shows the power control process.

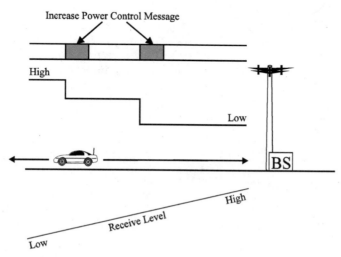

Figure 2.19, RF Power Control

Roaming

A home system identifier code is stored in the mobile telephone's memory which allows it to compare the home system identifier code to the system identifier code which is transmitted on the serving control channel. If they do not match, it means the subscriber is operating in a visited system and the mobile telephone will provide a ROAM indicator. The ROAM indicator is typically used by the subscriber to determine if billing rates have changed. Visited systems often charge a premium for service usage.

Signaling

Signaling is the transferring of control messages between two points. There are basic two parts of signaling: the physical transport of the message and the actual content of the message. Control messages are sent on radio control channels, radio voice channels, and between the network parts of the cellular and telephone system.

Radio Control Channels

Most cellular systems have dedicated control channels that carry several types of messages to allow the mobile telephone to listen for pages and compete for access. These messages include:

o <u>overhead messages</u> which continuously communicate the system identification (SID) number, power levels for initial transmissions, and other important system registration information
o <u>pages</u> which tell a particular mobile telephone that a call is to be received
o <u>access information</u> which is the information exchanged between the mobile telephone and the system to request service
o <u>channel assignment commands</u> that establish the radio channels for voice communications.

The control channel sends information by Frequency Shift Keying (FSK). To allow self-synchronization, the information is usually Manchester encoded which forces a frequency shift (bit transition) for each bit input [10]. Orders are sent as messages composed of one or more words.

To help coordinate multiple mobile telephones accessing the system, busy idle indicator bits are typically interlaced with the other message bits. Before a mobile telephone attempts access to the system, it checks the busy/idle bits to see if the control channel is serving another mobile telephone. This system is called Carrier Sense Multiple Access (CSMA) and it helps to avoid collisions during access attempts.

When a mobile telephone begins to listen to a control channel, it must find the beginning of messages so they can be decoded. Messages are preceded with an alternating pattern called a dotting sequence which is easy to sense and identifies a message will follow. Following the dotting sequence, a unique sequence of bits call a synchronization word is sent which allows the mobile telephone to match the exact start time of the message.

Radio channels can have rapid signal level fades which introduce errors, so the message words are repeated several times to ensure reliability (except the NMT system). Of the repeated words, the mobile telephone can use a majority vote system to eliminate corrupted messages. Message and signaling formats on the control channels vary between forward and reverse channels. The forward channel is synchronous and the reverse channel is asynchronous.

Forward Control Channel

On the forward control channel, several message words follow a dotting and synchronization word sequence. Each word has error correction/detection bits that are included so the data content can be verified and possibly corrected if received in error.

Reverse Control Channel

On the reverse control channel, words follow a dotting and synchronization word sequence. Because the reverse channel is randomly accessed by mobile telephones, the dotting sequence in the reverse direction is typically longer than the dotting sequence in the forward direction. Each reverse channel word has error detection and correction bits. Messages are sent on the reverse channel in random order and typically coordinated using the Busy/Idle status from a forward control channel.

Radio Voice Channels

After a mobile telephone is assigned a voice channel, voice and control information must share the same radio channel. Brief control messages that are sent on the voice channel include:

o <u>handoff messages</u> that instruct the mobile telephone to tune to a new channel
o <u>alert</u> message tells the mobile telephone to ring when a call is to be received
o <u>maintenance</u> command messages monitor the status of the mobile telephone
o <u>flash</u> requests a special service from the system (such as 3 way calling)

The analog voice channel typically transfers voice information between the mobile telephone and the base station. Signaling information must also be sent to allow base station control of the mobile telephone. Signaling on the voice channel can be divided into in band and out of band signaling. In-band signaling occurs when audio sig-

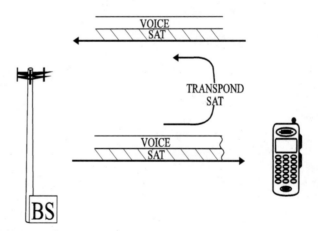

Figure 2.20, Transponding SAT

nals between 300-3000 Hz either replace or occur simultaneously with voice information. Out-of-band signals are above or below the 300-3000 Hz range, and may be transferred without altering voice information. Signals sent on the voice channel include a pilot or Supervisory Audio Tone (SAT), Signaling Tone (ST), Dual Tone Multi-Frequency (DTMF), and blank and burst FSK digital messages.

The supervisory tone provides a reliable transmission path between the mobile telephone and base station, and is transmitted along with the voice to indicate a closed loop. The tone functions are much like the current/voltage functions used in land line telephone systems to indicate that a phone is off the hook[11]. The supervisory tone may be one of the several frequencies (around 6 kHz for AMPS/TACS and 4 kHz for NMT) and this tone is different for nearby cell sites. If the supervisory tone is interrupted for longer than about 5 seconds, the call is terminated. In some systems such as NAMPS or NTACS, the SAT tone is replaced with a digital supervisory signal which is sent out of band on a sub-band digital signaling channel. Figure 2.20 shows how a supervisory tone is transponded back to the base station.

The use of different supervisory tone frequencies in adjacent cell sites is also used to mute the audio when co-channel interference occurs. Interfering signals have a different supervisory frequency than the one designated by the system for the call in progress. The incorrect supervisory tone alerts the mobile telephone to mute the audio from the interfering signal.

Re-transmission of the supervisory tone can also be used to locate the mobile telephone's position. An approximate propagation time can be calculated by comparing the phase relationship between the transmitted and received supervisory tones. This propagation time is correlated to the distance from the base station. However, multipath propagation (radio signal reflections) makes this location feature inaccurate and only marginally useful [12]. Only the re-transmission of the supervisory tone as a pilot tone is critical to operation.

A signaling tone (ST) is used in some systems to indicate a call status change. It confirms messages sent from the base station and is similar to a land line phone status change of going on or off hook [13]. Similar to the digital supervisory signal, in some systems such as NAMPS or NTACS, the signaling tone is replaced with a digital signaling tone which is sent out of band on a sub-band digital signaling channel.

Touch-tone (registered trademark of AT&T) signals may be sent over the voice channel. DTMF signals are used to retrieve answering machine messages, direct automated PBX systems to extensions, and a variety of other control functions. Bellcore specifies frequency, amplitude, and minimum tone duration for recognition of DTMF tones [14]. The voice channel can transmit DTMF tones, but varying channel conditions can alter the expected results. In poor radio conditions and a fading environment, the radio path may be briefly interrupted, sometimes sending a multiple of digits when a key was depressed only once.

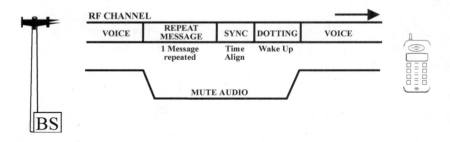

Figure 2.21, In Band Voice Channel Message

Blank and Burst Messages

When signaling data is about to be sent on the voice channel, audio FM signals are inhibited and replaced with digital messages. This interruption of the voice signal is normally so short (less than / second) that it is often not noticed by the mobile telephone user. Like control channel messages, these messages are typically repeated a multiple of times and a majority vote is taken to see which messages will be used.

To inform the receiver that a digital signaling message is coming, a bit dotting sequence is sent preceding the message. After the dotting sequence gets the attention of the receiver, a synchronization word follows which identifies the exact start of the message. Figure 2.21 shows how a voice channel message is sent.

Blank and burst signaling differ on the forward and reverse voice channels. On the forward voice channel, messages are repeated more times to ensure control information is reliable even in poor radio conditions. It is likely that messages will be sent in poor radio conditions as handoffs messages often occur when the signal is very weak.

Sub-Band Digital Audio Signaling.

A unique signaling feature used by some cellular radio systems is the sub-band digital audio signaling. In most analog cellular phones, an audio bandpass filter blocks the audio channel's lower range, but in mobile telephones which have sub-band digital signaling capability, a low speed digital signal replaces the lower audio range (below 300 Hz) with digital information. Figure 2.22 shows how sub-band digital and audio signals are combined with standard audio.

Network

Signaling commands must be passed between base stations, MCS, and the telephone network. These commands are for the maintenance, control, and administration of the network.

Signaling between the base station and MSC is performed on one of the multiple channels connected between base stations and the switching center. This control channel is designated exclusively as a control (data) channel. The MSC uses this data link to send commands to the base station and receive information about the calls in progress. The MSC uses this capability to switch calls between cell sites (handoff) and the public telephone network. Base stations also use software which is sometimes changed when new features or test software become available. The data link is used to download this new software.

Early cellular systems connected to the public switched telephone network in a way that was similar to a standard telephone line. Unfortunately, this type of connection does not provide much information about the call status. Most modern cellular systems connected to the public telephone network in a similar way as the standard public (landline) telephone switch connects to other telephone switches. This allows a cellular system to receive and send control messages which contain detailed information about the calls in process.

Advanced Mobile Phone Service (AMPS)

When the current U.S. cellular system was introduced in 1983, it was termed Advanced Mobile Phone Service (AMPS), now defined by the Electronics Industries Association (EIA) specification EIA-553, Base Station to Mobile Station Compatibility Standard. To work in the U.S. system, mobile and base station units must be manufactured to this specification.

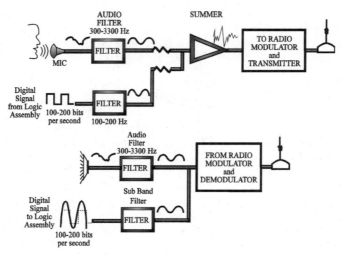

Figure 2.22, Sub Band Signaling

AMPS systems operate in over 72 countries [15]. The AMPS standard continues to evolve to allow advanced features such as increased standby time, narrowband radio channels, and anti-fraud authentication procedures.

History

In 1971, AT&T proposed a cellular radio telephony system [16] that could meet the FCC's requirement to serve a large number of customers (called subscribers) with a limited amount of radio frequencies. The AT&T proposal was the backbone of the U.S. commercial cellular system that started service in Chicago, October 1983. Since that time, cellular radio telephony has evolved into the Analog FM (EIA-553) cellular system we have today. This system was originally called the Advanced Mobile Phone Service (AMPS) system. Although the name of the analog system has changed, the EIA-553 system is commonly referred to as the AMPS system.

To provide for competition in the United States, the FCC allocated each cellular service area radio spectrum which is divided between two cellular companies, called A and B carriers. Today, there are 734 cellular service areas in the U.S., each with an A and a B carrier. The A carrier does not have a controlling interest in the local telephone company; the B carrier (often a Bell operating company) can have a controlling interest in a local telephone company.

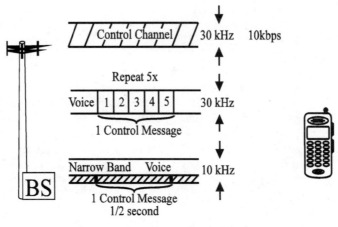

Figure 2.23, AMPS Radio Channels

System Operation

The AMPS radio system has dedicated control channels and voice channels. Mobile telephones scan a tune to one of 21 dedicated control channels to listen for pages and compete for access. The control chan-

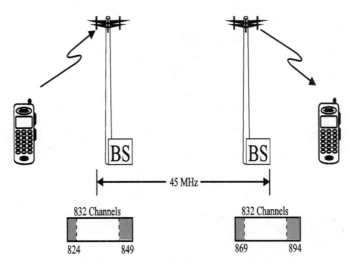

Figure 2.24, AMPS Frequency Allocation

nel continuously send system identification information and access control information. Although the control channel data rate is 10 kbp/s, messages are repeated 5 times which reduces the effective channel rate to below 2 kbp/s. This allows a control channel to send 10 to 20 pages per second.

Signaling on the radio voice channel is done by a 10 kbp/s blank and burst signaling. One of three supervisory tones (called SAT) are sent which are approximately 6 kHz. Recently, narrowband radio channels (from the NAMPS standard) have been added to the AMPS standard. Other recent changes include discontinuous reception which increases the battery life of portable phones and authentication which decreases fraudulent use. Figure 2.23 shows the different types of AMPS radio channels.

System Parameters

In 1974, 40 MHz of spectrum was allocated for cellular service [17] which provided only 666 channels. In 1986, an additional 10 MHz of spectrum was added to facilitate expansion [18] which expanded the system to 832 channels. Figure 2.24 shows the AMPS cellular frequency allocation:

The AMPS cellular system is frequency duplex with its channels separated by 45 MHz. The control channel and voice channel signaling is

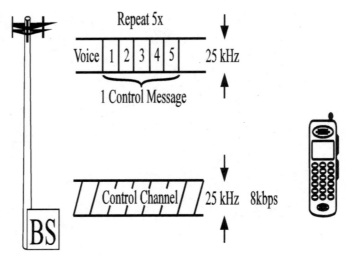

Figure 2.25, TACS Radio Channels

transferred at 10 kbp/s. AMPS cellular phones have three classes of maximum output power. A class 1 mobile telephone has a maximum power output of 6 dBW (3 Watts), class 2 has a maximum output power of 2 dBW (1.6 Watts), and the class 3 units are capable of supplying only -2 dBW (0.6 Watts). The output power can be adjusted in 4 dB steps and has a minimum output power of -22 dBW (approximately 6 milliwatts).

Total Access Communications Systems (TACS)

TACS is very similar to the US EIA-553 system with changes to the radio channel frequencies, bandwidths, and signaling rates.

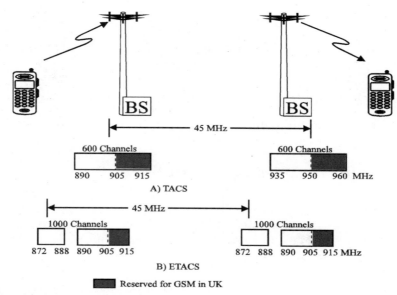

Figure 2.26, TACS Frequency Allocation

History

The total access communication system (TACS) was introduced to the U.K. in 1985. Since its introduction in the UK in 1985, over 25 countries offer TACS service. The introduction of the TACS system was very successful and the system was expanded to add more channels in which is called Extended TACS (ETACS).

System Operation

The TACS system was created by enhancing the AMPS cellular system specification. These enhancements allowed a smaller radio channel bandwidth, reduced speed signaling channel, and other new features. Figure 2.25 shows the basic radio channels in a TACS system.

System Parameters

The TACS system was initially allocated 25 MHz although 10 MHz of the 25 MHz was reserved for future pan-European systems in the UK. An additional 16 MHz of radio channel bandwidth was added to allow for Extended TACS (ETACS). Figure 2.26 shows the TACS and ETACS frequency allocations.

The ETACS system is frequency duplex with its channels separated by 45 MHz. The control channel and voice channel signaling is transferred at 8 kbp/s. There are 4 power classes for ETACS mobile telephones. Class 1 mobile telephones have a maximum output of 10 Watts, class 2 has 3 Watts, class 3 has 1.2 Watts, and class 4 has 0.6 Watts. Similar to AMPS, mobile telephones can be adjusted in 4 dB steps and have a minimum transmit power level of approximately 6 milliwatts.

Nordic Mobile Telephone (NMT)

There are two Nordic Mobile Telephone (NMT) systems; NMT 450 which is a low capacity system and NMT 900 which is a high capacity system. The NMT 450 system uses a lower frequency and higher maximum transmitter power level which allows a larger cell site coverage areas while the NMT 900 system uses a higher frequency and a lower maximum transmitter power which increases system capacity. NMT 450 and NMT 900 systems can co-exist which permits them to use the same switching center [19]. This allows some NMT service providers to start offering service with an NMT 450 system and progress up to a NMT 900 system when the need arises.

History

The Nordic mobile telephone (NMT) system was developed by the telecommunications administrations of Sweden, Norway, Finland, and Denmark to create a compatible mobile telephone system in the

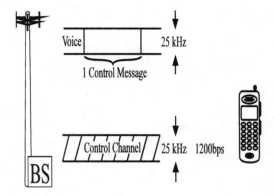

Figure 2.27, NMT Radio Channels

Nordic countries [20]. The first commercial NMT 450 cellular system was available at the end of 1981. Due to the rapid success of the initial NMT 450 system and limited capacity of the original system design, the NMT 900 system version was introduced in 1986. There are now over 40 countries that have NMT service available. Some of these countries use different or reduced frequency bands.

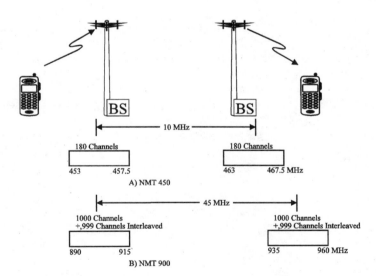

Figure 2.28, NMT Frequency Allocation

System Operation

Some operations of the NMT systems are very different than most other cellular systems. When NMT mobile telephones access the cellular system, they can either find an unused voice channel and negotiate access directly and begin conversation without the assistance of a dedicated control channel. Because scanning for free voice channels can be very time consuming, the NMT 900 system does allow for the use of a dedicated control channel called the calling channel. The NMT 900 system also allows discontinuous reception which increases the standby time of the portable phones. Figure 2.27 shows the different type of radio channels that are used in the NMT systems.

System Parameters

The NMT 450 system if frequency duplex with 180 channels (except Finland which only has 160 channels) [21]. The radio channel bandwidth is 25 kHz and the frequency duplex spacing is 10 MHz The NMT 900 system has 1000 channels for + 999 interleaved channels. Figure 2.28 shows the frequency allocation for the NMT systems.

Signaling on the NMT systems is performed at 1200 bp/s on the control (calling) channel (NMT 900) and voice channel. Because of the slow signaling rate and robust error detection/correction capability, no repeated messages are necessary.

NMT 450 base stations can transmit up to 50W. This high power combined with the lower 450 MHz frequency allows cell site size of up to approximately 40 km radius. NMT 900 base stations are limited to a maximum of 25W which has a maximum cell size radius of up to approximately 20 km [22].

There are three power levels (high, medium, and low) for NMT mobile phones and two power levels (high and low) for portables. NMT 450 mobile telephone power levels are: High 15W, Medium 1.5W, Low 0.15W. NMT 450 portable telephones; High 1.0W, Low 0.1W. NMT 900 mobile telephones: High 6.0W, Medium 1.0W, Low 0.1W and NMT 900 portable telephones: High 1.0W, Low 0.1W.

Four different pilot signals (similar to SAT tone) are assigned at 3955, 3985, 4015, and 4045 Hz.

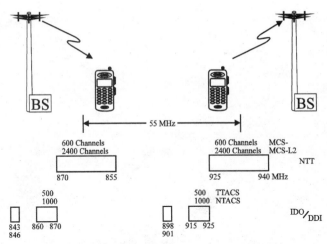

Figure 2.29, Japan Cellular System Frequency Allocation

The NMT system has various types of anti-fraud protection. NMT mobile telephone's hold a three digit password which is stored in the telephone and cellular switching center and is unknown to the customer. This password is sent to the cellular system during system access along with the mobile telephone number. The NMT system has also added a Subscriber Identity Security (SIS) system that provides additional anti-fraud protection. Not all NMT telephones have SIS capability.

Other Cellular Systems

Many of the existing cellular systems have been modified or adapted to the rest of the world. These changes may be frequency or different control messaging processes. Because there is significant savings when manufacturing production quantities are high, the trend is to use existing technology.

Japanese Cellular Systems

Japan had the first commercial cellular system in 1979. Because this system had achieved great success, several different types of cellular systems have evolved in Japan. These include the MCS-L1, MCS-L2, JTACS, and NTACS systems. Figure 2.29 shows the frequency bands allocated for the Japanese cellular systems.

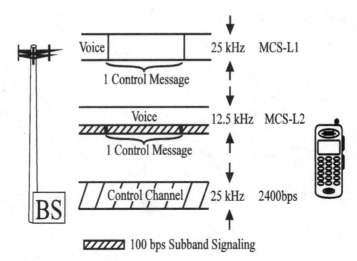

Figure 2.30, MCS-L2 Radio Channels

The MCS-L1 was the first cellular system in Japan which was developed and is operated by NTT. The system operates in the 800 MHz band. The channel bandwidth is 25 kHz and the signaling is at 300 bp/s. The control channels are simulcast from all base stations in the local area. This limits the maximum capacity of the MCS-L1 system.

Because the MCS-L1 system could serve a limited number of customers, the MCS-L2 system was developed. It uses the same frequency bands as the MCS-L1 system. The radio channel bandwidth was reduced from 25 kHz to 12.5 kHz with 6.25 kHz interleaving. This gives the MCS-L2 system 2,400 channels. The control channels transfer information at 2,400 bp/s and the voice channels can use either in-band (blank and burst) signaling at 2,400 bp/s or sub-band digital audio signaling at 150 bp/s. MCS-L2 mobile telephones have diversity reception (similar to diversity receive used in base stations). While this increases the cost and size of the mobile telephones, it increases the performance and range of the cellular system.

The competing cellular service providers in Japan use the JTACS and NTACS systems. The Japan TACS system is a modified TACS system from the UK. The only significant changes were the frequency bands and number of channels. The Narrowband TACS (NTACS) system reduced the channel bandwidth from 25 kHz to 12.5 kHz and changed the in-band 8 kbp/s signaling on the voice channel to 100 bp/s subband digital signaling. Figure 2.29 shows the radio channel types for the MCS-L1 and MCS-L2 systems in Japan.

Narrowband AMPS (NAMPS)

Narrowband Advanced Mobile Phone Service (NAMPS) is an analog cellular system which was commercially introduced by Motorola in late 1991 and is currently being deployed worldwide. Like the existing AMPS technology, NAMPS uses analog FM radio for voice transmissions. The distinguishing feature of NAMPS is its use of a "narrow" 10 kHz bandwidth for radio channels, a third of the size of AMPS channels. Because more of these narrower radio channels can be installed in each cell site, NAMPS systems can serve more subscribers than AMPS systems without adding new cell sites. NAMPS also shifts some control commands to the sub-audible frequency range to facilitate simultaneous voice and data transmissions.

In 1991, the first NAMPS standard, named IS-88, evolved from the US AMPS specification (EIA-553). The IS-88 standard identified parameters needed to begin designing NAMPS radios, such as radio channel bandwidth, type of modulation, and message format. During development, the NAMPS specification benefited from the narrowband JTACS radio system specifications. During the following years, advanced features such as ESN authentication, caller ID, and short messaging were added to the NAMPS specification.

CNET

The CNET system is a third generation German mobile telephone system which is used in Germany, Portugal, and South Africa [23]. The first CNET system started operation in 1985. The primary objective of the CNET system was to bridge the gap of cellular systems in Germany until to the digital European system could be introduced [24].

The CNET system Operates at 450 MHz with 4.44 MHz transmit and receive bands. The Frequency bands are 461.3 to 465.74 MHz and 451.3 to 455.74. The primary channel bandwidth is 20 kHz with 10 kHz channel interleaving.

The CNET system continuously exchanges digital information between the mobile telephone and the base station. Every 12.5 msec, 4 bits of information are sent during compressed speech periods [25]. CNET mobile telephones also use an Identification Card (IC) which slides into the telephone to identify the customer. This allows customers to use any compatible CNET telephone.

MATS-E

The MATS-E system is used in France and Kuwait [26]. The MATS-E system combines many of the features used in different cellular systems. MATS-E uses the standard European mobile telephone frequency bands; 890-915 MHz and 935-960 MHz. The channel bandwidth is 25 kHz which provides 1,000 channels. The MATS-E is a frequency duplex system separated by 45 MHz. Each cell site has at least one dedicated control channel with a signaling rate of 2400 bp/s. Voice channels use FM modulation with sub-band digital audio signaling with a data rate of 150 bp/s.

References:

1. Balston, D.M., "Cellular Radio Systems," Artech House, MA, 1993, p. 135.
2. Mehrotra, Asha, "Cellular Radio, Analog and Digital Systems," Artech House, MA, 1994, p. 177.
3. Balston, D.M., "Cellular Radio Systems," Artech House, MA, 1993, p. 113.
4. SuperPhone conference, Institute for International Research, New York NY, January 22-24, 1997.
5. Ibid.
6. Lee, William, "Mobile Cellular Telecommunications Systems," McGraw Hill, NY, 1989, p.2.
7. Harte, Lawrence, "Dual Mode Cellular, P.T. Steiner Publishing, PA, 1991, p. 7-3.
8. FCC Regulations, Part 22, Subpart K, "Domestic Public Cellular Radio Telecommunications Service," 22.903, (June 1981).
9. Balston, D.M., "Cellular Radio Systems," Artech House, MA, 1993, p. 89.
10. The Bell System Technical Journal, Vol. 58, No. 1, American Telephone and Telegraph Company, Murray Hill, New Jersey, January, 1979.
11. Ibid, p. 47.
12. Bohaychuk, Ron, personal interview, Ericsson Radio Systems, 7 October, 1990.
13. The Bell System Technical Journal, Vol. 58, No. 1, American Telephone and Telegraph Company, Murray Hill, New Jersey, January, 1979, p.47.
14. Bellcore, "LSSGR; Signaling, Section 6," TR-TSY-000506, Rev 1, December, 1988.
15. Harte, Lawrence, "Cellular and PCS/PCN Telephones and Systems," APDG Publishing, NC, 1996, pp. 377-381.
16. Calhoun, George, "Digital Cellular Radio," Artech House, MA. 1988, pp. 50-51.
17. Lee, William, "Mobile Cellular Telecommunications Systems," McGraw Hill, NY, 1989, p. 5.
18. Ibid, p. 265.
19. Balston, D.M., "Cellular Radio Systems," Artech House, MA, 1993, p. 74.
20. Mehrotra, Asha, "Cellular Radio, Analog and Digital Systems," Artech House, MA, 1994, p. 177.
21. Ibid, p. 177.
22. Balston, D.M., "Cellular Radio Systems," Artech House, MA, 1993, p. 77.
23. Mehrotra, Asha, "Cellular Radio, Analog and Digital Systems," Artech House, MA, 1994, p. 193.
24. Ibid, p. 186.
25. Ibid, p. 188.
26. Ibid, p. 193.

Chapter 3
Digital Cellular and PCS

Digital Cellular and PCS Systems

This chapter presents a brief history of the technologies, reviews the requirements for advanced cellular services, and explains the planning and development phases of the technology. Following this overview, each new technology will be introduced. Frequency Division Multiple Access (FDMA), Time Division Multiple Access (TDMA), and Code Division Multiple Access (CDMA) systems are briefly discussed and explained. Semi-technical descriptions of advanced digital services common to all digital technologies are provided which will summarize potential new services and how they may be implemented.

History

Shortly after analog cellular systems were introduced, it became apparent that robust, high capacity, and lower cost wireless systems were needed to better service customers. In the early to mid 1980's, several industry associations, such as the Cellular Telecommunications Industry Association (CTIA) and the Conference of European Posts and Telecommunications (CEPT), developed requirements for the next generation of cellular technology. Manufacturers and cellular service providers then became partners working together to create the industry specifications that would fulfill these new requirements.

Requirements

As the customer base continues to expand in a cellular system, the system must then expand to accommodate the growth. If cellular systems can expand by adding new cells, why change cellular technology? The answer is twofold. First, the new technologies may provide less costly ways to expand than the existing analog cellular systems and second, the analog systems lack the capability to support many of the new advanced services.

CTIA Requirements

In 1988, the Cellular Technology Industry Association (CTIA) Advanced Radio Technology Subcommittee (ARTS) was established to identify technology requirements that would lead to the timely introduction of cost-effective technologies and new features. They created a User Performance Requirements (UPR) document which provided the goals for the new technology [1]. The UPR document assembled inputs from a Booz-Allen marketing study, assessments of cellular carrier needs, and consultations with manufacturers. Cellular service providers indicated that they wanted a new technology product life cycle of eight to ten years. The manufacturing industry then proceeded to work with CTIA to define a series of specific milestones to be achieved and to introduce new technology products by 1991. The result was the product development timeline contained in the UPR. The UPR document did not focus on any one technology, but specified the following customer service requirements for all technologies:

* a tenfold increase in system capacity compared to AMPS
* dual mode during transition
* provide for new features (for example, short message services)
* ensure that equipment would be available by 1991
* set standards for high quality service

The Telecommunications Industry Association (TIA) was asked to create a specification based on the UPR requirements. As a result, Interim Standard 54 (IS-54) which uses a Time Division Multiple Access (TDMA) digital technology was released in early 1991 [2]. As a result the TDMA equipment was demonstrated and tested in mid-1991 in Dallas and Sweden.

However, the TDMA IS-54 standard did not meet all UPR goals, and soon after IS-54 was created, several companies proposed new technologies to meet industry needs. As a result, during the past several years, many new industry standards have been accepted. These standards include IS-136 for a new generation of TDMA [3], IS-95 for Code Division Multiple Access (CDMA) [4] and IS-88 for Narrowband AMPS (NAMPS) [5]. Each of these has inherent advantages over AMPS technology, and each could result in savings for the industry.

Regardless of the technology used, the FCC requires all cellular systems to maintain AMPS service [6]. Because AMPS service will remain universally available, the CTIA desired that the next-generation of cellular radios operate with both the new technology and the existing AMPS system, automatically defaulting to AMPS mode where new technology is currently unavailable. This dual capability, called "dual-mode," has become the standard, although it is not required by the FCC. Dedicated single mode phones may be smaller and cheaper, but they lack ROAMING capability in systems without digital service.

Figure 3.1 illustrates how a dual-mode cellular system provides both AMPS and at least one new digital cellular service. Dual-mode cellular systems consist of dual-mode mobile telephones, base stations, and a Mobile Switching Center (MSC). Dual-mode mobile telephones have both analog and digital capability. When digital channels are available, the cellular system prefers to assign the mobile telephone to a digital channel. However, if no digital channel is available, such as when the mobile telephone is in a system that does not have digital capability, an analog channel will be assigned.

Dual-mode cell site base stations must also have both analog and digital channels. In addition to the new digital radio channels, dual-mode base stations require modifications to their scanning receiver, communications interface, and RF amplifiers.

The Mobile service Switching Center (MSC) performs the same function as the Mobile Telephone Switching Office (MTSO). Modifications to the MTSO to upgrade it to digital service capability include software, echo cancel hardware, and other communications interface changes.

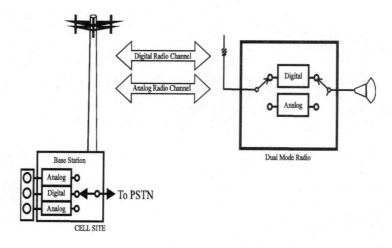

Figure 3.1, Dual Mode Cellular System

CEPT Requirements

In the 1980's, the early success of European cellular systems demonstrated the need for a updated version of the cellular system. Unlike the United States, which enjoyed a standard AMPS cellular system, countries in Europe had created a variety of unique systems such as TACS, NMT-450, NMT-900, CNet and MATS-E. As a result, a new system was required that could meet the long term system service and cost objectives for European cellular service providers. In 1982, the Conference of European Posts and Telecommunications (CEPT) held a meeting to begin the standardization process for a single next generation European cellular system. This meeting established the Groupe Special Mobile (GSM) standards body. The GSM technology developed by this group is now referred to as Global System for Mobile communication (GSM)

During 1985, the Consultative Committee of International Telegraph and Telephones (CCITT) created a list of technical recommendations that would be used to create the specifications for an updated cellular system. From these requirements, an action plan was created to coordinate the creation of the GSM specifications.

Increased Capacity

As a cellular service provider of an analog cellular systems continually adds new customers, more radio channels will be needed. If the cell sites are not filled to capacity (approximately 50) with radio channels, the cellular service provider simply adds more radio channels. If cell sites are filled with radio channels, more cell sites must be added to accommodate the increased capacity. Either way, to expand by using analog cellular technology, cellular service providers must add radio channels.

Digital cellular technologies will allow for capacity increases in a different way: by allowing more subscribers to share the same radio channel spectrum. The intensified use of radio spectrum is accomplished either by allowing more subscribers to share the same radio channel, or by packing more radio channels into a single cell site. To simultaneously serve multiple subscribers on the same radio channel, new technologies either assign time slots or unique codes to separate the calls. Ultimately, all of these techniques reduce the amount of radio spectrum needed, and allow more subscribers to use a cellular service area. In this way, the new technologies should also reduce the average system equipment cost per customer.

To upgrade a cellular system for digital service, digital radio channels replace the analog radio channels or are added to the equipment racks at the base station. Digital cellular system manufacturers typically offer digital channels that simply replace an analog radio channel. Some manufacturers offer base transceiver units which can operate in either the analog or the digital mode. This permits a more flexible capability to deal with different analog and digital traffic loads at different times or days.

Cellular service providers can evaluate the potential system capacity factors of the new cellular technologies by reviewing several types of efficiency: radio channel efficiency and infrastructure efficiency. Radio channel, or spectral, efficiency is measured by the number of conversations (voice paths) that can be assigned per frequency bandwidth. Geographic spectral efficiency is measured by the number of conversations per frequency bandwidth per unit of service area. Infrastructure efficiency is measured by the cellular system equipment and operating costs, calculated on a per-subscriber basis, or per available channel per unit of service area. Chapter 11 discusses the economic impact of these factors.

Figure 3.2 illustrates how cellular systems allow users to share a cell site. The number of subscribers who can share a cell site is much greater than the number of available radio channels but, because not everyone places calls at exactly the same time (except during traffic jams), many users can share a single radio channel. For analog systems that can allow only one conversation per radio channel, a typical analog system may add 20-32 customers to the system for each available voice/radio channel [7]. If an analog cell site has 50 radio channels installed, this has enough capacity to serve about 1000 customers. The new digital technologies claim to multiply this number by 3-20 times. If all of an analog cellular system cell site's available channels were converted to digital, it could serve 3,000 to 20,000 customers.

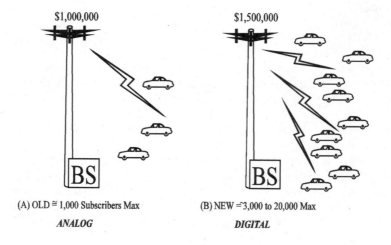

Figure 3.2, Serving More Users

New Features

All of the proposed new technologies can provide similar features (for example, calling number identification), although all of the possible features may not be included in current or proposed industry standards for each technology. The new digital technologies also allow for some advanced features not possible with analog technology, such as simultaneous voice and data transmissions. The following new features will result in a larger revenue potential. The new features include:

* calling number identification (similar to paging)
* short message transmission (similar to dispatch or alphanumeric paging)
* voice-activated control
* priority access (for emergency services)
* extension phone service
* lighter and more portable units
* enhanced voice privacy
* vehicle location
* ESN security (fraud protection)
* imaging (video) service.

Radio Technology Basics

There are three basic types of cellular technology: Frequency Division Multiple Access (FDMA), Time Division Multiple Access (TDMA), and Code Division Multiple Access (CDMA). Digital cellular systems fall into these categories and many systems use a combination of these technologies. There are also variations in the way radio technologies allow duplex operation, called Frequency Division Duplex (FDD),and Time Division Duplex (TDD).

Frequency Division Multiple Access (FDMA)

FDMA systems allow for a single mobile telephone to call on a radio channel. Typically FDMA systems use analog FM radio modulation but occasionally will use digital phase modulation (such as CT-2). Figure 3.3 illustrates an FDMA system. FDMA systems typically have a control channel which coordinates radio channel assignment to a voice channel. After the mobile telephone coordinates its access on the control channel, the cellular system assigns it to a voice channel, however;. each voice channel can communicate with only one mobile telephone at a time.

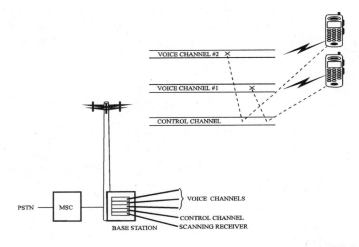

Figure 3.3, FDMA Cellular System

While FDMA systems do not allow more mobile telephones to share a single radio channel, it is possible to increase the number of channels that can be placed in a cell site by reducing the radio channel bandwidth. One characteristic of FM radio signals is the degree to which they are allowed to deviate or shift the radio frequency as they transmit information. The more FM signals that are allowed to deviate, the wider the radio channel and the better they resist noise, fading, and interference. The modulation allowed for very narrow radio channels is more susceptible to co-channel interference. However, there are techniques that are used to allow analog cellular systems to improve their quality while reducing the radio channel bandwidth. By using radio signal interference reporting by the mobile telephone, interfering signals will be detected early and the radio channel frequency changed to another radio channel which has less interference. This compensates for the decreased deviation , and a report from Motorola states that overall, subscribers which use mobile reported interference NAMPS systems experience less co-channel interference than subscribers on wider radio channel systems [8].

FDMA mobile telephones typically cost less than digital mobile telephones because they are relatively simple to design and as a result large quantities are being produced. However, with the continued integration of digital circuits, the cost of FDMA mobile telephones may eventually be equal to or even more expensive than digital mobile telephones.

All of the other technologies described below combine FDMA with another technology, since all of them use assigned carrier frequencies for each signal. However, it is customary in the industry to describe the others only in terms of the distinctive multiplexing or coding technology alone. The term FDMA is used specifically for a system in which there is only one voice conversation per radio carrier frequency, normally using a very narrow bandwidth.

Time Division Multiple Access (TDMA)

TDMA systems allow several mobile telephones to communicate simultaneously on a single radio carrier frequency. These mobile telephones share the radio frequency by dividing their signal into time slots. Time slots can then be dedicated or dynamically assigned.

TDMA systems divide the radio spectrum into radio carrier frequencies typically spaced 30 kHz to 200 kHz apart. This spacing between the carrier frequencies is the nominal or effective bandwidth of the total multi-channel multiplexed signal. TDMA systems typically have narrowband radio voice or traffic channels but can have wideband radio signals. The DECT system (discussed in chapter 7) is a wideband TDMA system with 1.7 MHz wide radio signals. In an FDMA system, the bandwidth of a radio signal waveform is the same thing as the bandwidth of a radio channel, since there is only one channel using the entire signal waveform. In TDMA and CDMA systems, a channel is not the same thing as an entire signal waveform, but is instead only a part of it. Many documents which refer only to analog or FDMA systems (and unfortunately also for some TDMA or CDMA systems as well) use the word "channel" without distinguishing between the two, and the reader must know enough about the subject to understand which meaning of the word "channel" is implied by the author.

The distinguishing feature of TDMA systems is that they employ digital techniques at the base station and in the cellular radio to subdivide the time on each channel into time slots. Each time slot can be assigned to a different mobile telephone. Voice sounds and access information are converted to digital information which is sent and received in bursts during the time slots. The bursts of digital information can be encoded, transmitted, and decoded in a fraction of the time required to produce the sound. The result is that only a fraction of the air time is used by one channel, and other subscribers can use the remaining time on the radio channel.

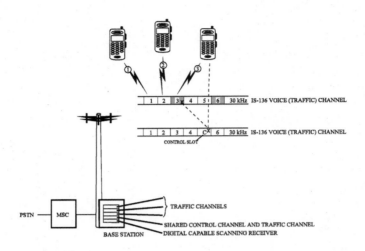

Figure 3.4, TDMA Cellular System

Most TDMA mobile telephones accesses the system through a dedicated control channel slot. This digital control channel (DCC) has the capability for many new features. Some TDMA systems (such as the IS-54 system) can use the existing analog control channel to assign a mobile telephone to a digital traffic channel. Figure 3.4 illustrates a TDMA cellular system.

TDMA mobile telephones are much more complex than FDMA mobile telephones. When implemented in a digital signal processing system (DSP), a typical analog mobile telephone has signal processing capability for .5 to 1 million instructions per second (MIPS) while TDMA mobile telephones have over 50 MIPS processing [9]. This added complexity, limited demand, and lower production numbers of digital phones have resulted in higher costs for TDMA mobile telephones. However, TDMA systems allow more customers to simultaneously use radio channels in the cell site, reducing the total number of required cell sites and radio channels, and ultimately reducing the cellular system infrastructure cost.

Code Division Multiple Access (CDMA)

CDMA technology differs from TDMA technology in that it divides the radio spectrum into wideband digital radio signals with each signal waveform carrying several different coded channels. Each coded channel is identified by a unique Pseudo-random Noise (PN) code. Digital receivers separate the channels by correlating (matching) signals with the proper PN sequence and enhancing the correlated one without enhancing the others. The CDMA RF signal waveform uses some of its coded channels as control channels. The control channels include a pilot, synchronization, paging, and access channel.

Figure 3.5 illustrates a CDMA system, and reveals several changes from an analog system. The base station uses a wide CDMA RF signal waveform which provides for many different coded channels. Some of these coded channels are used for control and access coordination and others are used for voice communications. The CDMA mobile telephone accesses the system either through an analog control channel or coded channel on the CDMA RF signal waveform.

When the mobile telephone obtains access on the CDMA system, the CDMA control channel responds by assigning the CDMA mobile telephone to a new coded channel. This is typically on the same RF carrier frequency. Qualcomm, the company which invented cellular CDMA, plans to use the same RF carrier frequency in all cells of the system. In a CDMA system of this type, because neighboring cell sites

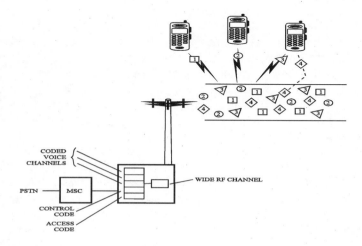

Figure 3.5, CDMA Cellular System

use the exact same frequency, CDMA mobile telephones can simultaneously connect two (or more) cell sites by using the same coded channel information transmitted by each of the base transmitters, and using the same decoding process at all the base receivers for the mobile set's transmitted signal. This allows a process called soft hand-off.

CDMA mobile telephones are similar to TDMA phones in complexity. The CDMA system uses similar digital voice compression, but the channel coding process can easily apply variable rate speech compression. Variable rate speech compression increases the amount of compression as the speech activity decreases. This decreases the average number of bits representing the voice signal so that more users share a radio channel.

Duplex Operation

To allow apparently simultaneous transmitting and receiving (no need to push to talk), a mobile telephone uses Frequency Division Duplex or Time Division Duplex (TDD) systems.

FDD systems allow a transmitter and receiver to work simultaneously at different frequencies. The FDD systems must separate the transmitter energy from overpowering the receiver. This is made possible by either using 2 antennas separated by a distance and/or having a filter connected to the receiver and transmitter to block the transmitter frequencies from being connected to the receiver.

Time Division Duplex (TDD) uses separate transmit and receive time slots that do not overlap. When the transmitter is operating, the receiver is off. TDD systems may be designed to use the same radio frequency in both directions (such as the CT-2 system) This simplifies the design and reduces the number of components in the radio transceiver. It also permits the base unit to use a sophisticated and complex adaptive equalizer to optimize performance of radio transmission in both directions, without the need to put a similar capability into the mobile set. The disadvantage of using the same frequency for transmit and receive is the coordination of time slot periods. If the TDD system allows the mobile telephone to transmit directly after it has received a time slot, the time delay could cause an overlap with other time slots. The time delay for short range systems (such as cordless telephones) is not a challenge. However, in systems with large cell sizes (for example, above 2 to 3 km), the time delay could cause slots to interfere with each other. One way to address this problem is to design the mobile set to have an adjustable transmit burst timing advance. Short range TDD systems do not normally have this capability.

Most of the long-range TDMA cellular technologies combine FDD and TDD operation. This keeps a similar frequency allocation which is used for analog cellular systems. Figure 3.6 shows the different types of duplex systems.

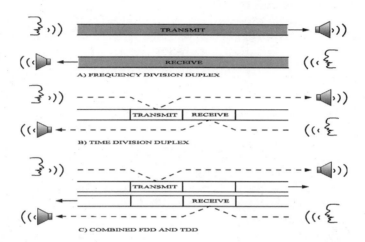

Figure 3.6, Duplex Systems

Digital Services

Customers do not desire to purchase mobile telephones or data devices instead they purchase the desired benefits and services these devices should provide. Digital technologies provide the potential for more benefits and services, and in so doing, will likely capture more and more of the market over time. There are digital services common to all digital technologies. While some digital services may have the same features as analog cellular features, its method of operation can be very different.

Digital cellular systems are generally newer than their analog predecessors, and have benefited from the recent emphasis on user features and services. The new digital technologies are inherently better at providing advanced services because all the different types of information which flows through a digital system is coded as digital data. Digital data transmissions merge voice, data, and video signals in one common communications channel, enabling messaging, private services, and improvements in performance. Through a combination of digital transmission and new control messages which have been added to the design specifications for the radio communication channel, digital systems can offer advanced applications that target both human and non-human services. Furthermore, digital services can be implemented on most of the older technologies, including analog cellular systems such as AMPS, TACS, or NMT (nearly all cellular data transmissions today are over analog connections). Although purely analog based systems offer data transmission and some advanced features such as short message services, the future of data transmission at higher rates is through digital systems. This chapter describes the benefits and services of the new digital cellular systems.

Multi-media is the mixing of voice, data, and video services. Digital information is a universal medium capable of transporting digital voice, data files, and digitized video. The new digital cellular technologies will therefore add multi-media capabilities to wireless systems which today primarily deliver real-time voice information.

Ideally, the digital communication channel transfers information unchanged between originator and receiver. This idea would allow transmission of any combination of digitized voice, data files, and digital video, provided the total amount of information did not exceed the digital channel capacity. Unfortunately, connecting digital radio with the public switched telephone network (PSTN) limits digital transmission capacity, and restricts some of the services that could be offered.

The cellular network today primarily transports voice information from one caller to another. Because information is only transported by the network, this is called a bearer service. The cellular telephone network is the bearer of voice information.

Advanced services can be offered using the cellular system as a bearer service. If a subscriber transfers a fax message on a mobile telephone, the cellular system is only transporting the fax information through a standard communications channel. If a caller number identification is transported via special tones without changes to the cellular system, this would also be a bearer service.

Some of the significant changes in the digital cellular systems include tele-services. Tele-services process information inside the network. For example, if the caller number identification is received by the cellular system from the landline telephone company, and the cellular system converts this to messages which contain the digits and adds the name of the caller, the cellular system provides a tele-service.

The new digital cellular systems also have new control messages that allow the system to provide new features. These new control messages can be sent while the mobile telephone is idle (in standby) or while a conversation is in progress.

Caller Identification

Caller identification on a cellular system works much as it does with landline telephone caller ID systems, by providing the user a description of the calling party prior to answering a call. This identification can be either a phone number, a name, or both. The identifying information appears on the user's phone display during the ring sequence. The subscriber then has the option of answering the call or not.

A caller ID phone number may be transferred in two different ways from a caller to a mobile telephone. Most common is for the phone company to send the calling number to the cellular system. The alternative is for the calling person to enter the digits from a touch tone phone, similar to the method used for sending phone numbers to numeric pagers.

Figure 3.7 shows the sequence of events that occur when caller ID data is sent. In step 1, when the mobile telephone has been dialed, the caller ID data is forwarded from the telephone company to the MSC. If the connection between the MSC and the PSTN uses traditional PBX signaling, this information is conveyed by means of a burst of modem information between the first two ringing cycles. If the connection uses the more sophisticated PRI ISDN signaling, the information is conveyed via a digital packet message associated with the call setup message that initiates the call attempt. Next, the cellular system decodes the calling number ID, changes its format, and sends it as a message to the mobile telephone via the base station. The message containing the caller number identification arrives after the mobile telephone is alerted that a call is incoming (paging). The mobile telephone decodes the message and the phone number appears on the display. As an option, the mobile telephone may examine its internal phone book data to see if the phone number matches any previously stored number. If a match is found, the text string (normally a name) can be displayed along with the incoming call phone number.

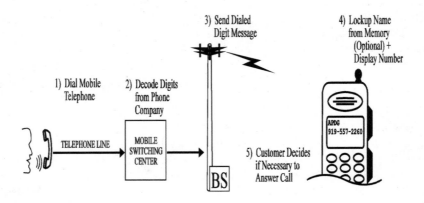

Figure 3.7, Caller Identification

Privacy concerns have led most systems to provide a method of allowing users to block the transmission of their phone number when placing a call. When the caller number identification has been blocked by the calling person, the caller ID information received by the mobile telephone is marked as private. A message such as PRIVATE or BLOCKED, ANONYMOUS or UNAVAILABLE is displayed to indicate this to the user.

Short Message Services (SMS)

Short message service (SMS) gives cellular subscribers the ability to send and receive text messages. Short messages usually contain about one page of text, or approximately two thousand bytes of information or less. Some systems limit the short message to 160 alphanumeric characters They can be received while the mobile telephone is in standby (idle), or in use (conversation). While the mobile telephone is communicating both voice and message information, short message transfer takes slightly longer than it does while the mobile telephone is in standby.

SMS can be divided into three general categories: Point-to-point, Point-to-multi-point, and broadcast. Point-to-point SMS sends a message to a single receiver. Point to multi-point SMS sends a message to several receivers. Broadcast SMS sends the same message to all receivers in a given area. Broadcast SMS differs from point to multi-point because it places a unique "address" with the message to be received. Only mobile telephones capable of decoding that address receive the message.

Much like voice mail or beeper type systems, point to point SMS sends a short message from one source to one receiver. An example of this type of message would be "You Have Won the Lottery" or "Change time of our appointment to 10:30."

Point to Point Messaging

Much like voice mail or beeper type systems, a point to point message system sends a short message from one source to one receiver.

Figure 3.8 illustrates point to point SMS transfer. Initially, the message goes to the Mobile Switching Center (step 1) to be routed and stored in the message center (step 2). The cellular system searches for the mobile telephone (step 3) and alerts the mobile telephone that a message is coming. The mobile telephone tunes to the voice or control channel where the message will be sent. The system then attempts to send the message (step 4). As the message is being sent, the system waits for acknowledgment messages (step 5) to confirm accurate deliv-

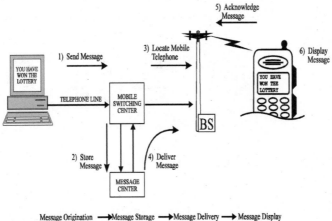

Figure 3.8, Point to Point Short Message Delivery

ery of each part of the message. If the transmission is successful, the message may be removed from MSC message center. If unsuccessful, the system attempts delivery again.

Messages can also be sent from a mobile station to the message center for delivery, but since most mobile phones have very few keys, most messages will probably go from the wireless system to mobile stations. However, keyboards and telemetry monitoring equipment may be connected to mobile phones, allowing mobile telephones to transfer messages to the cellular system. Another alternative is to use predefined messages that can be selected from a list by the subscriber and transmitted.

Point-to-Multipoint Messaging

Point-to-multipoint messaging is a process of sending a message to a group of mobile stations. An example of Point-to-multipoint messaging might be a message to a corporate sales team indicating "Sales meeting on Tuesday is canceled."

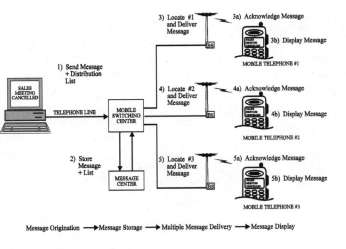

Figure 3.9, Point-to-Multipoint Short Message Delivery

Figure 3.9 illustrates how a message is transferred to a group of users. Like the point-to-point message service, the message first goes to the MSC message center (step 1) where the message center determines that this message is designated for multiple units. The designated list of message recipients may be pre-arranged (such as a sales staff) or it may be included with the originating message. In either case, the message center stores the message and recipient list (step 2). To complete message delivery. the MSC then searches for each mobile telephone in the list. The cellular system then individually alerts each mobile telephone that a message is coming (steps 3-5). The mobile telephones tune to the message channel (voice or control), the system sends the message to each unit, and the units receive and store it. If the transmission is successful, the mobile telephone typically sends an acknowledgment, and the message may be removed from MSC message center. If the message transmission to one or more of the receivers was not successful, the cellular system attempts delivery again later.

Broadcast Messaging

Broadcast messaging sends messages to a all mobile telephones that have been pre-set to monitor a specific messaging channel. Every base station transmits each broadcast message along with enough information to accurately decode the message.

Figure 3.10 illustrates how broadcast SMS messages are delivered to users. Like the point-to-point SMS service, the message first goes to the MSC message center (step 1) where the message center determines that this message is designated for all mobile units with a

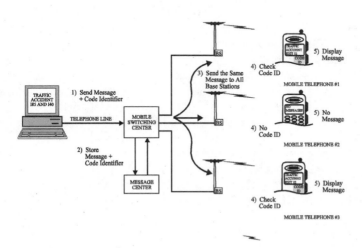

Figure 3.10, Broadcast Short Message Delivery

unique code. The delivery code may be pre-arranged (such as a traffic report code) or included with the originating message. The MSC then broadcasts on a designated message channel, which may be part of a control channel. in every cell site where the broadcast message is designated to be received.

Unlike point-to-point and point-to-multipoint messages, mobile telephones do not acknowledge receipt of broadcast SMS. If a mobile telephone is off or is not tuned to the message channel, it misses the broadcast message. To address this limitation, messages may be broadcast several times, and mobile telephones that have already received message may ignore repeats.

Private Systems

Both public and private systems are in use today. Public systems, including today's cellular systems, are available to anyone who pays a service fee to a public telephone company. Private systems are used by corporations or other private owners to interconnect employees or private subscribers without routing calls through a public telephone company. Private systems lease radio channels which are FCC licensed to a local public cellular provider. Figure 3.11 illustrates how private systems operate within public ones. To entice companies to set up a private system for employees to use at work and home, public providers may offer localized flat billing rates or other financially advantageous price plans for private systems.

Private system radio equipment may be located, for example, in a company building. Alternatively, small private systems can exist without the expense of a separate infrastructure. Several independent systems may exist within a single RF channel, so that a new system can be created simply by identifying it to the phone users. In such cases, multiple systems have a separate billing rate as a key part of their functions. For example, a business with several buildings near each other could have its own cellular system. Employees using cellular phones at work would be billed at a different rate than if they used the same phones in the public system. The private system exists at the same time and by means of the same equipment as the public system. The distinction between the two systems is provided by the different call processing messages supported in the network and the different price and billing arrangements.

New digital technologies enable typical private systems to offer advanced services such as localized billing rates, telephone directory listings, three or four digit dialing, and other features. Private mobile telephones may also be designed to access company databases for directory listings and other services via a digital communication channel.

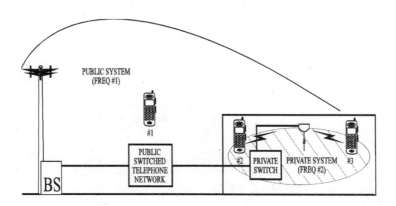

Figure 3.11 Private Cellular System

Data Services

Sending and receiving FAX data over a cellular link is an important key to a truly mobile office. Unfortunately, wireless FAX data has problems similar to those of cellular modem transmissions. However, while modem algorithms have been adapted to work well with sometimes poor cellular connections, FAX algorithms have not. Most FAX transmission schemes do not recover or correct errors well. Depending on the error type and location, some or all of a wireless FAX page can be lost, or the FAX page may defaced with black or white streaks as a result of data transmission errors. With newer digital data transmission protocols, this need not be the case. Digital data transmission protocols correct errors through an encoding scheme and through retransmission of missing or incorrect data. Assuming no dropped calls, wireless FAX transmissions could be as reliable as on a wire line.

Figure 3.12 illustrates digital FAX transmission from a landline fax machine to a mobile telephone capable of receiving a FAX. The landline FAX machine dials the mobile telephone's FAX telephone number or a universal number that allows FAX message detection (step 1). The cellular system converts the FAX modem signal into a digital signal (step 2). Next, the mobile telephone is alerted that a FAX is to be received (step 3). The mobile telephone receives the digital FAX. During the digital FAX message delivery, each bit group of digital information is checked by means of some extra error detection bits which are used only over the radio link, and re-sent if it is received in error. If the mobile telephone is connected to an analog FAX (standard FAX machine), a special FAX adapter in the mobile telephone will convert the digital FAX signal back into an analog FAX signal (step 4).

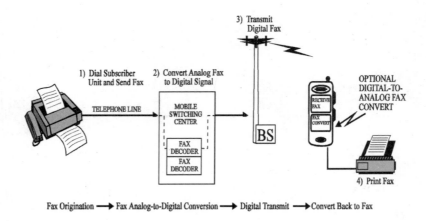

Fax Origination ⟶ Fax Analog-to-Digital Conversion ⟶ Digital Transmit ⟶ Convert Back to Fax

Figure 3.12, Fax Over Digital Cellular

Voice Privacy

Digital systems are inherently more secure than analog systems because they can easily use an encrypted mode of operation. This encrypted mode of operation "scrambles" voice data before it is sent to the base station. The encryption uses a mask value calculated from some of the authentication data. When the voice data is received at the base station it is decrypted using the same mask value that was used to encrypt it. The use of a mask in this way is similar to adding a pre-arranged secret number to the data value before transmitting it, and then subtracting the same secret number at the receiver end of the radio link. The result is the correct data value, but an interceptor cannot learn the true data value unless the secret number which was added to it is also known. Figure 3.13 shows how voice privacy mode can be requested by a subscriber.

In addition to encrypting the voice data, most digital systems have a message encryption mode that encrypts the signaling between the mobile phone and the base station. While this is not a direct benefit to the user of the phone, it does help to prevent unauthorized access to the cellular system.

Extended Battery Recharge Life

The battery recharge life of a portable mobile telephone is the amount of time a fully-charged battery can operate the unit in use before it must be recharged. The simplest way to increase battery recharge life is to use a larger battery, but this is often impractical. Battery recharge life affects two important elements of phone use: the length of time that a mobile telephone can be in receive-only mode waiting for messages to be sent, and the length of time that a mobile telephone

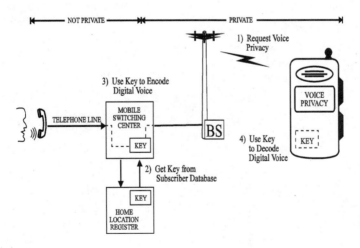

Figure 3.13, Digital Voice Privacy

can continuously transmit on a radio channel. Phone manufacturers spend considerable effort to maximize battery recharge life.

The length of time that a mobile telephone can be in receive-only mode (waiting for messages such as a page to be sent) is called standby time. The best way to increase standby time is for the phone to "sleep" between paging messages. When a mobile set is sleeping, all the electronic portions of the set are powered down except an electronic clock which controls the sleep-awake cycle. The initial digital specifications strongly emphasized maximizing sleep times. These standards create pre-scheduled sleep periods using paging frame class groups which allow a mobile telephone to sleep while pages not designated for that phone are broadcast. During the times when pages for that phone are broadcast, the phone is "awake." Figure 3.14 shows how a mobile telephone sleeps between groups of pages. Typical delays vary from seconds to a minute or two. When a mobile telephone operates as a portable telephone, the maximum delay is typically 2-4 seconds. A longer sleep period might cause callers to hang up before the mobile telephone woke up to answer. Much longer delays are acceptable when a mobile telephone is used as an alpha-numeric pager or other message device. The longer the sleep period, the longer the standby time, and the longer the battery recharge life.

The length of time a mobile telephone can continuously transmit on a single battery charge is called talk time. Digital systems' talk time is

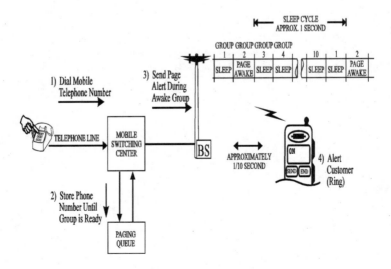

Figure 3.14, Enhanced Standby Time

inherently longer than that of analog systems because digital systems transmit short bursts of data and are idle between bursts, whereas analog systems transmit (and use power) continuously. IS-136 TDMA transmitters are only on 1/3rd of the time, and GSM transmitters operate only 1/8th of the time. IS-95 CDMA mobile telephones also have a lower average transmit power although they transmit continuously while the user is speaking. Figure 3.15 shows why digital cellular offers lower average RF transmit power and longer talk time.

Most of the digital specifications support some type of discontinuous transmission protocol (DTx or VOX). VOX protocols reduce the mobile telephone's transmit power when no voice information is being sent. Since many conversations involve listening more than half the time, VOX protocols greatly increase talk time.

Unfortunately, VOX algorithms detect voice activity much like a speaker phone, and they can cut off the first syllable of speech (an effect called "clipping"), which is unacceptable for some users.

Newer digital specifications have added new lower power classes that allow the mobile telephone to transmit at lower average power levels when it is very near a cell site, resulting in longer talk time. The lower output power classes are needed in systems with microcells.

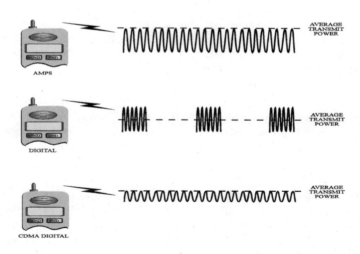

Figure 3.15, Digital Cellular Offers Lower Average Transmit Power

Total battery life (as opposed to recharge life) is measured by the number of times the battery can be successfully recharged before it must be discarded. Each recharge cycle causes chemical changes inside the battery which reduce its total energy capacity. Recent advances in battery design have led to rechargeable batteries which can be recharged effectively hundreds to even thousands to times, and thus will last for a year or more in a cellular or PCS application.

Anti-Fraud Control

Authentication is a way to prevent fraudulent access to the system by validating cellular users. All US digital standards have adopted the same basic authentication standard. The North American Systems (IS-54 TDMA, IS-91 AMPS, IS-136 TDMA, and IS-95 CDMA) use an authentication process based on the CAVE algorithm. The GSM system uses an authentication process based on the A3 algorithm. The North American and GSM authentication processes are very similar.

Figure 3.16 illustrates the authentication process for the North American systems. During setup, each cellular phone is issued a number called an A-KEY. The A-KEY is issued much as Swiss banks issue a secret account number to identify secretive clients. This A-KEY value is never disclosed to others. The subscriber enters it into the phone via the keypad. The phone uses the A-KEY to calculate and store a shared secret data (SSD) key. The network also performs the same calculations to create and store the SSD. During each call, the SSD key creates an authentication response code, and during access, the phone transmits only the authentication response code. The authentication response changes during each call because the system sends a random number which is also used to create the authentication response code value. A criminal who intercepts an authentication transaction over the air has no clue regarding the correct value of the SSD key, and cannot repeat the response given in one authentication transaction to try to fool the system in another authentication transaction.

Authentication is the process of validating a mobile subscriber to determine if it is fraudulent, and if so, to deny access to the cellular system. Authentication functions by transferring secret information between the mobile telephone and the system. The North American (NAMPS, TDMA, and CDMA) authentication processes differ slightly from the European (GSM) authentication process.

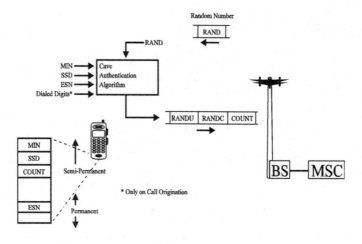

Figure 3.16. North American Authentication

A secret algorithm is at the heart of the authentication process. The algorithm defines a mathematical manipulation of data so that if two processors have the same initial values, they produce the same answer. The answer from the authentication algorithm is used to determine if a subscriber seeking access to the system is a valid registered unit.

The authentication process in North American Systems (NAMPS, IS-54/IS-136 TDMA, and CDMA) is called the CAVE authentication algorithm. The CAVE algorithm operates on a group of data bits called the shared secret data (SSD). The SSD is contained in both the mobile telephone and cellular system. If either the mobile telephone or cellular system have an incorrect value of the shared secret data, authentication fails. The SSD is 128 bits of data divided into two halves called SSD-A and SSD-B. SSD-A is used by the authentication process, and SSD-B by message encryption and voice privacy processes.

The key to authentication process is not the secrecy of the CAVE algorithm, but the initial values used when running the algorithm. Each subscriber receives a secret number called the A-KEY (or authentication key). The cellular subscriber enters the A-KEY on the keypad after typing A-K-E-Y (letters on the keypad), then pressing the function key twice. The A-KEY is entered into the mobile set *one time only* by the subscriber, and can then be forgotten. The subscriber does *not* need to remember and use it repeatedly, like the PIN number used with some bank cards and in some analog cellular phone backup authentication methods still in use today. The mobile telephone does not use the A-KEY itself to authenticate the mobile set, but instead creates and stores a secret key called shared secret data (SSD). After the A-KEY is entered, it is known only to the subscriber and the network Home Location Register (HLR).

The cellular system begins the authentication process by sending an AUTH bit over the control channel in the continuous System Parameter Overhead Message (SPOM). When the mobile unit receives the AUTH information, it is set so that it will always send the authentication response information in addition to other values such as the mobile's ESN and dialed digits when starting a call.

Mobile telephones add other data in addition to the authentication response value computed by the CAVE algorithm. One extra data element is a code derived from the random challenge value sent from the base station. The purpose of this is to ensure that the mobile and base stations are using the same random challenge value in their calculation to produce the authentication response. The other extra data element is the CALL COUNT value which counts all calls made by the mobile set.

After receiving the results of the mobile's authentication process, the base station compares the answer to its own calculations. If the values match, the call processing continues. Once a voice channel is assigned, the base station may update the mobile's SSD with a new value to be used in future transactions.

Authentication successfully prevents "cloning" of mobile phones because the A-KEY is never sent over the air. Even if the A-KEY were captured and duplicated, it would be very difficult for an invalid mobile to receive all the SSD updates and COUNT value increments that the "valid" mobile telephone was receiving.

In addition to being used for authentication, the CAVE algorithm is also used for message encryption and voice privacy. Message encryption "scrambles" non-voice messages sent between the mobile telephone and the base station. The base station controls which messages are encrypted.

The GSM system uses the A3 authentication algorithm. The GSM A3 authentication algorithm is contained in a removable subscriber identity module (SIM) chip or card. Unlike the CAVE authentication algorithm, which is standard for all mobile telephones, the GSM A3 authentication process has several versions for use in different countries [10]. The SIM card contains a microprocessor which can store different authentication programs, so that different system operators can use different authentication algorithms.

The GSM algorithm processes data (RAND) with shared secret data (called Ki) to create a signed result (SRES). The Ki is stored in both the mobile telephone and cellular system. After receiving the results of the mobile's authentication process, the cellular system compares the answer to its own calculations. If the values match, the call processing continues. If either the mobile telephone or cellular system have an incorrect piece of the shared secret data the authentication process fails. The Ki key has a maximum length of 128 bits of data. Ki is also used to create the key used for voice privacy encryption.

Figure 3.17 shows the authentication process used in the GSM system. A random number (RAND) is sent on the broadcast control channel as part of the secret key processing. This random number changes periodically. The random number, the Ki secret data, and other information in the mobile telephone are processed by the A3 authentication algorithm to create an Signed Response (SRES).

The GSM system uses a different algorithm for message encryption and voice privacy. The A5 algorithm creates a message encryption mask for voice privacy. The encryption mask uses a Kc key, which is

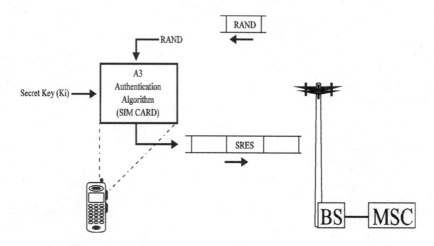

Figure 3.17, GSM Authentication Process

created at the beginning of each call, with an A8 encryption algorithm. Throughout the call, the A5 algorithm uses the Kc key to scramble voice data sent to and from the mobile telephone. Since the cellular system has access to the same set of secret information, it generates the same encryption mask as the mobile telephone and uses it to unscramble the voice data before sending it to the land line network.

Digital Data Services

Data transmission is the transfer of digital information from one location to another. Because cellular and landline telephone systems have not had direct digital connections, analog modems on analog voice channels have provided the only means for public data transmission.

Most early analog cellular modems used the same protocols as modems using public wire lines. The only requirement was an adapter cable to carry the analog data signal between the phone and the modem. Cellular radio systems experience radio distortion, and so do not provide continuous high quality signals like landline telephone systems. Cellular modems had to re-send information due to errors, greatly slowing cellular data transmission rates in comparison to land line rates. Special cellular telephone modems were created to overcome the slow data transmission rate, but unfortunately, both ends of the data communications required the special modems.

Digital systems which transmit everything as data improve cellular data transmissions. Data can be sent directly between a mobile telephone and another data device in two ways: either on a continuous circuit or by packets. A continuous circuit (circuit switched) transmission allows the network to route continuous data to a single location. A packet transmission system allows packets of information to be directed to any location.

Circuit Switched Data

At the beginning of transmission, the network identifies a single destination for the data, and routes circuit switched data to it. The sender provides the receiver with a destination address such as a phone number or internet address, and then begins a continuous transmission. An advantage of circuit switched data is that there is little overhead communication associated with it. A disadvantage is that it requires dedicated communications equipment (such as a radio channel) even when no data is being sent.

Cellular systems must convert digital information from the mobile telephone to a form suitable for telephone (or other) networks. Therefore, to send circuit switched data on digital cellular systems, the mobile telephone must tell the cellular system what type of data is to be sent.

Figure 3.18 illustrates how digital subscribers send data on digital cellular systems. The mobile telephone obtains access to the cellular system (step 1), then informs the system that circuit switched data is to be sent (step 2) and provides the destination phone number or internet address. The base station then sends a request for data transmission to the switching center (step 3) and the Mobile Switching Center routes the data to a modem (step 4) which converts the digital information to suitable form (FAX or data modem) for transfer through the telephone network. The network then transfers the data to the destination.

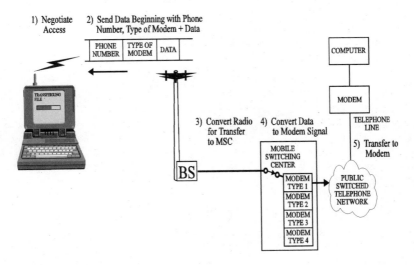

Figure 3.18, Direct Digital Data Transmission

Packet Switched Data

Packet switched data is sent in small pieces or packets, which contain all of the information to be sent, or a piece of a larger block of data. When blocks of data are sent by packets, each packet that is sent contains a sequence number which allows the recreation of the block of data after all of the packets have been received successfully. If a packet of data is lost in transmission, its replacement can be requested by using the sequence number.

Each packet of data contains some address information (called a header) indicating the packet's destination. Packet switched data is especially useful where the transmission path may change during the transmission, or where the sender needs to transmit to different addresses or communication with multiple receivers is desired.

Figure 3.19 illustrates digital cellular transmission of packet data. First, the mobile telephone gains access to the cellular system (step 1), then sends a packet of information containing both the destination address and data (step 2). The base station converts the packet to a form which can be sent to the Mobile Switching Center (step 3). According to the address sent with the packet, the Mobile Switching Center then routes the data directly to a packet data network, typically internet or a public packet data network. The packet network uses the address with the packet to route it to its destination.

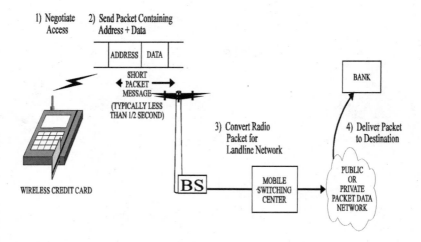

Figure 3.19, Packet Data Transmission

Digital Technology

Most of the new cellular technologies use digital voice technology to achieve the goals of the next generation cellular technologies. Digital technology increases system efficiency by voice digitization, speech compression (coding), channel coding, use of spectrally efficient modulation.

Digital Signal Regeneration

As a radio signal passes through the air, distortion and noise enter the signal. A digital signal can be processed to enhance its resistance to distortion in three ways: signal regeneration, error detection, and error correction. Signal regeneration removes the added distortion and noise by creating a new signal without noise from a noisy one (see figure 3.20). Error detection determines if the channel impairments have exceeded distortion tolerances. Error correction uses extra bits provided with the original signal to recreate correct bits from incorrect ones.

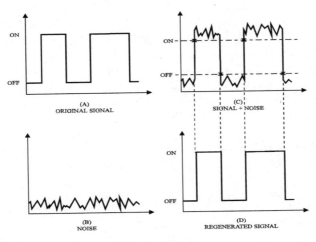

Figure 3.20, Digital Signal Regeneration

Figure 3.20 shows how noise 2.3 (b) is added to a digital signal time waveform 2.3 (a). By using ON/OFF threshold detection and conversion 2.3 (c), the original signal can be regenerated 2.3 (d). Provided that the signal is sufficiently stronger than the noise, it can be received almost error-free despite the presence of noise. In certain types of radio signals (particularly FM and phase modulation) this recovery of the error-free signal in the presence of noise is called the capture phenomenon.

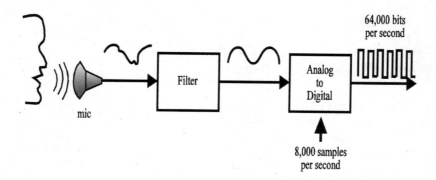

Figure 3.21, Voice Digitization

Signal Digitization

Figure 3.21 illustrates the conversion from an analog signal to a digital one. Speech into the microphone creates an analog signal. An audio bandpass filter remove high and low frequencies that interfere with digitization. The filtered signal is then sampled 8,000 times per second. This sample rate is standard in the telephone industry. For each sample, an 8 bit digital value is created. The resulting 64,000 bits per second represent the voice.

About 64 kilobits per second (kbp/s) of data are required to reasonably digitize an analog voice waveform. Because transmitting a digital signal via radio requires about 1 Hz of radio bandwidth for each bp/s, an uncompressed digital voice signal would require more than 64 kHz of radio bandwidth. Without compression, this bandwidth would make digital transmission less efficient than analog FM cellular, which uses only 25-30 kHz. Therefore, very high speech compression is necessary to increase cellular system capacity. Speech compression removes redundancy in the digital signal and attempts to ignore data patterns that are not characteristic of the human voice. The result is a digital signal which represents the voice audio frequency spectrum content, not a waveform.

The low bit-rate speech coder analyzes the 64 kbp/s speech information and characterizes it by spectrum, pitch, volume, and other parameters. Figure 3.22 illustrates the low bit-rate speech compression process. The speech coder examines a 20 millisecond time window of the speech. By means of mathematical analysis of this sample of the waveform, which contains many cycles of the speech waveform, it produces several numbers called the prediction coefficients. These numbers are used in a mathematical process called predictive coding. When a pulse (in digital numerical representation form) is put into this predictive coder as input, the resulting output is a waveform (also represented in digital form) which has a similar audio frequency spectrum to the original speech. That is, it contains high and low power in the corresponding parts of the audio frequency spectrum where the original speech has correspondingly high and low power. As the speech coder characterizes the input signal, it looks up codes in a code book table which represent various pulse patterns (rather than just a single pulse) to chose the pattern which comes closest to matching the output of the predictive coder to the original time window voice signal.The coding process requires a large number of mathematical calculations because the predictive coder output must be re-calculated many times at a rate faster than real time in order to find the best match. The coder system then sends the coefficients used in the predictive coder and the entry from the code book table to a decoder at the other end of the radio link. The compression process may be fixed or variable. Taking the 64,000 bit/second rate of a standard landline telephone coder, we see that the digital coders used in various cellular and PCS systems all compress or reduce the bit rate. For the TDMA system, the compression is 8:1. For CDMA, the compression varies

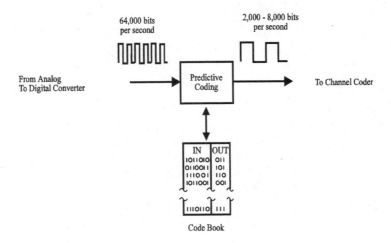

Figure 3.22, Speech Coding

from 8:1 to 64:1 depending on speech activity. GSM Systems Compress the voice by 5:1.

High bit rate speech coders (small amount of compression, such as 32 kb/s ADPCM) typically convert the waveform into a representative digital signal. Low bit rate speech coders (high amount of compression such as VSELP or QCELP) analyze the waveform for key characteristics. In essence, low bit rate speech coders model the source of the waveform. This process makes low bit rate speech coders more susceptible to distortion from background noise and bit errors, poorer voice quality from a poor coding process model, and echoes from the speech coder processing time.

When there is a significant amount of background noise, distortion in the coding process occurs. Because the speech coder attempts to characterize the waveform as a human voice, the background noise is not in its code book. The speech coder will find the code that comes closest sound that matches the combined background noise and the human voice. The result is usually distortion

All low bit-rate coders are more sensitive to bit errors than higher bit-rate coders, since any error in a bit constitutes a greater fraction of the total information transmitted at a low bit rate.

As a general rule, with the same amount of speech coding analysis, the fewer bits used to characterize the waveform, the poorer the speech quality. If the complexity (signal processing) of the speech coder can be increased, it is possible to get improved voice quality with fewer bits. The speech coders used in the digital cellular phones typically require 8 million instructions per second (MIPS) to process the voice signal. It has been estimated that it will take 4 times the amount of processing to reduce the amount of bits by additional factor of two [11].

Voice digitization and speech coding take processing time. Typically, speech frames are digitized every 20 msec and input to the speech coder. The compression process, time alignment with the radio channel, and decompression at the receiving end all delay the voice signal. The combined delay can add up to 50-100 msec. Although such a delay is not usually noticeable in two-way conversation, it can cause an annoying echo when a speakerphone is used, or the side tone of the signal is high (so the user can hear themselves). However, an echo canceller can be used in the MSC to process the signal and remove the echo.

Digital Radio Channel Coding

Once the digital speech information is compressed, control information bits must be added along with extra bits to protect from errors that will be introduced during radio transmission. Three types of error protection coding are used in digital cellular and PCS systems: block coding, cyclic redundancy check (CRC) codes, and convolutional (continuous) coding. (Only the GSM system and its derivatives use block coding, and only for call processing messages at that). Control messages (such as power control) must be combined with speech information. Control messages are either time multiplexed (simultaneous) or they replace (blank and burst) the speech information.

Block coding and CRC append extra bits to the end of a data block of information. These bits allow the receiver to determine if all the information has been received correctly. Convolutional coding produces a new and longer string of bits by combining the data with another predetermined string of bits in a process analogous to numerical multiplication. Unlike ordinary arithmetic there is no borrow or carry between the columns of binary bits. At the receiving end of the radio link, the received data is processed by a process analogous to dividing the received (binary) numerical value by the same pre-determined number. The result is two (binary) numbers analogous to the quotient and the remainder in ordinary arithmetic division. If the remainder is zero, there was likely no error. When the remainder is not zero, there was some error. Depending on the values chosen for the predetermined number, a limited number of errors can be corrected since these particular errors correspond to certain known non-zero remainder values. When there are many bit errors, it is not possible to correct them because more than one pattern of errors is known to correspond to the same non-zero remainder value. However, in certain cases it is useful to know that there are an errors, even if they cannot be corrected. This coding allows for quick checking and correction of information. Figure 3.23a shows how error detection and correction bits are added to the compressed speech. In all, error detection and correction bits add approximately 50% to the total number of bits used per subscriber. Error detection/correction reduces the number of bits available to users and decreases the system capacity.

Convolutional coders are described by the relationship between the number of bits entering and leaving the coder. For example, a 1/2 rate convolutional coder generates two bits for every one that enters. The larger the relationship, the more redundancy and better error protection. A 1/4 rate convolutional coder has much more error protection capability than a 1/2 rate coder.

CRC parity generation divides a given binary data value by a pre-defined binary number. However, the division is not like ordinary arithmetic division, because there is no carry or borrow operation. The remainder resulting from this division is appended to the data to allow comparison when received. The division-like process is repeated on the data at the receiver and the quotient is compared. If the quotients do not match, one can infer that an error has occurred, and certain limited patterns of error can be corrected. Figure 3.23b shows a CRC generator. An electronic circuit consisting of modules called a shift register and exclusive OR gates allow the division of the binary data values.

As digital specifications were originally being developed, great emphasis was placed on a layered approach to their design. A key to this layered design was the concept of logical channels. A logical channel is simply a specification structure that separates different types of data from each other. This contrasts with the physical channel separation that is used on other analog systems such as EIA-553. The advantage of a logical channel separation is that it permits a natural division of the software development tasks during the design of the system.

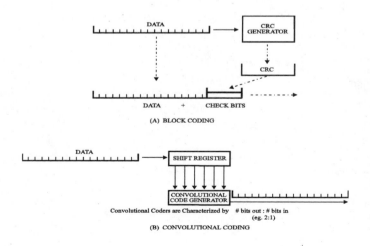

Figure 3.23, Block Error Detection and Convolutional Coding

An example of logical channels is the control signals that are mixed with voice signals to direct the phone's communications with the base station. There are two types of control signaling, fast and slow. Slow signaling typically sends continuous channel quality measurements such as signal strength and a report stating the number of bits received in error in the last few frames. Fast signaling primarily sends channel assignment messages which must be acted on quickly. Fast messages replace the speech information for brief periods. Slow messages are transmitted slowly at a low bit rate and are multiplexed into the bit pattern of the speech frames. Figure 3.24 shows the process of fast and slow message signaling.

Modulation

Analog cellular typically uses Frequency Modulation (FM). Frequency modulation is a process of shifting the radio frequency in proportion to the amplitude (voltage) of the input signal. Digital technologies use both FM (used for GSM and its derivative systems) and phase modulation (used for IS-54, IS-136, etc., and for IS-95 CDMA), a process that converts digital bits into phase shifts in the radio signal. Phase modulation is a result of time advancing or retarding the carrier frequency waveform to introduce phase changes at specific points in time. Figure 3.25 displays a basic digital modulation circuit. The dig-

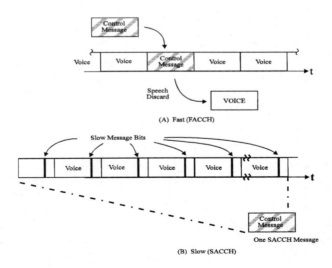

Figure 3.24, Fast and Slow Message Signaling

ital signal is supplied to a pulse shaper. The pulse shaper adjusts (smoothes) the edges so the abrupt changes of the digital pulses do not force the modulator to produce energy outside the allowable bandwidth. The modulator converts these pulses to a low level RF frequency which changes in phase which represent one or more digital pulses.

RF Amplification

The RF amplifier increases the low level RF signal from the modulator to a high power RF signal ready for transmission through the antenna. A major difference between certain digital and analog technologies is that the DQPSK digital modulation used in IS-54 and IS-136 requires a linear RF amplifier. It is acceptable for analog cellular technologies to use a very efficient RF amplifier of a type called Class C, which may add some signal distortion. Most digital cellular technologies cannot accept this type of distortion and therefore they use a

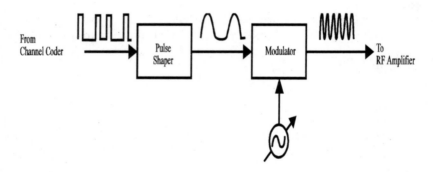

Figure 3.25, Radio Channel Modulation

more linear amplifier. Unfortunately, the battery-to-RF energy conversion efficiency for linear amplifiers is 30-40% compared with 40-55% RF amplifiers in most analog cellular phones [12]. Linear amplifiers require more battery power to produce the same RF energy output during transmission. Digital technologies overcome this limitation either by transmitting for shorter periods, or by precisely controlling power to transmit at lower average output power.

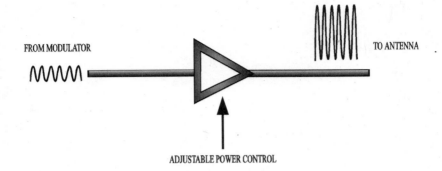

FROM MODULATOR

TO ANTENNA

ADJUSTABLE POWER CONTROL

Figure 3.26, RF Power Level Control

Mobile telephones adjust their transmitter output as a result of commands they receive from the cellular system. For CDMA mobile telephones, their output power is adjusted by a combination of the receive signal strength and fine adjustment messages from the cell site. Figure 3.26 shows that the RF power amplifier can vary its output power.

References:

1. CTIA, User Performance Requirements, Issue 1, September 8, 1988.
2. Electronic Industries Association, EIA Interim Standard IS-54, Rev 0, "Dual-Mode Mobile Station - Base Station Compatibility Standard,," 1990.
3. Electronic Industries Association, EIA Interim Standard IS-136, Rev 0, "Dual-Mode Mobile Station - Base Station Compatibility Standard,," 1995.
4. Electronic Industries Association, EIA Interim Standard IS-95Rev 0, "Dual-Mode Mobile Station - Base Station Compatibility Standard,," 1994.
5. Electronic Industries Association, EIA Interim Standard IS-89Rev 0, "Dual-Mode Mobile Station - Base Station Compatibility Standard,," 1990
6. FCC Regulations, Part 22, Subpart K, "Domestic Public Cellular Radio Telecommunications Service," 22.903, June 1981.
7. SuperPhone conference, Institute for International Research, New York NY, January 22-24, 1997.
8. Special Seminar, NAMPS Panel, CTIA Next Generation Cellular: Results of the Field Trials, Washington, DC, December 4-5, 1991.
9. NAMPS conference, Telecommunications Industry Association, Chicago, IL, 1991.
10. Mouly, Michel, The GSM System for Mobile Communications, (M. Mouly et Marie-B. Pautet), pp 478,479.
11. SuperPhone conference, Institute for International Research, New York NY, January 22-24, 1997.
12. Chorney, Paul, personal Interview, MA/COM, October 11, 1994.

Chapter 4

North American TDMA

North American TDMA

The North American TDMA system is based on the IS-136 industry specification. IS-136 uses Time Division Multiple Access (TDMA) system combines new digital TDMA radio channels and features with Advanced Mobile Phone Service (AMPS) functionality. TDMA systems differ from FDMA (analog) systems by the dividing of a single radio carrier waveform into time slots which allow multiple users to share it. The IS-136 system is sometimes referred to as Digital AMPS (DAMPS) or North American digital cellular (NADC). It is the next generation beyond IS-54 dual-mode analog-digital cellular.

History

In 1988, the Cellular Telecommunications Industry Association created a development guideline for the next generation of cellular technology. This guideline was called the User Performance Requirements (UPR) and the Telecommunications Industry Association (TIA) used this guideline to create a TDMA digital standard, called IS-54. This digital specification evolved from the original EIA-553 AMPS specification. The first revision of the IS-54 specification (Rev 0) identified the basic parameters (for example, time slot structure, type of radio channel modulation, message formats) needed to begin designing TDMA cellular equipment. Unfortunately, IS-54 Rev 0 lacked some basic features that were introduced in the first commercial TDMA phones, and IS-54 Rev A was soon introduced to correct errors and add essential basic features (such as caller ID) to the TDMA standard. In 1991, IS-54 Rev B added features such as authentication, voice privacy, and a more capable caller ID with greater benefit to the user. Digital TDMA technology has continued to evolve beyond the current IS-54 Revision B to the digital control channel system covered by the IS-136 specification.

The IS-136 specification concentrates on features that were not present in the earlier IS-54 TDMA system. These include longer standby time, short message service functions, and support for small private or residential systems that can coexist with the public systems. In addition, IS-136 defines a digital control channel to accompany the Digital Traffic Channel (DTC). The digital control channel allows a mobile telephone to operate in a single digital-only mode.

Revision A of the IS-136 specification now supports operation in the 800Mhz range for the existing AMPS and DAMPS systems as well as the newly allocated 1900MHz bands for PCS systems. This permits dual band, dual mode phones (800 MHz and 1900 MHz for AMPS and DAMPS). The primary difference between the two bands is that mobile telephones cannot transmit using analog signals at 1900MHz.

During development of IS-136, many new features were influenced by or borrowed from the GSM (Global System for Mobile Communications) specification (covered in chapter 6). The overall control channel signaling processes are very similar to GSM control channel structure.

System Overview

The IS-136 cellular system allows for mobile telephones to use either 30 kHz analog (AMPS) or 30 kHz digital (TDMA) radio channels. The IS-136 TDMA radio channel allows multiple mobile telephones to communicate on the same frequency, apparently at the same time, by sharing brief 6.6 millisecond time slots on the same radio channel. This time division of radio channel is called Time Division Multiple Access (TDMA). The TDMA system includes many of the same basic subsystems as other cellular systems, including a switching network, base stations (BS), and mobile telephones. IS-136 systems can serve mobile telephones of three types: AMPS only, dual mode IS-54 (original TDMA) and single digital mode IS-136 (TDMA). Figure 4.1 shows an overview of an IS-136 radio system.

A primary feature of the IS-136 systems is their ease of adaptation to the existing AMPS system. Much of this adaptability is due to the fact that IS-136 radio channels retain the same 30 kHz bandwidth as AMPS system channels. Most base stations can therefore replace TDMA radio units in locations previously occupied by AMPS radio units. Another factor in favor of adaptability is that new dual mode mobile telephones were developed to operate on either IS-136 digital traffic (voice and data) channels or the existing AMPS radio channels as requested in the CTIA UPR document. This allows a single mobile telephone to operate on any AMPS system and use the IS-136 system whenever it is available.

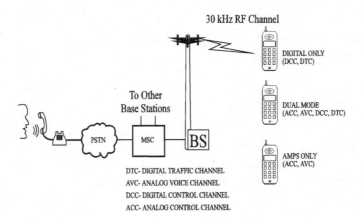

Figure 4.1, Overview of an IS-136 System

All IS-136 TDMA digital traffic frequency signals are divided into frames with 6 time slots. Every communication frequency signal consists of two 30 kHz wide carrier frequency signals, a forward or downlink waveform (from the cell site to the mobile telephone) and a reverse or uplink waveform (from the mobile telephone to the cell site). Each time slot on one of these waveforms forms a forward or reverse part of a communication channel, respectively. The time slots used for the correspondingly numbered forward and reverse channels are time-related so that the mobile telephone does not simultaneously transmit and receive. The entire repeating pattern of time slots is called a frame. In IS-136 there are 6 time slots per frame. In a TDMA radio system, the name "channel" is sometimes used for the entire carrier frequency signal waveform, and sometimes for just one time slot.

When you read different documents about TDMA systems, you must be careful to understand the proper meaning of this word in each context.

The IS-136 system has defined a new type of control channel which is part of a digital traffic channel. This digital control channel (DCC) carries system and paging information, and coordinates access in a way similar to the analog control channel (ACC). The DCC has many more capabilities than the ACC such as extended sleep mode, short message service (SMS), private and public control channels, and others. SMS is a service which uses the alpha-numeric display of the mobile telephone to show a number or an alphabetic message to the user. It is similar to alpha-numeric paging service, but does not use a separate system or pager unit. Because the DCC uses the DTC slot structure, a DCC may co-exist on the same carrier frequency waveform with DTC traffic channels used for voice, each one on a different time slot.

IS-136 systems allow several users to share each radio carrier frequency by dedicating a specific time slot from each frame to individual users. Voice channels can be either full rate or half rate. Full rate IS-136 systems assign two time slots per frame of 6 slots to each user , allowing 3 users to simultaneously share a radio channel. Half rate IS-136 systems assign one time slot every frame to allow up to 6 users to share a single radio channel. At the time this book goes to press, research is underway for development of a half-rate digital speech coder, and the IS-136 system is designed to use that coder when it is available, but it is not ready yet.

Full Rate IS-136

Time intervals on IS-136 channels are divided into 40 msec frames with six time slots on two different radio frequencies. One frequency is for transmitting from the mobile telephone (called reverse or uplink); the other is for receiving to the mobile telephone (called forward or downlink). During a voice conversation in IS-136, in every sequence of three time slots, one time slot is dedicated for transmitting, one for receiving, and one remains idle. The mobile telephone uses the idle time slots to measure the signal strength of carrier frequencies in surrounding cells. These measurements assist in target cell selection and hand-off. The total bit rate of the carrier frequency waveform is 48.6 kbit/s. This time sharing and the use of some bits for synchronization and low bit-rate data transmission results in a user-available data rate of 13 kb/s. Some of the 13 kb/s are used for error detection and correction, so only 8 kb/s of data are available for digitally coded speech.

Subscribers talk and listen at the same time, so the mobile telephone must function as if it is simultaneously sending and receiving (called full duplex). When in conversation mode (called dedicated mode), IS-136 mobile telephones do not transmit and receive simultaneously, but only appear to do so. Speech data bursts alternate briefly between transmitting and receiving, and when received, the digitally coded speech bursts are expanded in time to create a continuous audio signal.

Figure 4.2 shows how IS-136 full duplex radio carrier frequency waveforms are divided in time. IS-136 digital radios transmit on one frequency and receive on another frequency 45 MHz higher, but not at the same time. The mobile telephone transmits a burst of data on one frequency, then receives a burst on another frequency, and is briefly idle before repeating this cycle process.

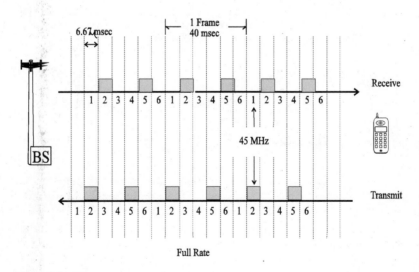

Figure 4.2, Full Rate Channel Sharing

Half rate IS-136

The number of simultaneous users that can share a single radio carrier frequency waveform can be doubled by using only half the number of slots per frame, which is called a half rate channel. Half rate channels use 1 of six slots to transmit and 1 to receive, leaving 4 idle. Using only one of the six time slots results in a user-available data rate of about 6.5 kb/s. A half rate system supports up to six simultaneous users per radio channel. Figure 4.3 shows how half rate systems share the IS-136 radio carrier frequency waveform.

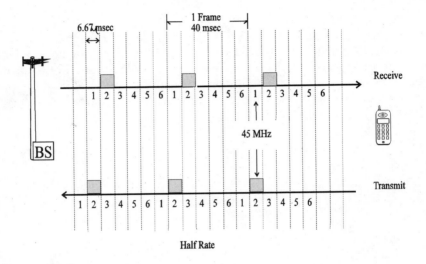

Figure 4.3, Half Rate Channel Sharing

System Attributes

The IS-136 digital cellular system uses the same 30 kHz RF bandwidth as the AMPS cellular system. This allows for a simplified conversion from AMPS radio channels to a new digital channel with the same bandwidth. The key attributes of the IS-136 system is the use of digital transmission, a new control channel structure, and advanced features.

IS-136 mobile telephones have acquired some of the system control functions that were part of the AMPS system features. While IS-136 systems use a Mobile-service Switching Center (MSC) to centralize control of mobile telephones moving through the system, the addition of new types of control channels permit for some of the switching to be performed locally (for example, a wireless office switching system) without the use of a central system.

IS-136 Radio Channels

The IS-136 cellular system has three basic types of radio channels: analog control channels (ACC), analog voice channels (AVC), digital traffic channels (DTC). The analog radio channel is either a control channel or a voice channel. The TDMA radio carrier waveform is divided into several different types of control and voice channels by the use of different time slots or shared portions of time slots.

Control channels continuously provide information to mobile telephones which are operating in the system and coordinate their access. Typically, one radio channel per base station will be an analog control channel and one digital radio channel in a base station will have at least one time slot dedicated as a digital control channel. Each 800 MHz cellular system operator is allowed up to 416 carrier frequencies. The analog control channel frequency is assigned by the system operator from one of the 21 radio frequencies authorized only for analog control channels. The radio channel which has the digital control channel time slot can be assigned to any one of that system operator's other 395 cellular radio channels. Mobile telephones must scan for available control channels before attempting to access the system. The scanning process can be time-consuming, which is why analog control channel frequencies are limited to only 21 of the 416 authorized radio channels. In the IS-136 system, the DCC might be located on any one of 395 radio channels, so a means was designed to help the mobile telephone to quickly find the control channel. To speed up this search, a DCC locator message can be continually sent on the AMPS control channel, parts of the previous IS-54 radio channel, and as part of a release message so at the end of the call which allows a mobile telephone to find the nearby DCC after a call is completed.

The IS-136 radio channel multiplexes several users onto a single radio carrier frequency through use of distinct time slots. A single digital radio carrier frequency waveform is capable of transferring 48.6 kb/s of information. Time slots are the smallest individual time period available to each mobile telephone. The TDMA frame is a 40 msec time interval. The frame is divided into 6 time slots, each normally containing 324 bits of information. Of the 324 bits, some are dedicated as data, and others are dedicated for control or synchronization.

Sixteen TDMA frames are combined to form a traffic channel super-frame. The traffic channel superframes are defined for convenience in synchronizing a low bit rate channel which uses the bits from each DCC time slot in a pre-scheduled manner. These low bit-rate channels carry call control information (for example, power level control). Figure 4.4 shows the basic IS-136 digital radio channel structure.

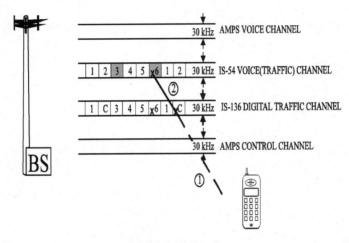

Figure 4.4, IS-136 Digital Radio Channel Structure

Traffic Channel Structure

The IS-136 digital traffic channel multiplexes (time shares) control, voice, and data channels on a single radio carrier frequency waveform. To accomplish this, the digital traffic channel is symbolically divided into several different logical channels, and these channels' informa-tion format (for example, control or voice) is defined. Control channels transfer broadcast, paging, and access control. Traffic channels trans-fer voice and data (for example, fax) information. Figure 4.5 shows the different types of control channels that are part of a traffic channel.

Several messaging and control functions are part of a control channel. Some of the bits in each control channel time slot have been dedicat-ed for this capability. These new logical control channels consist of bits combined from a period of 32 associated slots to form a single superframe. The superframe is divided into sub channels which are simply logical groupings of the time slots. The groupings are obtained by combining predefined data and by using information contained on the DCC.

The downlink control channel (from base station to mobile telephone) consists of three components: 1) short message, paging and access channels (SPACH), 2) a broadcast control channel (BCCH), and 3) system control field (SCF) channels. The SPACH combines the short message, paging, and access channels. The BCCH broadcast channel provides system identification and access information. The SCF helps coordinate access to the system.

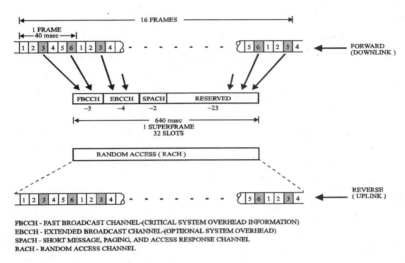

FBCCH - FAST BROADCAST CHANNEL-(CRITICAL SYSTEM OVERHEAD INFORMATION)
EBCCH - EXTENDED BROADCAST CHANNEL-(OPTIONAL SYSTEM OVERHEAD)
SPACH - SHORT MESSAGE, PAGING, AND ACCESS RESPONSE CHANNEL
RACH - RANDOM ACCESS CHANNEL

Figure 4.5, IS-136 TDMA Traffic Channel Structure

The uplink control channel provides an access point for mobile telephones to randomly attempt to access the cellular system. This control channel is call a random access channel (RACH). This is the channel which mobile telephones use to transmit their first radio burst to the base station when they begin to set up a call. The RACH is a shared, point-to-point unacknowledged channel, although the SCF sends an acknowledgment message on the downlink channel when the system recognizes that the mobile telephone has attempted contact to the system.

The broadcast channels (BCCH) contain general system information needed by all mobile telephones. The BCCH sends information on fast and extended channels. The fast channel sends information that the mobile station needs to obtain quickly such as the system identification information. An extended channel carries information that the mobile station uses after locking on (or camping on) to the DCC. The SMS, paging and access channel (SPACH) contains data for the mobile telephone about paging messages (which are the first message

111

used when starting a mobile destination call) and short messages, as well as access responses. The shared control feedback (SCF) flags are part of the access channel, and provide information to control access to the channel by multiple mobile telephones. The uplink channel (from mobile telephone to base station) labeled with the same time slot number used for SPACH downlink messages is the random access channel (RACH) used for random access attempts by the mobile telephone to the system.

A SMS point-to-point paging and access response channel (SPACH) contains information for specific mobile telephones. As the name suggests, this channel contains three logical sub channels to carry point-to-point SMS messages, page messages and access responses.

A portion of the control channel is used for shared control feedback (SCF) flags. These flags are used as an acknowledgment of access attempts on the RACH channel. The phone monitors the SCF channel to determine if it is permitted to attempt access on the RACH, and to determine if the base station received the access attempt.

Mobile telephones spend most of their time while they are tuned to the DCC waiting for a page. To conserve battery power, paging messages for telephones with similar last digits are scheduled to occur at or about the same time in a cyclic time pattern. The mobile telephones are designed to "sleep" (operate with power turned off to most of the set except for a timer) during the time intervals when paging for their particular group will not be scheduled. Most portable battery-powered phones use power conservation algorithms that turn off unused portions of the hardware. The IS-136 paging channel was designed to maximize the sleep time available to a portable phone, and it is a primary reason for increased standby times of IS-136 sets compared to analog sets. The paging channel is a multiplexed channel comprised of many paging channel groups. Each group has a different period during which it is "active," or transmitting page information. A mobile telephone finds the proper paging channel through an algorithm based on its mobile telephone identification number and information found on the BCCH.

The access response channel is a shared channel that all mobile telephones use for sending information to the base station to request service. The ARCH is a contention based access system, but some control flags on the downlink help to arbitrate the access. Contention based access is a system by which several mobile telephones can talk to the base at approximately the same time , and avoid information loss or repeated radio interference due to simultaneous transmissions.

IS-136 systems can be set up as small private systems within the existing public systems. The small private systems are controlled by two new types of system identifiers: PSIDS (private system id's) and RSIDS (residential system ID's). This separate control channel allows the billing system to charge users different rates depending on which system they access.

Capacity Expansion

There are three types of capacity expansion addressed in the IS-136 system; voice channel capacity, paging channel capacity, and network switching capacity. The basic method of voice channel capacity expansion in a IS-136 TDMA system is to allow more users to share a single radio channel than in the AMPS system. Because IS-136 TDMA channels which provide for 3 voice channels (and 6 in the future when half-rate is available) can replace an AMPS radio channel that only provide 1 voice channel, this increases the capacity of the system. Additional capacity expansion was anticipated through the ability to reuse frequencies geographically closer to each other. If more frequencies can be reused in nearby cells, system capacity is increased. There are theoretical reasons to expect that the IS-54 and IS-136 TDMA systems will operate at higher interference levels because of their digital error correction capabilities. Error correction does improve voice quality in high interference, but in real application, TDMA radio channels appear to tolerate only about the same carrier to interference ratio (17 dB) as analog cellular systems, thus allowing frequency reuse similar to that of AMPS [1]. Other improvements such as better adaptive equalizers and base receiver diversity systems may also be able to reduce the minimum C/I ratio, and thus allow closer placement of the same re-used radio frequency and therefore higher geographical traffic capacity.

If the cellular system increases the number of available voice channels, the number of paging messages delivered is likely to increase. AMPS control channels can only provide a limited capacity for paging (approximately 19 paging messages per second). The IS-54 TDMA standard allows a new optional band of secondary dedicated analog control channels which increases the paging channel capacity through the addition of new control channels. These secondary dedicated control channels are also sometimes used when a new digital system must co-exist with an older AMPS-only cellular system. When the AMPS system cannot coordinate digital radio channel assignment, dual mode mobile telephones can access the system through the secondary dedicated control channels. IS-136 digital control channels have paging channels which increases the paging capacity of the system. The IS-136 system also uses a temporary mobile station identifier (TMSI) to increase paging capacity. A TMSI, which is shorter than a MIN, is assigned to a mobile telephone by means of a radio message from the base station. A TMSI is has fewer digits, so that up to five paging messages can be packed into a single frame of data. Another large capacity improvement in IS-136 systems is the ability to use any channel as a control channel. The increase in the number of available control channels solves many of the frequency reuse issues by simply using more control channels when more paging or short message traffic occurs. In practice, however, control channels cannot be added indefinitely without running out of voice and data channels.

Through the use of public and private control channels, some of the switching for an IS-136 system can be accomplished by private systems. This reduces the switching requirements and interconnections to the cellular network.

Signaling

Signaling is the physical process of transferring control information to and from the mobile telephone from the cellular system. Signaling on the analog control channel and analog voice channel is almost identical to the AMPS system signaling. Signaling on a digital radio traffic channel has been divided into two signaling channels: the fast associated control channel (FACCH) and the slow associated control channel (SACCH). The fast channel replaces speech with signal data when needed. The slow channel uses a few dedicated bits within each time slot. Messages sent via the SACCH channel have long delays of 440 msec compared with 40 msec required to send a FACCH message. To save time on certain time-critical messages such as power control adjustment , messages may be sent as FACCH messages instead of using the SACCH channel.

The AMPS control channel signaling is almost unchanged in dual mode IS-136 systems. A few new messages were defined: The AMPS control channel was modified to allow assignment to a TDMA radio channel. AMPS voice channel signaling was modified to allow handoff to a digital channel. This change involves adding a time slot and type of channel (full rate or half rate) to the channel (frequency) assignment message. The IS-136 system also has the capability to indicate it has digital capability through the use of sending a protocol capability indicator (PCI) bit in the standard overhead messages sent by the system. The PCI bit lets the mobile telephone know that the control channel it is monitoring comes from a base station which has digital traffic channel capability.

Slow Associated Control Channel (SACCH)

SACCH is a continuous data stream of signaling information sent beside speech data. For historical reasons this is sometimes called out-of-band signaling, but it is in the same radio frequency and band as the speech, and just uses a few separate digital bits in the same time slot as the digitally coded speech. SACCH messages are sent by dedicated bits in each slot, so SACCH messages do not affect speech quality. However, the transmission rate for SACCH messages is slow. For rapid message delivery, messages are sent via the FACCH channel instead of using the SACCH channel. The SACCH and FACCH system was designed to maximize the number of bits devoted to speech and minimize the number of bits devoted to continuous signaling. Figure 4.6 illustrates the SACCH signaling process. SACCH mes-

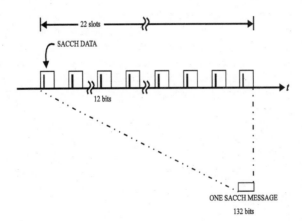

Figure 4.6, SACCH Signaling

sages are divided up into small parts and transmitted over a sequence of 12 time slots. This interleaving of the data bits over a long interval of time helps the error correction code do a more effective job, since the probability of a large number of consecutive data bit errors, after all the data bits are de-interleaved and re-assembled into the SACCH message, is very slight. The convolutional error protection code can identify and correct bit errors if there are not too many consecutive bit errors. As we will see below, digitally coded speech and FACCH, in distinction to SACCH, is interleaved also but only over two time slots rather than 12.

Fast Associated Control Channel (FACCH)

FACCH control messages replace speech data with signal messages. For historical reasons this is called in-band signaling). The digital speech coder used in IS-136 has the capability to bridge over frames of speech data which are missing (due to replacement by a FACCH message) or unusable due to very bad bit error levels. It will repeat the last good frame (20 milliseconds) of speech when the current frame is missing or unusable. If several consecutive speech frames are omitted, the coder will reduce the loudness for each repetition and after 5 repetitions it will produce silence. Under most FACCH or channel error conditions, the occasionally repetition of a speech frame goes un-noticed by the user.

FACCH data is error-protected by a 1/4 rate convolutional coder. That means that fl of the bits in the transmission are error protection codes, and only / of the bits are the actual FACCH message. Most other error protection codes used in IS-136 use a fi rate error protection code. The / rate code is used for the FACCH because some of its most important control messages, used for handoff, are often sent in poor radio conditions at the outer boundary of the cell. Figure 4.7 illustrates the FACCH signaling process. Notice that a FACCH message replaces speech slots, and each FACCH message is interleaved between two slots.

In IS-136 there are no special bits or bit fields in a slot to identify if the traffic channel has speech or FACCH signaling. The receiver discovers this by a sequence of error protection decoding steps, based on the fact that the FACCH and the digital voice coding use different types of error protection codes. Each time slot of data is de-interleaved and the data is then first error protection decoded as if it were digitally coded speech. If the result is good (free of un-corrected errors), the data is passed to the speech decoder to be converted into audio. If this decoding indicates it is *not* a valid speech slot, the receiver will then attempt to error protection decode it as a FACCH message. If this second error protection decoding calculates correctly, it is a FACCH message and is passed to the internal control processor to be interpreted and acted on. If the second error protection calculation decodes incorrectly, then it is probably data which has errors due to radio channel fading or interference, and it is discarded. (If you are familiar with GSM and PCS-1900, you will note that those systems, in distinction with IS-136, merely use some reserved bits in each time slot to indicate to the decoder whether the other bits are carrying digitally coded speech versus FACCH).

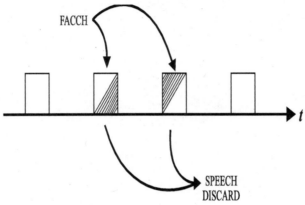

Figure 4.7, FACCH Signaling

When a mobile telephone is operated in the discontinuous transmission (DTX) mode to conserve battery power, the SACCH channel is not operating continuously. Therefore, all messages are sent as FACCH rather than SACCH.

Speech Coding

To allow several users to share a single 30 kHz wide radio channel, voice signals are digitally coded , before transmission. Because the bit rate of the IS-136 coder is significantly lower than the bit rate traditionally used in the PSTN, this particular type of coding is sometimes called compressed coding. Figure 4.8 illustrates IS-136 speech coding. At the mobile set microphone, the analog voice signal is first sampled 8,000 times each second and digitized. Each sample is uniformly coded using 12 or 13 digital bits, for use in the digital signal processor (DSP). On the MSC end of the connection, the PSTN has already digitally coded the speech by a similar process: the speech is sampled or measured 8,000 times each second. However, in the PSTN only 8 bits are used to encode each sample, using a non-uniform code called Mu-law, which uses larger voltage differences between code values near the high end of the scale, and smaller differences near zero volts. To get the higher accuracy needed for use in the DSP operations, each 8-bit sample data value is converted into a 12 or 13 bit value. Every 20 msec the digital audio information representing a 20 msec window of the speech waveform is supplied to a speech coder. The IS-136 speech coder uses vector sum linear prediction coding (VSELP) to convert audio to 8 kb/s compressed voice. Because fading radio channels can cause digital bit errors, additional error protection data bits are also included before transmission, to help correct the most important information of the compressed digital audio in case of errors. These additional error protection bits increase the total data rate to 13 kb/s. To

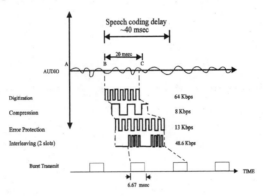

Figure 4.8, Speech Data Compression

help protect the data from short fractional millisecond intervals when the radio signal is weak (called fast or rapid fading), the 13 kb/s error-protected data is interleaved (distributed) over 2 adjacent slot periods. These slot periods are assigned to 2 radio transmit bursts from the mobile telephone.

Dynamic Time Alignment

Dynamic time alignment is a technique that allows the base station to receive digital mobile telephones' transmit bursts in an exact time slot, even though not all mobile telephones are the same distance from the base station. Time alignment keeps different digital subscribers' transmit bursts from colliding or overlapping. Dynamic time alignment is necessary because subscribers are moving, and their radio waves' arrival time at the base station depends on their changing distance from the base station. The greater the distance, the more delay in the signal's arrival time.

The base station adjusts for the delay by commanding mobile telephones to alter their relative transmit times based on their distance from the base station. The base station calculates the required offset from the mobile telephone's initial transmission of a shortened burst in its designated time slot. A shortened burst initial transmission is only necessary in cells where propagation time is unknown before the first transmission). To account for the combined receive and transmit delays, the required timing offset is twice the path delay. The mobile telephone uses a received burst to determine when its burst transmission should start. The mobile telephone's default delay between

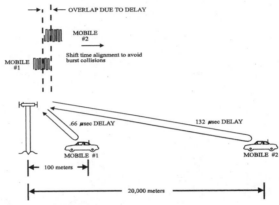

Figure 4.9, Dynamic Time Alignment

receive and transmit slots is 44 symbols (approximately 1.8 msec), which can be reduced in 1/2 symbol (or 1 bit time) increments to 15 symbols. This standard allows mobile telephones to operate at a maximum of 72 miles from the base station without slot collisions. Figure 4.9 illustrates dynamic time alignment.

Duplex Channels

To prevent a mobile telephone's receive and transmit messages from interfering, IS-136 systems separate transmit and receive channels by the same 45 MHz as IS-54 Rev B systems on the 800 MHz cellular band, and by 80 MHz on the 1.9 GHz band. In addition to separating transmit and receive frequencies, IS-136 voice (traffic) and control channels can separate transmission and reception in time. The time separation, (sometimes called Time Division Duplex — TDD)* opera-

1. The term TDD is more often used for a system which transmits and receives on the same frequency but different time slots, like DECT.

tion, simplifies the design of transmitters and receivers. Figure 4.10 shows IS-136 system time and frequency separations between receive and transmit channels [2].

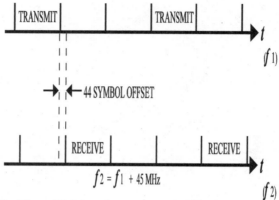

Figure 4.10, Time Division Duplex Radio Channels

In the 800 MHz cellular band, the transmit band for the base station is 869-894 MHz and the transmit band for the mobile telephone is 824-849 MHz. The 30 kHz AMPS channel bandwidth has been maintained. Figure 4.10 illustrates the 44 symbol offset between the forward and reverse channel that allows for dynamic time alignment.

Paging Classes

As portables took over a larger share of the market, the length of a battery's operating or recharge cycle time became a concern. To help increase the battery life, the system can allow the phone to "sleep" during times when messages are not expected. Paging channels (a logical channel, not a physical RF channel) are multiplexed reduce power consumption during receive mode. Cellular telephones in 1994 had an average battery life of 15 to 20 hours with standard sized batteries [3]. The new specification allows for sleeping between paging groups, and IS-136 phones should double or even triple the average standby time. If a mobile user can tolerate a slightly longer delay when a page is being received, that user can be assigned to a paging class which sends pages at a slower rate. The phone knows when the next frame containing page messages will occur, so it can sleep to conserve battery power until the next frame for its paging class. Delays in receiving pages for classes 1 - 3 range from 1.2 to 3.84 seconds, a range that is probably acceptable for normal use. However, paging class 8 has a maximum delay of over 120 seconds, too long for a caller to wait for

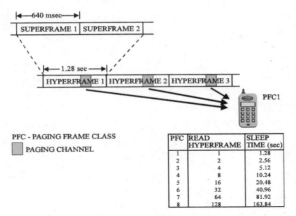

Figure 4.11, Paging Classes in an IS-136 System

the subscriber to answer. This paging class is targeted for remote control applications where maximum battery life is needed and a quick response to a page is secondary. Figure 4.11 shows the paging class process used in the IS-136 system.

RF Power Classification

A new class IV power class of mobile telephone has been added. The Class IV mobile telephone output power is identical to Class III, but its minimum power is 12 dB lower. The lower minimum power allows systems to reduce the minimum cell site radius. Table 4.1 shows the power classification types for the IS-54 radio system.

RF Power	Class I	Class II	Class III	Class IV	Class V-VI
Maximum Power	4 Watts	1.6 Watts	.6 Watts	.6 Watts	Reserved
Average Power (Full Rate TDMA)	1.333 Watts	.533 Watts	.2 Watts	.2 Watts	Reserved
Minimum Power	6 mW	6 mW	6 mW	.5 mW	Reserved

Table 4.11, IS-136 RF Power Classification

Basic Operation

An IS-136 mobile telephone initializes, when it is first turned on, by scanning for a control channel and tuning to the strongest one it finds. During initialization, it also determines if the system is digital-capa-

121

ble. If the system is not digital-capable, the mobile telephone looks for an optional secondary control channel. After initialization, the mobile telephone enters idle mode and waits to be paged for an incoming call or for the user to place a call (access). When a call is to be received or placed, the mobile telephone enters system access mode to try to access the system via a control channel. When it gains access, the control channel sends commands to the mobile telephone to tune to an analog or digital traffic channel. The mobile telephone responds by tuning to the designated channel and entering conversation mode.

Access

Access to an IS-136 system can occur through an analog control channel or digital control channel. Prior to accessing an IS-136 system on an analog control channel, a mobile telephone observes the broadcast data bits transmitted by the base station control channel for a protocol capability indicator (PCI) flag to determine if it is digital-capable. When the mobile telephone attempts access to a system that is digital-capable, the system indicates its dual mode capability with the mobile protocol capability indicator (MPCI). If the access attempt succeeds, the system assigns the mobile telephone an analog or digital voice channel. If it assigns an analog channel, the system sends an initial voice channel designation (IVCD) message which contains the voice channel number. If the system assigns a digital channel, it sends an initial traffic channel designator (ITCD) message containing the channel number and time slot.

Figure 4.12 shows the access procedures for when a mobile telephone access the cellular system through a digital control channel. When two or more mobile telephones contend for access at the same time, the base station eventually recognizes one of them and begins communicating with it. While monitoring the DCC, a mobile telephone attempts access by first examining the to the SCF digital bit field to see if the channel is busy. If it is not busy, the mobile telephone sends its message. After starting to send its message, the mobile telephone looks for the base station to indicate that the channel status has changed to busy. If the channel becomes busy within a specific period of time, the phone assumes that the base station has received its message. The base station also sends other information, called the partial echo (PE) field, to indicate whether it has received the mobile telephone's message or a message from another mobile telephone. The PE field is a value obtained from the mobile telephone's message echoed back from the base station.

This type of access system works well, except when two mobile telephones see the same idle slot and initiate transmissions simultaneously. If the base station correctly receives one of the messages, both mobile telephones detect the received flag, but only one can be served. Such collisions are prevented with the PE field. The PE field is a value that allows mobile telephones to identify which access attempt the

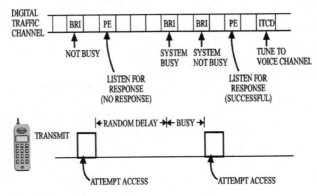

Figure 4.12, TDMA System Access

base station is responding to. If a mobile telephone sees the received flag without the correct PE value, it stops the access attempt and retries after a random delay. This protocol greatly reduces the occurrence of unintentional simultaneous transmissions by multiple mobile telephones and thus increases the number of good messages per second, compared to the simpler and less effective protocol used on the analog control channel.

Paging

In response to a page message, but prior to accessing the IS-136 system, a dual mode mobile telephone determines if the system has digital capability. If the system has digital capability (the PCI bit is set), the mobile telephone responds to the page message by attempting to access the cellular system. During the access attempt, the mobile telephone indicates that it is responding to a page message and that it has digital capability (MPCI).

Dual mode cellular systems can also send calling number identification. After the page message is received and the voice channel or traffic channel assigned, the system can transfer the calling number ID (CNI). The CNI is transferred on a the voice or traffic channel prior to the subscriber answering the call. Figure 4.13 illustrates the IS-136 paging process.

Pages which are sent on the digital radio channel are part of the SPACH channel. A mobile telephone monitors its paging channel for any messages from the base station. An algorithm using the MIN of the mobile telephone determines which paging channel to monitor. Once a mobile telephone has locked on to its paging channel, it wakes up for designated superframes to read a slot of paging information. If

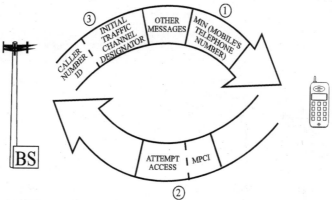

Figure 4.13, TDMA System Paging

the mobile telephone detects a page message, it sends a page response to the base station. The base station responds with an initial traffic channel designation. Multiple paging channels can exist on the same cell, so the mobile telephone must use its MIN and information found on the BCCH channel to first find the correct paging channel wake up time.

Handoff

When a dual mode mobile telephone is operating on a digital channel, the hand-off process is different from AMPS. Long before a hand-off, the base station typically commands the mobile telephone to measure nearby radio channels' signal strength. This is part of the mobile assisted handoff (MAHO) process. MAHO is a system in which the mobile telephone assists the hand-off decision by sending radio channel quality information back to the base station. In pre-existing analog systems, hand-off decisions were based only on measurements of mobile telephones' signal strength made by receivers at the base station. IS-136 systems use two types of radio channel quality information: signal strength of multiple neighbor cell's channels and an estimated bit error rate of the operating channel. The bit error rate is estimated using the result of the forward error correction codes for speech data and call processing messages. Having the mobile telephone report quality information also allows for measurements of the forward or downlink quality that are not possible from the base station.

Figure 4.14 illustrates the MAHO process. The system sends the mobile telephone a MAHO message containing a list of radio channels from up to 12 neighbor cells (usually only 6 neighboring cells are measured). During its idle time slots, the mobile telephone measures the signal strength of the channels on the list including the channel it is currently operating on. The mobile telephone averages the signal strength measurements over one second, then continually sends MAHO channel strength reports back to the base station every second. The system combines the MAHO measurements with its own information (such as knowing which cells have available idle voice or digital traffic channels) to determine which radio channel will offer the best quality, and it initiates hand-off to the best channel when required.

Once the mobile telephone is on an analog voice channel or a digital traffic channel, it may hand off to another voice or traffic channel.

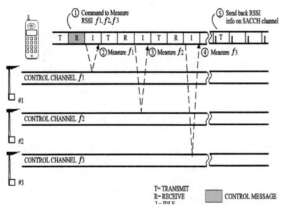

Figure 4.14, Mobile Assisted Hand-off

Initially, the mobile telephone is commanded by the base station to begin monitoring the radio signal of a nearby base station. (time 1). The mobile telephone continually sends the radio channel quality information back to the base station using the SACCH channel (time 2). Use of the SACCH channel leaves voice quality unaffected. When the base station determines that the mobile telephone can be better served by another cell site, it commands the unit to tune to a new radio channel (time 3) When the handoff is to a new cell which differs in radius from the old one by more than 3 km (2 miles), the audio is now muted on the digital traffic channel. When the mobile telephone has tuned to the new radio channel, it sends shortened bursts if the handoff was between two cells which differ by more than 2 miles (time 4) to avoid potential collisions with bursts sent from other mobile telephones. When the base station determines the necessary time adjustment, it commands the mobile telephone to time align (time 5). The mobile telephone then unmutes the audio and begins voice communications (time 6). When the two cells are about the same size, no muting or shortened burst are used. This is also the case for a sector to sector handoff within the same cell. Then there is no gap in the speech, providing a so called "seamless" TDMA handoff.

IS-136 systems may hand off between different system ID's. For example, a mobile telephone could hand off from a private system to the public system, or between two adjacent public systems. The mobile telephone experiences no difference when handing off between systems, but within the network, some additional information is required for billing. Rates on private systems are not necessarily the same as the public system, even when they occupy the same cell site and RF channel.

References:

1. Calhoun, George, "Digital Cellular Radio," Artech House, MA, 1988.
2. SuperPhone conference, Institute for International Research, New York NY, January 22-24, 1997.

Chapter 5
CDMA (IS-95)

Code Division Multiple Access

The Code Division Multiple Access (CDMA) system which originated in the United States is based on the IS-95 industry specification. IS-95 CDMA combines new digital spread spectrum CDMA and Advanced Mobile Phone Service (AMPS) functionality into one dual-mode cellular telephone on the 800 MHz band, and can use a CDMA-only handset on the 1.9 GHz PCS band. CDMA systems primarily differ from FDMA (analog) and TDMA systems through the use of coded radio channels. In a CDMA system, users can operate on the same radio channel simultaneously by using different coded sequences.

History

Spread spectrum radio technology has been used for many years in military applications. CDMA is a particular form of spread spectrum radio technology. In 1989, CDMA spread spectrum technology was presented to industry standards committee but it did not meet with immediate approval by the standards committee since they had just resolved a two-year debate between TDMA and FDMA and were not eager to consider another access technology. CDMA cellular service began testing in the United States in San Diego California during

1991. In 1995, IS-95 CDMA commercial service began in Hong Kong and now several CDMA systems are operating throughout the world, including a 1.9 GHz all-digital system in the US since November, 1996.

The development of CDMA was partially inspired by an attempt to satisfy the goals of the Cellular Telecommunications Industry Association (CTIA) User Performance Requirements (UPR) objectives for the next generation of cellular technology, particularly the goal of increasing capacity to 10 times that of analog cellular technology. In response to these objectives, radio specifications were created that CDMA proponents claim to satisfy the requirements. A proprietary specification was presented by Qualcomm to the Telecommunications Industry Association (TIA) which modified and accepted it as the IS-95 CDMA specification.

System Overview

The IS-95 CDMA system allows for voice or data communications on either a 30 kHz AMPS radio channel (when used on the 800 MHz cellular band) or a new 1.23 MHz CDMA radio channel. The IS-95 CDMA radio channel allows multiple mobile telephones to communicate on the same frequency at the same time by special coding of their radio signals. This different type of radio channel is called Code Division Multiple Access (CDMA). The CDMA system includes many of the same basic subsystems as other cellular systems, including a switching network, base stations (BS), and mobile telephones. IS-95 systems can serve mobile telephones of three types: AMPS only, IS-95 Dual Mode, or IS-95 (digital only). Figure 5.1 shows an overview of an IS-95 radio system.

The IS-95 CDMA cellular system has three basic types of radio channels: analog control channels (ACC), analog voice channels (AVC), digital traffic channels (DTC). There are two types of analog radio channels: either a control channel or a voice channel. The CDMA radio channel is divided into several different types of control and voice channels by the use of different codes and by time scheduling

CDMA radio channels carry control, voice, and data functionality by dividing a single traffic channel (TCH) into different sub-channels. Each of these channels is identified by a unique code. When operating on a CDMA radio channel, each user is assigned to a code for transmission and reception. Some codes in the TCH transfer control channel information, and some transfer voice channel information.

The control channel which is part of a digital traffic channel on a CDMA system has new advanced features. This digital control channel (DCC) carries system and paging information, and coordinates access similar to the analog control channel (ACC). The DCC has many more capabilities than the ACC such as a precision synchronization signal, extended sleep mode, and others. Because each CDMA radio channel has many codes, more than one control channel can exist on a single CDMA radio channel and the CDMA control channels co-exist with other coded channels that are used for voice.

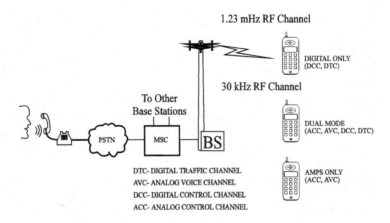

1.23 mHz RF Channel

DIGITAL ONLY
(DCC, DTC)

30 kHz RF Channel

To Other
Base Stations

DUAL MODE
(ACC, AVC, DCC, DTC)

AMPS ONLY
(ACC, AVC)

PSTN   MSC   BS

DTC- DIGITAL TRAFFIC CHANNEL
AVC- ANALOG VOICE CHANNEL
DCC- DIGITAL CONTROL CHANNEL
ACC- ANALOG CONTROL CHANNEL

Figure 5.1, Overview of an IS-95 CDMA Radio System

System Attributes

The IS-95 CDMA cellular system has several key attributes that are different from other cellular systems. The same CDMA radio carrier frequencies may be optionally used in adjacent cell sites, which eliminates the need for frequency planning, the wide-band radio channel provides less severe fading, which the inventors claim results in consistent quality voice transmission under varying radio signal conditions. The CDMA system is compatible with the established access technology, and it allows analog (EIA-553) and dual mode (IS-95) subscribers to use the same analog control channels. Some of the voice channels are converted to CDMA digital transmissions, allowing several users to be multiplexed (shared) on a single RF channel. As with other digital technologies, CDMA produces capacity expansion by allowing multiple users to share a single digital RF channel.

The CDMA system has moved some system control functions to the mobile telephone. Like other cellular technologies, CDMA uses a Mobile Switching Center (MSC) to centralize control of mobile telephones moving through the system.

IS-95 Radio Channels

The IS-95 cellular system has three basic types of radio channels: analog control channels (ACC), analog voice channels (AVC), and digital traffic channels (DTC). The analog radio channel is either a control channel or a voice channel. The CDMA radio channel is divided into several different types of control and voice channels by the use of different channel codes.

The IS-95 CDMA radio channel divides the radio spectrum into wide 1.23 MHz digital radio channels. CDMA radio channels differ from other technologies in that it multiplies (and therefore spreads the spectrum bandwidth of) each signal with a unique pseudo-random noise (PN) code that identifies each user within a radio channel. CDMA transmits digitized voice and control signals on the same frequency band. Each CDMA radio channel contains the signals of many ongoing calls (voice channels) together with pilot, synchronization, paging, and access (control) channels. Digital mobile telephones select the signal they are receiving by correlating (matching) the received signal with the proper PN sequence. The correlation enhances the power level of the selected signal and leaves others un-enhanced.

Each IS-95 CDMA radio channel is divided into 64 separate logical (PN coded) channels. A few of these channels are used for control, and the remainder carry voice information and data. Because CDMA transmits digital information combined with unique codes, each logical channel can transfer data at different rates (for example, 4800 b/s, 9600 b/s).

CDMA systems use a maximum of 64 coded (logical) traffic channels, but they cannot always use all of these. A CDMA radio channel of 64 traffic channels can transmit at a maximum information throughput rate of approximately 192 kb/s [1], so the combined data throughput for all users cannot exceed 192 kb/s. To obtain a maximum of 64 communication channels for each CDMA radio channel, the average data rate for each user should approximate 3 kb/s. If the average data rate is higher, less than 64 traffic channels can be used. CDMA systems can vary the data rate for each user dependent on voice activity (variable rate speech coding), thereby decreasing the average number of bits per user to about 3.8 kb/s [2]). Varying the data rate according to user requirement allows more users to share the radio channel, but with slightly reduced voice quality. This is called soft capacity limit.

Figure 5.2 shows how CDMA channels share each radio channel. Digital signals are coded to produce multiple chips (radio energy) for each bit of information to be transmitted. In the jargon of CDMA, a chip is a name for one of the bits in the PN bit sequence which is generated at a higher bit rate than the data rate to be coded. The receiver has an internal chip generator which can produce exactly the same PN chip sequence as the one used for encoding at the transmitter. The time delay of this chip generator in the receiver is adjustable to allow for the time the radio signal requires to travel from the transmitter to the receiver. The receiver shifts the PN chip pattern in time until it matches the coded pattern. The particular chip pattern is illustrated symbolically by a combination of pictures such as circles, squares, and diamonds in figure 5.2. Chips on the forward radio channel (from the Base Station to the mobile telephone) are selected to collide only infrequently with chips from other users. A chip collision occurs when the binary sequence of a chip pattern (such as 011010111000) matches that of another chip pattern for a short interval of time. When two chip sequence patterns are designed so that there is no complete collision between the two during one data bit interval, this is known as orthogonal coding. The 64 chip or PN patterns used in IS-95 CDMA are perfectly orthogonal to each other, but each pattern is combined

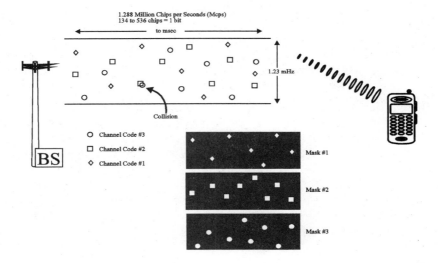

Figure 5.2, CDMA Radio Channel

with a pattern which is unique to the mobile telephone used for that channel, and some codes are used in adjacent cells. As a result, chips patterns from many subscribers may produce short term collisions on the reverse channel.

Several chips are created for each bit of user information (speech or data), so if some of them encounter interference form of code collisions shown in figure 5.2, most of the remaining chips will still be received successfully. CDMA channels are designed using error protection codes so they can operate with some limited amount of interference among users (chip collisions), so CDMA radio channels can tolerate a limited amount of interference without significantly reducing voice quality.

The CDMA radio channel is constructed of coded signals on the same frequency, so if adjacent cells use different codes they may reuse the same radio frequency. Interference from neighboring cells appears only as chip collisions when the PN chip codes for the two conversations are not completely orthogonal. The effect of chip collision interference is to reduce system capacity.

The IS-95 system adds several types of information to the transmitted data. Error protection bits are added to protect the digitized speech from errors created during radio transmission. Control messages are inserted to coordinate operation between the Base Station and the mobile telephone, facilitating soft hand-off and other channel maintenance functions.

Capacity Expansion

The IS-95 system increases voice channel capacity and the paging channel capacity of a cellular system significantly compared to analog cellular systems. The basic method of voice channel capacity expansion in a IS-95 CDMA system is to allow more users to share a single radio frequency than in the AMPS system. In a typical cellular system, a single IS-95 CDMA radio channel which can provide for many voice channels (approximately 10 to 20 [3] are proposed by CDMA inventors) replaces two AMPS radio channels that only provide 2 voice channels. Additional voice capacity can be added by reducing interference from nearby cells or by slightly decreasing the voice quality intentionally. Interference can be reduced from nearby cells by using directional antennas or focused repeaters. Voice channel capacity can also be increased by reducing the average data rate available to the users in the system. This divides the total channel capacity to more users. A reduced data rate typically results in reduced audio quality.

Although a CDMA system can page mobile telephones through an analog control channel, each CDMA radio channel has its own paging channel. When a mobile telephone registers on a CDMA radio channel, pages can be sent to the CDMA paging channel and not to the analog control channel. This increases the total system paging channel capacity.

Traffic Channel Structure

Each CDMA 1.23 MHz bandwidth radio frequency pair contains its own digital control channel (DCC). A CDMA DCC is composed of four types of channels, each identified by a unique code: pilot, synchronization, paging, and access.

Each CDMA Base Station transmits a pilot signal on a unique pilot channel (shown in figure 5.3). The pilot channel is distinguished from other PN coded channels by a particular PN chip code. The pilot signal is the reference timing signal for demodulating the signal and for estimating received signal strength to indicate which cell site can best communicate with the mobile telephone. CDMA mobile telephones simultaneously measure the pilot signal strengths of all neighboring Base Stations in the system.

After the mobile telephone determines the strongest pilot channel, it demodulates the synchronization channel. The synchronization channel contains information parameters that allow the mobile telephone to synchronize to other CDMA channels (paging, access, and voice). These include system parameters, access parameters, channel list information, and a neighboring radio channel list.

The CDMA paging channel continuously sends system parameter and paging information intended for a single mobile telephone or group of mobile telephones. After the mobile telephone has initialized, it continuously listens to the strongest paging channel to determine if a call is to be received.

When system attention is required, the mobile telephone competes for access on the access channel. This is a random process in which the mobile telephone continually increases its access power until the base

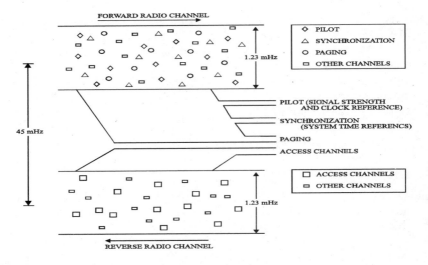

Figure 5.3, Digial Control Channels

station responds. If the Base Station does not respond within an allotted time, or the mobile telephone exceeds a maximum power level, the access attempt is aborted. This procedure avoids interference between mobile telephones during access attempts. When access is complete, the voice information is transmitted over separate voice channels. Figure 5.3 shows that each CDMA radio channel contains a pilot channel, synchronization channel, paging channel, and several different traffic channels.

Signaling

Signaling is the physical process of transferring control information to and from the mobile telephone. When a dual mode IS-95 set operating on the AMPS channels, signaling is performed using the blank-and-burst method of EIA-553. When operating on the CDMA channel, signaling is sent by blank-and-burst or by dim-and-burst. Dim-and-burst signaling sends control information in unused bit locations during periods of low speech activity. Variable rate speech coding varies the coding rate so that both voice and control messages may be sent during each 20 msec frame, thereby allowing for fast or slow dim-and-burst signaling.

The AMPS control channel signaling is unchanged in dual mode CDMA systems. The AMPS control channel was not modified to allow assignment to a CDMA radio channel because each CDMA radio channel has its own control channel which can assign either an AMPS or CDMA radio channel. The CDMA system adds a new control channel composed of a pilot, synchronization, paging, and access channel. CDMA control messages can be sent using either one of two alternative techniques. One technique, called blank-and-burst, momentarily blanks the audio and replaces it with a burst of control information. An alternative technique, called dim-and-burst, reduces the number of digitized speech bits available to characterize the voice, and adds a short burst of control information bits. Any control message can be sent by either blank-and-burst or dim-and-burst signaling.

Blank and Burst

Blank-and-burst signaling replaces speech data with signal messages. For historical reasons, this is called "in-band signaling." Blank-and-burst message transmissions degrade speech quality because they replace speech frames with signaling information. The QCELP speech coder used in IS-95 will repeat the sound generated by the previous good frame of digitally coded voice if the current frame has been replaced by a blank and burst message. The quality degradation for only one isolated replacement is almost imperceptible. Figure 5.4 illustrates blank-and-burst signaling.

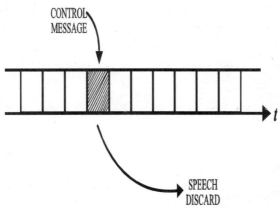

Figure 5.4, CDMA Blank-and-Burst Signaling

Dim and Burst

Dim-and-burst inserts control messages when speech activity is low. The QCELP digital speech coder is designed to produce a lower bit rate when the voice is silent, between syllables or when there is only background noise, for example. As the speech coder changes bit rate, some of the unused bits are re-assigned as control message bits. Because all of the speech bits are not available for the control message, several frames are needed to send the message. The number of required frames varies according to speech activity. The degrading effect of dim and burst messages on speech is less perceptible (in fact, almost no degradation is perceptible at all) than blank and burst, but the message requires somewhat more time to be transmitted.

The gross user data rate available is 9600 b/s. The data rate is reduced by cycling the transmitter off for several 1.25 msec bursts (called power groups) during each 20 msec frame. The user data rate is determined by checking the frame quality bits (CRC). If the frame quality bits do not check for one data rate, decoding at another data rate will be attempted.

The mix between voice data and signaling data is determined by a mixed mode flag bit sent at the beginning of each frame. Additional flag bits include a burst format bit which indicates if the message is being sent via blank-and-burst or dim-and-burst, a traffic type which identifies primary or secondary channel, and a traffic mode bit which sets the proportional mix between voice and signaling data. Figure 5.5 illustrates a dim-and-burst message.

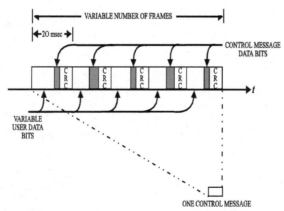

Figure 5.5, CDMA Dim-and-burst Signaling

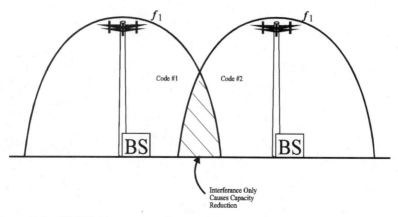

Figure 5.6, CDMA Frequency Reuse

Frequency Reuse

Figure 5.6 shows that CDMA allows the same frequency (f1) to be reused in all adjacent cells (This is known in the cellular and PCS industry as a frequency plan with N=1). In the shaded area where interference is significant, chip collisions from adjacent cells and other subscribers are more frequent, but this only reduces the number of users that can share the radio channel. Reusing the same frequency in every cell eliminates the need for frequency planning in CDMA systems.

Each CDMA radio channel occupies 1.23 MHz of spectrum (a bandwidth corresponding to about 40 AMPS radio channels). However, one CDMA radio channel typically replaces 2 to 6 AMPS radio channels in a single cell site or sector. This is because AMPS radio channels in each cell or sector are placed 7 to 21 channels apart (210 to 630 kHz frequency separation) to allow for n=7, 12, 13, or 21 frequency planning. In an n=7 cell frequency plan with 3 120° sectors, the frequency separation in one sector alone is 21 channels or 630 kHz. However, in most cases a CDMA system would be used in the entire cell (3 sectors) rather than only one sector.

Soft Hand-off

In AMPS cellular systems, hand-off occurs when the Base Station detects a deterioration in signal strength from the mobile telephone. As AMPS subscribers approach hand-off, signal strength may vary abruptly, and the voice is muted for at least 200 milliseconds in order to send control messages and complete the handoff. In contrast, CDMA uses a unique "soft hand-off," which is nearly undetectable and loses few if any information frames. CDMA inventors also claim that CDMA's soft hand-off is much less likely to lose a call during hand-off.

Soft hand-off allows the mobile telephone to communicate simultaneously with two or more cell sites to continuously select the best signal quality until hand-off is complete. The CDMA mobile telephone measures the pilot channel signal strength from adjacent cells and transmits the measurements to the serving Base Station. When an adjacent Base Station's pilot channel signal is strong enough, the mobile telephone requests the adjacent cell to transmit the call in progress. The serving Base Station also continues to transmit as well. Thus, prior to complete hand-off, the mobile telephone is communicating with both Base Stations simultaneously. Using two base stations with the same frequency and the same PN chip code simultaneously during hand-off maintains a much higher average signal strength throughout the process. During soft hand-off, the base receivers choose the best frames of digitally coded speech from either Base Station by sending both frames to the MSC via digital links and having the MSC evaluate the errors disclosed via the error protection coding used with the digitally coded speech. The mobile telephone uses its RAKE receiver (described below) to add or optimally combine the two base transmitter signals, even if they do not arrive at the mobile telephone in synchronism due to different distances from the two base stations. Soft hand-off produces almost no perceptible interruption in voice communications, and does not lose any data bits for modems, credit card machines, and other services transferring digital data. It is similar to, but slightly better than, the seamless handoff used in various TDMA systems in this regard. Figure 5.7 shows how CDMA systems use two base stations simultaneously during hand-off.

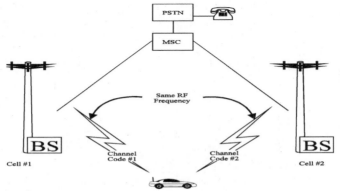

Figure 5.7, CDMA Soft Hand-off

Variable Rate Speech Coding

CDMA speech coding occurs at a variable rate. The coding process begins with an analog-to-digital converter digitizing the user's voice at a fixed sample rate of 8,000 samples per second 64 kb/s. The digitized voice is supplied to a speech coder, which encodes the speech by characterizing the speech into voice parameters. When voice activity is low, the variable-rate CDMA speech coder represents the speech by a signal with fewer bits. This added coding efficiency increases CDMA system capacity by a factor related to the ratio of silence to sound intervals in the speech.

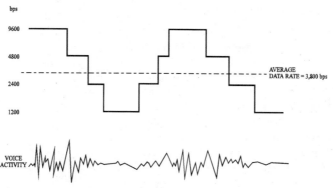

Figure 5.8, Variable Rate Speech Coding

Figure 5.8 shows how the speech coder compression rate varies with speech activity. The speech signal is divided into 20 msec intervals. The speech coder produces data rates in the range of 9600 down to 1200 b/s. As the speech activity decreases, the bit rate decreases.

Discontinuous Reception (Sleep Mode)

Discontinuous reception (DRX) enables mobile telephones which are not engaged in conversation to power off non-essential circuitry during periods (sleep) when pages will not be received. To provide for this sleep mode, the paging channel is divided into paging sub-channel groups.

Figure 5.9 shows the DRx (sleep mode) process. When the mobile telephone registers on a CDMA radio channel, it informs the system of its sleep mode capability. CDMA paging channels are divided into 200 msec slots (paging groups) which allow the mobile telephone to sleep during unwanted groups. Each paging group is composed of 10 frames (200 msec). The system can dynamically assign up to 640 paging

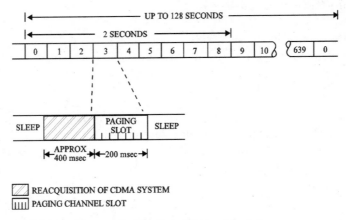

Figure 5.9, Discontinuous Reception (Sleep Mode)

groups to allow a maximum sleep period of 2 minutes and 8 seconds. For normal operation, about 10 groups will be used for a maximum delay of about 2 seconds. Approximately 400 msec before the end of a sleep period, the mobile telephone wakes up to allow re-acquisition with the control channel.

Soft Capacity

A cellular system is in a condition of over capacity when more subscribers attempt to access the system than its radio interface can support at a desired quality level. CDMA allows the system to operate in a condition of over capacity by accepting a higher-than-average bit error rate, or reduced speech coding rate. As the number of subscribers increases beyond a threshold, voice quality begins to deteriorate, but subscribers can still gain access to the system.

Figure 5.10 shows that as more users are added to the system, voice quality deteriorates. When voice quality falls below the allowable minimum (usually determined by an acceptable bit error rate), the system is over capacity. Allowing more subscribers on the system by trading off voice quality (or the subscribers data rate) creates a soft capacity limit.

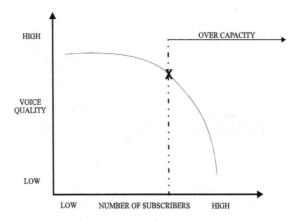

Figure 5.10, IS-95 Soft Capacity Limit

RF Power Control

To effectively separate the coded channels, the received signals at the Base Station from all mobile telephones must be at almost the same level. If one received signal were much more powerful than the others, the receiver could not effectively decode the weaker ones, making it much less sensitive to weaker channels. To accommodate this requirement for uniform signal levels, the CDMA system precisely controls mobile telephone power. The power control system performs two simultaneous operations; open loop control and closed loop control. The open loop control is a coarse adjustment and the closed loop control is a fine adjustment. The power control system maintains received signals within (1 dB (33%)) of each other. Demonstrations have also shown that a strong interfering signal reduces the number of users per radio channel in a serving cell site. When interference is too great, mobile telephones are handed off to another cell [4].

A CDMA mobile telephone's coarse (open loop) RF amplifier adjustment is controlled by feedback from its receiver section. The mobile telephone continuously measures the radio signal strength received from the Base Station to estimate the signal strength loss between the Base Station and mobile telephone. Figure 5.11 shows that as the mobile telephone moves away from the Base Station, the received signal level decreases. When the received signal is stronger, the mobile telephone reduces its own RF signal output; conversely, when the

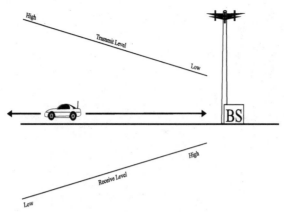

Figure 5.11, CDMA Open Loop RF Power Control

mobile received signal level is weaker, the mobile telephone increases the amplification of its own RF signal output. The end result is that the signal received at the Base Station from the mobile telephone remains at about at the same power level regardless of the mobile telephone's distance.

Because the open loop power adjustment does not adequately control received signal level by itself, the Base Station also fine-adjusts the mobile telephone's RF amplifier gain by sending power level control commands to the mobile telephone during each 1.25 msec time interval. The commands are determined by the Base Station's received signal strength. The power control bit communicates the relative change from the previous transmit level, commanding the mobile telephone to increase or decrease power from the previous level.

Figure 5.12 illustrates closed loop power control. As the received signal power increases, the power control bit signals the mobile telephone to reduce transmit power level. When the received signal is lower than desired, the power control bit commands the mobile telephone to increase power. The closed loop adjustment range (relative to the open loop) is 24 dB minimum. The combined open and closed loop adjustments precisely control the received signal power at the Base Station.

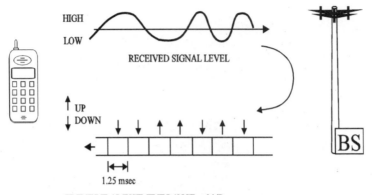

HIGH

LOW

RECEIVED SIGNAL LEVEL

↑ UP
↓ DOWN

1.25 msec

FINE TUNE ADJUSTMENT RANGE: ±24dB

Figure 5.12, CDMA Closed Loop RF Power Control

Mobile telephones also reduce their average power by transmitting only in bursts when the channel data rate is reduced. Figure 5.13 shows how a 4800 bps channel transmits only 1/2 of the time.

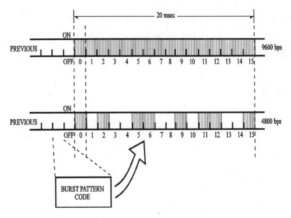

Figure 5.13, CDMA Mobile telephone Transmit Power Bursts

Rake Reception

When multiple reflected (multipath) signals are received at slightly different times, CDMA can combine multipath signals, adding several weak multipath signals to construct a stronger one. The gains from this process are similar to those obtained from antenna diversity. This process is called RAKE reception. The result is better voice quality and fewer dropped calls than would otherwise be available.

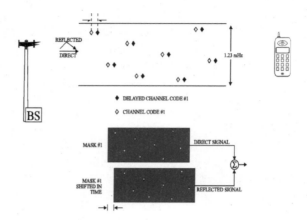

Figure 5.14, CDMA Rake Reception

Figure 5.14 shows how a multipath signal can be added to the direct signal. The radio channel shows two code sequences. The shaded codes are time delayed because the original signal was reflected and received a few microseconds later. The original signal is decoded by mask #1. The mask is shifted in time until it matches the delayed signal. The output of each decoded channel is combined to produce a better quality signal.

IS-95 system parameters are significantly different from the EIA-553 AMPS system. The IS-95 system parameters include a new wide RF radio channel, 45 MHz duplex channel separation (on the 800 MHz band) or 80 MHz duplex channel separation (on the 1.9 GHz PCS band), digital phase modulation, and new RF output power levels.

Guard Bands

While the 25 MHz x 2 cellular band allocation on the 800 MHz band remains the same, the frequency allocation for use by CDMA systems has been divided so that 9 or 10 of the 1.23 MHz CDMA channels can be allocated in the A or B frequency bands. Because CDMA channels have more bandwidth than analog channels and require a guard band at the ends of the spectrum, the allowable channel assignments for CDMA channels are 1013 through 1023 and 1 through 777. The center frequency is used for the CDMA channel assignment. Figure 5.15 shows the CDMA channel guard bands.

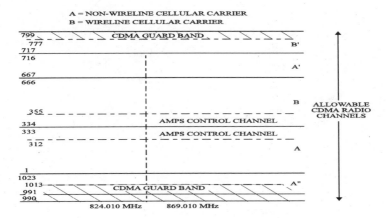

Figure 5.15, CDMA Channel Guard Bands

Duplex Channels

Although the CDMA radio channel is digitized, the transmission continues to be frequency duplex. The separation of forward and reverse channels on the 800 MHz cellular band is 45 MHz. The transmit band for the Base Station is 869-894 MHz. The transmit band for the mobile station is 824-849 MHz.

Figure 5.16 illustrates the time offset between the forward and reverse channel, called time alignment. Time alignment advances or retards transmit bursts in 1.25 msec steps. Stepping the transmit time up or back effectively reduces interference to nearby cells. Unlike other technologies, CDMA systems do not need to receive frames at a precise time relative to transmit frames. This ability to shift mobile telephones' time alignment allows CDMA systems to prevent all the mobile telephones from transmitting at the same time, thus reducing the average interference level received by the Base Station.

Frequency Diversity

The CDMA radio channel spreads the signal over a wide 1.23 MHz frequency range, making it less susceptible to radio signal fading that occurs only over a specific narrow frequency range. As a result, radio signal fades affect only a portion of the CDMA signal bandwidth, and most of the information gets through successfully. With only a small portion of information corrupted, digital information transmissions over a CDMA radio channel are relatively robust.

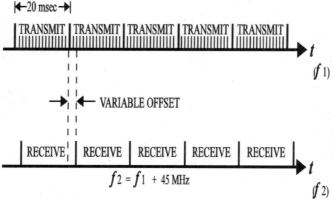

Figure 5.16, CDMA Duplex Channel

RF Power Classification

Maximum RF output power for AMPS and CDMA mobile telephones is the same. The requirements for precise power control demand that CDMA mobile telephones transmit results in an average transmit power which is 10% of analog FM mobile telephones [5].

Basic Operation

When a IS-95 mobile telephone is first powered on, it initializes by scanning for control channels and tuning to the strongest one. In this initialization mode, it scans all the specific CDMA carrier frequencies to determine if the system is CDMA digital capable. If the system is not CDMA capable, the mobile telephone locks on to an AMPS control channel. If it is CDMA capable, the mobile telephone locks onto the CDMA control channel and registers with the system. After initialization, the mobile telephone enters idle mode and waits to be paged for an incoming call or for the user to place a call (access). When a call is to be received or placed, the mobile telephone enters system access mode to try to access the system via a control channel. When it gains access, the control channel commands the mobile telephone to tune to an analog or digital traffic channel. The mobile telephone tunes to the designated channel and enters conversation mode.

Access

Prior to accessing an IS-95 system, a mobile telephone listens for a CDMA pilot channel to determine if it is digital capable. If the system is digital capable, the mobile telephone attempts access via the CDMA radio channel. If the access attempt succeeds, the system assigns the mobile telephone an analog or digital voice channel. If it assigns an analog channel, the system sends an initial voice channel designation (IVCD) message which contains the voice channel number. If the system assigns a digital channel, it sends an initial traffic channel designator (ITCD) message with the channel number and channel code.

The CDMA access channel is on the reverse radio link (mobile telephone to base station). Each CDMA radio channel can have up to 32 separate (coded) access channels. Access channel messages are grouped into 20 msec frames of 88 information bits. The gross channel rate for the access channel is 9600 b/s. However, access channel messages are repeated twice, reducing the effective channel rate to 4800 b/s.

The mobile telephone accesses the CDMA channel by increasing its power level stepwise until the Base Station acknowledges the request. The power level increases continue up to a maximum power limit designated by the control channel before the access attempts begins. If the first access attempt (a complete sequence of power level increases) is unsuccessful, mobile telephone waits for a random period before attempting access again. Figure 5.17 illustrates the CDMA system access.

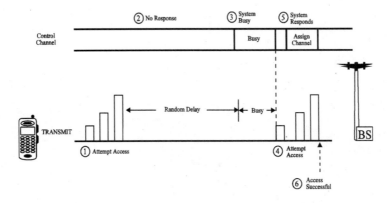

Figure 5.17, CDMA System Access

Paging

Paging is the process of sending a page message to the mobile telephone to indicate that a call is to be received. The IS-95 mobile telephone listens for pages on either the CDMA paging control channel or AMPS analog control channel (ACC), the preferred choice being the CDMA paging channel. CDMA paging channels can page in groups to allow sleep modes.

Because the CDMA mobile telephone can simultaneously decode more than one coded channel, it can monitor multiple paging channels. When the mobile telephone determines that a neighboring cell site has a higher quality paging channel, it initiates hand-off (registers) to the new paging channel.

Page messages are sent on the paging channel. Figure 5.18 illustrates the paging process. Initially (step 1), the mobile telephone monitors the paging channel for pages. When the mobile telephone is paged, it requests service from the cellular system (step 2) indicating that it is responding to a page message. The cellular system then assigns it a new channel code (step 3) where it will be authenticated (step 4) and the conversation may begin.

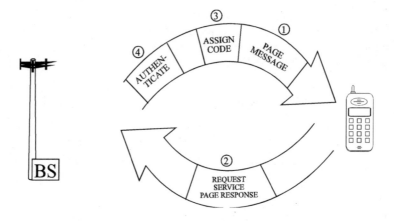

Figure 5.18, System Paging

The IS-95 cellular system can send calling line identification (CLI). After the page message has been received and the voice channel or traffic channel has been assigned, the system can transfer the calling number ID. The calling number ID can be transferred on a voice or traffic channel before the subscriber answers.

Handoff

The hand-off command sent on the AMPS voice channel does not support AMPS-to-CDMA hand-off. It is technically possible to hand-off from AMPS to CDMA through a hard hand-off (with a mute interval) like the standard AMPS hand-off, but this was deemed unnecessary [6]. Because there are no AMPS to CDMA hand-offs, the signaling format of the AMPS voice channel is unchanged.

The hand-off process for dual mode subscribers operating on a digital channel is different from that of AMPS. Figure 5.19 illustrates CDMA system digital channel hand-off. Before hand-off, the mobile telephone has received a list of neighboring cells' pilot channels that are candidates for hand-off from the serving cell (#1, time 1). The mobile telephone continuously measures the signal strength of the candidate radio channels (time 2). When the pilot channel of a neighboring cell #2 is sufficient for hand-off, the mobile telephone requests simultaneous transmission from that channel (time 3). The system then assigns the new channel to transmit simultaneously from cell #1 and cell #2 (time 4). The mobile telephone continues to decode both channels (different codes on the same frequency) using the channel with the best received quality. When the signal strength of the original channel falls below a threshold, the mobile telephone requests release of the original channel (time 5) and voice transmissions from cell #1 ends (time 6). The illustration shows the mobile telephone simultaneously communicating with only two Base Stations, but simultaneous communications with more than two Base Stations is possible.

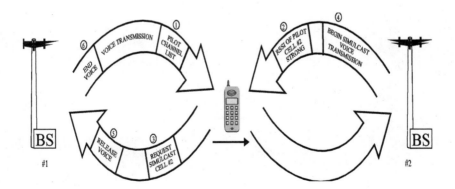

Figure 5.19, CDMA Hand-off

References:

1. Harte, Lawrence, "Cellular and PCS/PCN Telephones and Systems," APDG Publishing, NC, 1996, p. 127.
2. Cellular Telecommunications Industry Association, Winter Symposium, Results of the Field Trials," Washington, DC, 1991, December 4-5.
3. EMCI, "Digital Cellular, Economics and Comparative Analysis," Washington DC, 1993, p. 74.
4. Cellular Telecommunications Industry Association, Winter Symposium, Results of the Field Trials," Washington, DC, 1991, December 4-5.
5. EMCI, "Digital Cellular, Economics and Comparative Analysis," Washington DC, 1993, p. 75.
6. Cellular Telecommunications Industry Association, Winter Symposium, Results of the Field Trials," Washington, DC, 1991, December 4-5.

Chapter 6
European Digital Cellular

European Digital Cellular

The Global System for Mobile Communications (GSM) system is a digital radio system which uses Time Division Multiple Access (TDMA) technology. When communicating in a GSM system, users can operate on the same radio channel simultaneously by sharing time slots. The GSM specification or variants of the specification (particularly PCS-1900 in North America and DCS-1800 in the UK) is used in a variety of systems and frequencies including Personal Communications Services (PCS) and Personal Communications Network (PCN) systems.

History

Global System for Mobile Communications (GSM) is a digital cellular specification which was initially created to provide a single-standard pan-European cellular system. GSM began development in 1982, and the first commercial GSM digital cellular system was activated in 1991. Because it was created by representatives from many countries, it is accepted as the digital standard in more than 70 countries, including more than 30 countries outside Europe.

Before GSM, Europe had 9 different analog cellular systems, using 7 different incompatible technologies. Because most mobile telephones in the 1980's could only operate on a single type of cellular system, most customers in Europe could not roam to neighboring countries. Mobile telephones in Europe were also more expensive due to high cost of manufacturing small quantities of unique radio products.

At the 1982 Conference of European Posts and Telecommunications (CEPT), the standardization body, Groupe Spécial Mobile, was formed to begin work on a single European standard. The standard was later named Global System for Mobile Communications (GSM). This standard incorporated compatibility with Integrated Services Digital Network (ISDN) [1] and Signaling System Number 7 (SS7). In 1990, phase 1 of the GSM specifications were completed, including basic voice and data services. At that time, work began to adapt the GSM specification to provide service at 1800 MHz. This 1800 MHz standard, called DCS 1800, is used for the Personal Communications Network (PCN) in the UK and is planned for use in other European nations. Some 900 MHz band GSM systems began operating in 1991, and by 1992 most major European GSM commercial systems were operating. Phase 2 of the GSM and DCS 1800 specifications, which added advanced short messaging, microcell support services, and enhanced data transfer capability, are now complete. The North American PCS-1900 standard is essentially similar to the British DCS-1800 standard, except that different radio frequencies are used and smaller cells with lower power mobile and base station transmit power are used as well. PCS-1900 systems have been operating on the 1.9 GHz PCS band in the US since 1996, the first in the Washington DC area.

System Overview

The GSM cellular system allows 8 mobile telephones to share a single 200 kHz bandwidth radio carrier waveform for voice or data communications. The GSM radio channel structure allows multiple mobile telephones to communicate on the same frequency by using different time slots on the radio channel. The GSM system includes many of the same basic subsystems as other cellular systems, including a switching network, base stations (BS), and mobile telephones. The GSM system separates the base station controlling function from the base station and moves it into a base station controller (BSC). A single BSC can serve several GSM base stations. GSM systems can serve mobile telephones of two basic types: full rate and half rate. Figure 6.1 shows an overview of a GSM radio system.

A GSM conversation uses a type of radio channel called a traffic channel (TCH), which is organized into frames and time slot bursts. The traffic channel carries voice, data, and control information. A frame is

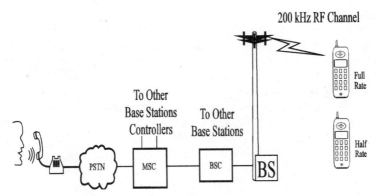

Figure 6.1, Overview of a GSM System

composed of 8 time slots used by 8 different conversations. From each frame, each user is assigned to a particular time slot burst for reception, and a particular corresponding burst for transmission. Some slots in the TCH transfer control channel information, and some transfer voice channel information.

GSM voice communication is conducted on two 200 kHz wide carrier frequency waveforms, a downlink or forward carrier (from the cell site to the mobile telephone) and an uplink or reverse carrier (from the mobile telephone to the cell site). The time slots between downlink/forward and uplink/reverse channels are related so that the mobile telephone does not simultaneously transmit and receive.

The GSM system has several types of control channels which are used in conjunction with the digital traffic channel. At least one carrier frequency in each cell or sector carries control channels. These digital control channels carry system and paging information, and coordinates access like the control channels on analog systems. The GSM digital control channels have many more capabilities than analog control channels such as broadcast paging, extended sleep mode, and others. Because the GSM control channels use only a portion (one or more slots), they typically co-exist on the same carrier frequency with DTC traffic channels that are used for voice communication. Some other carrier frequencies in the cell use only traffic channels and no control channels.

GSM systems allow several users to share each radio channel carrier frequency by dedicating a specific time slot from each frame to individual users. Voice channels can be either full rate or half rate. Full rate GSM systems assign one time slot per frame to each user , allowing 8 users to simultaneously share a radio channel. GSM is designed so that it can easily accommodate a future half-rate speech coder (a digital speech coder which produces good speech quality at half of the bit rate of the present speech coder) which is expected to emerge from the research laboratory in the next few years. Half rate GSM systems assign one time slot every other frame to allow up to 16 users to share a radio channel.

Full Rate GSM

Time intervals on full rate GSM channels are divided into frames with 8 time slots on two different radio frequencies. One frequency is for transmitting from the mobile telephone; the other is for receiving to the mobile telephone. During a voice conversation at the mobile set, one time slot is dedicated for transmitting, one for receiving, and six remain idle. The mobile telephone uses the idle time slots to measure the signal strength of surrounding cell carrier frequencies. These measurements assist in channel selection and hand-off. This time sharing results in a user-available data rate of 22.8 kb/s. Some of the 22.8 kb/s are used for error detection and correction, so only 13 kb/s of data are available for compressed speech data, or typically 12 kb/s for customer data or fax communication.

Subscribers talk and listen at the same time, so the mobile telephone must function as if it is simultaneously sending and receiving (called full duplex). When in conversation mode (called dedicated mode), GSM mobile telephones do not transmit and receive simultaneously when examined on a detailed time scale, but only appear to do so. Speech data bursts alternate between transmitting and receiving, and when received, the compressed speech bursts are expanded in time to create a continuous audio signal.

Figure 6.2 shows how GSM full duplex radio channels are divided in time. On the 900 MHz band, GSM digital radio channels transmit on one frequency and receive on another frequency 45 MHz higher, but not at the same time On the 1.9 GHz band, the difference between transmit and receive frequencies is 80 MHz. The mobile telephone transmits a burst of data on one frequency, then receives a burst on another frequency, and is briefly idle before repeating the process. Figure 6.2 shows how full rate systems share the GSM radio channel.

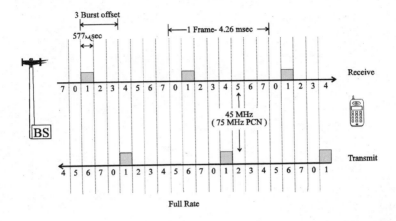

Figure 6.2, Full Rate Channel Sharing

Half rate GSM

A radio channel's capacity can be doubled by dedicating only one slot every other frame per subscriber, creating a half rate channel. Half rate channels use 1 of sixteen slots to transmit and 1 to receive, leaving 14 idle. Figure 6.3 illustrates the half rate GSM channel structure. Using only one of the sixteen time slots results in a user-available data rate of about 11.4 kb/s. A half rate system supports up to sixteen simultaneous users per radio channel. Introduction of half rate service is planned for the near future (revision 2+), after the standards committee has approved a suitable digital speech coder. Figure 6.3 shows how half rate systems share the GSM radio channel.

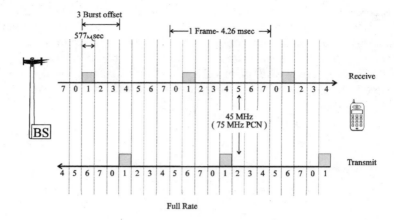

Figure 6.3, Half Rate Channel Sharing

System Attributes

The key attributes of a GSM system include digital only operation, a wide bandwidth TDMA radio channel, and advanced signaling/messaging capability. The GSM system is a digital-only system and was not designed to be backward-compatible with the established analog systems. The GSM radio band is shared temporarily with analog cellular systems in some European nations.

The GSM radio carrier frequency waveform is relatively wide when compared to other analog and digital standards. The 200 kHz wide radio carrier waveform allows for slightly reduced signal fading and thus has better performance for high data rates when required, compared to 25 or 30 kHz bandwidth signals. The GSM system includes a new set of control channels and features that allow the mobile telephone to perform some of the functions that had been performed by analog cellular systems.

GSM Radio Carriers (Channels)

GSM radio carriers (analogous to a single radio channel in an analog cellular system) use a single radio waveform which is divided into time periods called time slots to allows several users to share each radio carrier. Each GSM radio carrier is divided into several different types of control and voice channels by the use of different time slots or shared portions of time slots. Notice that in an analog cellular system, there is one carrier frequency for each voice channel, so the two terms are used synonymously. In a TDMA system such as GSM, the standard terminology established in the GSM standards documents distinguishes clearly between a carrier frequency waveform and the 8 channels which comprise it. The GSM control channels continually provide information to mobile telephones which are operating in the system but not engaged in a conversation, and coordinate their access. Typically, one radio carrier per cell or per sector has a single time slot dedicated as a control channel. This is called the beacon frequency in that cell or sector. In some cases, where there is a large amount of call setup or short message activity, other time slots in that same carrier frequency signal are also used for these types of activity in addition to the first dedicated channel.

The GSM radio system multiplexes several users onto a single radio carrier waveform through the use of distinct time slots. A GSM carrier transmits at a bit rate of 270 kb/s, but a single GSM digital radio channel or time slot is capable of transferring only 1/8 th of that, about 33 kb/s of information (actually less than that, due to the use of some bit time for non-information purposes such as synchronization bits). Some of the radio channels are used as control channels which transfer broadcast, paging, and access control information and channels carry voice or data information. Time slots are the smallest individual time period available to each mobile telephone. The frame is a 4.615 msec TDMA frame. The frame is divided into 8 time slots, each normally containing 148 bits of information. Of the 148 bits, some are used to carry data, and others are dedicated as control, and still others are used for synchronization. Different types of channels which are used for different purposes are designed so they cycle through a pre-scheduled sequence of operations, and different types of information are transmitted on each time slot during this cycle. There is a cyclic sequence of 26 TDMA frames for traffic channels which carry digitally coded speech, called a 26-frame multiframe. The traffic channel multiframes allow individual call control information (for example, power level control) to be transmitted at pre-scheduled times on the time slot which mostly carries digitally coded voice information. Frames 1 though 12 inclusive and and 14 through 25 inclusive are used to carry digitally coded speech. Frames 13 and 26 of the multiframes are *not* used to transmit digitally coded speech. Instead frame 13 is dedicated for slow control messages and frame 26 is unused for full rate speech coding applications. Figure 6.4 shows the basic GSM radio channel structure.

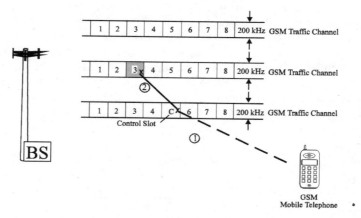

Figure 6.4, GSM Radio Channel Structure

Capacity Expansion

Capacity expansion in the GSM system is primarily a result of allowing several users to simultaneously share a single radio carrier frequency and through the use of multiple paging channels. A secondary improvement in capacity arises from the ability of GSM to use smaller cell clusters such as n=4, with consequently more carriers and channels per cell. The ability to reuse frequencies in more closely spaced cells is due to the robust digital error correction capability of a GSM system and the use of slow frequency hopping to average the interference of nearby cell sites. This ability to operate robustly in the presence of strong interference is described by the minimum carrier to interference (C/I) ratio. Analog cellular systems require an 18 dB C/I value, but GSM can operate at about 14 dB C/I value, and with future improvements in antennas, base antenna diversity methods, and adaptive equalizers, it may approach a theoretical C/I value of 9 dB. Another small capacity increase of the GSM system is the result of an interim assignment to a voice channel without completing the voice connection. While a call is waiting to be answered (ringing), a GSM mobile telephone can be assigned to a temporary channel which only consists of message signaling. Because time slots are not being used to transfer voice, this does not use a voice channel.

The GSM system has the potential to offer a significant amount of paging channel capacity. This is a result of localized paging and multiple paging channels. When a mobile telephone regularly registers on a GSM base station, the system knows a particular area where the mobile telephone is located. This allows the system to only send paging messages to local cell sites near where the mobile telephone last registered. Because the same paging message does not need to be broadcast throughout the entire system, this frees the paging channel capacity for paging channels in other cell sites. Paging channels can be located on any of the GSM radio channel frequencies and multiple paging channels can exist on a single cell site. The paging channel is also divided into groupings of classes and message types. Message types include paging messages and short messages.

Traffic Channel Structure

GSM uses a single of radio carrier frequency waveform to multiplex (time share) control, voice, and data channels. There can also be multiple carrier frequencies in each cell. The radio carriers are divided into time slots, and the time slots are used for different types of channels. The channels' information format (for example, fax or data) is defined in the GSM standards. Control channels transfer broadcast, paging, and access control. Traffic channels transfer voice and data (for example, fax) information.

Each cell has a broadcast channel. By examining the signal strength of each nearby broadcast channel, and using the error detecting codes which are incorporated into the digital transmission from that base station, the GSM mobile telephones which are not engaged in a conversation can measure the quality of nearby cell sites' radio channels to determine which is the optimal control channel to select. Once the mobile telephone has found the best broadcast channel, it continues to receive that frequency and time slot until there is a reason to chose another. The reasons to chose another include movement of the mobile telephone into another cell, which will cause the mobile telephone to move away from the old base station, so the signal strength will decrease and the bit error rate will increase. In that case, the mobile telephone again scans for the best broadcast channel all over again. There are other causes as well, which will not all be covered here.

The control channels have a 51 frame multiframe structure for purposes of scheduling certain types of information transmission. Several types of control channels use the 51-frame multiframe structure. Of the 51 bursts, some are used to broadcast system parameters, some are used to page and assign radio channels, and optionally, some are used to broadcast short messages. Each GSM system has an internal counter which counts sequences of frames, multiframes, and also two other longer sequences called a superframe and a hyperframe, respectively. A super-frame is composed of 51 of the 26-frame type of multiframes (6.12 seconds). One can also say that a superframe is composed of 26 of the 51-frame type of multiframes, as well. The hyper-frame is the largest time interval in the GSM system, and is composed of 2048 super-frames (approximately 3 1/2 hours). During a hyper-frame period, every time slot has a unique sequential number composed of frame number and time slot number. Figure 6.5 shows the basic GSM traffic channel structure. The hyperframe counter is used to synchronize several internal operations in the GSM system, including the sequential transmission of different types of pre-scheduled information at certain times, and the frequency hopping sequence (which is an optional feature) pattern, and the synchronization of the encryption process for privacy of subscribers' conversations. Once a mobile telephone receives broadcast channel transmissions from a base station, and also a particular scheduled transmission called the synchronizing burst message, it can synchronize its own internal hyperframe counter with the base system's hyperframe counter. Then all the equipment, both base and mobile, in the system is synchronized. It is *not* necessary that different systems in different cities synchronize their various hyperframe counters.

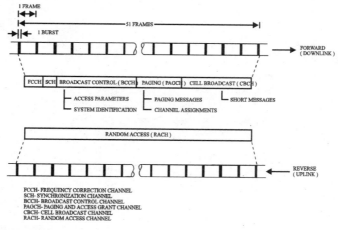

Figure 6.5, GSM Traffic Channel Structure

As mentioned above, the traffic channels (for digitally coded voice and data) have a 26 frame multiframe structure. In a single traffic channel multiframe, frame numbers 13 and 26 are used for control bursts. Time slots which occur during frame number 13 transfer control messages, and slots in frame 26 remain unassigned for full rate traffic channels. (the mobile telephone can use this idle burst time to measure the signal on a frequency transmitted by another nearby cell as part of the mobile assisted handover process). These pre-scheduled control bursts which occur once out of each 26 bursts are called the slow associated control channel (SACCH), and are used to transfer information that is sent continuously between the mobile telephone and the BTS, such as channel quality information. When it is necessary to send control messages quickly, some of the user data (speech bursts) can be discarded and replaced with control commands. This unscheduled replacement of the user speech bursts is called the fast associated control channel (FACCH). There are special bits in the transmission pattern to notify the receiver when the normal digitally coded speech data has been replaced by the FACCH bits, so they will be properly processed at the receiver. The speech coder is specially designed so that it can bridge over one data frame of missing digital speech data by repeating the last fully received frame of digital speech data. The result is usually hardly noticeable to the listener. For user data and fax transmission, the data can be held back when a FACCH frame burst is transmitted, so that there is no loss or corruption of the data.

During full rate voice conversation (dedicated mode), time slot number 26 is unassigned. Because control channel information time periods (51 frames per multiframe) are different from speech channel frame periods (with 26 frames per multiframe), eventually, other beacon frequency synchronization channels (SCH) will be monitored by the mobile telephone which is engaged in the conversation mode. Storing the relative synchronization information of other radio channels allows the mobile telephone to pre-synchronize to those channels prior to hand-over.

Signaling

Signaling is the physical process of transferring control information to and from the mobile telephone. Signaling in connection with the traffic channel is carried either via the the fast associated control channel (FACCH) and/or the slow associated control channel (SACCH). The FACCH channel replaces speech with signal data. The SACCH channel uses dedicated (pre-scheduled) frames within each burst.

Slow Associated Control Channel (SACCH)

SACCH is a continuous stream of signaling data sent at pre-scheduled times on the same time slot which is used at other times for the speech data, and, for historical reasons, the SACCH is called "out of band" signaling. The name is somewhat misleading, since the SACCH messages are transmitted in the same band, and in fact on the same frequency which is also used for digitally coded speech. SACCH messages are sent by dedicated slots in each traffic channel's multiframe sequence, so that sending SACCH messages does not affect the quality of speech. The transmission rate for SACCH messages is very low. For rapid message delivery, messages are sent via the FACCH channel instead of the SACCH channel. A balance was maintained to allow the maximum number of bits to be devoted to speech, and to allocate a minimum number of bits to continuous signaling. Figure 6.6 illustrates SACCH signaling.

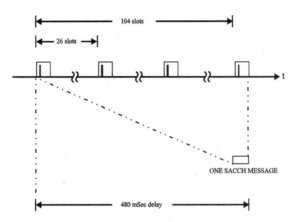

Figure 6.6, SACCH Full Rate Signaling

Fast Associated Control Channel (FACCH)

FACCH replaces speech data with signaling messages when required, a process called (for historical reasons) "in band" signaling. As many as one out of six speech frames may be stolen for FACCH messages, by replacing speech frame(s) with signaling information, so FACCH messages degrade speech quality. When a frame of coded speech is lost due to replacement by FACCH data, or when temporary fading or interference is severe, the GSM RELP speech coder can repeat the last good received frame of speech coder data. Under these circumstances, listeners in GSM systems hear a brief prolongation of a speech sound (for 20 milliseconds) rather than a silence, "click" or other gross disturbance of speech. FACCH messages result in non-linear degradation of speech quality as the number of stolen frames increases.

Because data is bit-interleaved for radio transmission over 8 consecutive channel bursts, the data bits for a FACCH message are transmitted piece by piece over 8 sequential channel bursts (57 bits per channel burst). Figure 6.7 illustrates FACCH signaling.

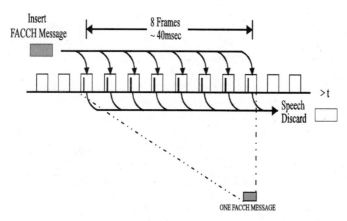

Figure 6.7, FACCH Signaling

Frequency Reuse

GSM uses a wider bandwidth radio carrier waveform than its analog predecessors, and it uses digital error protection coding and modulation to better reject interfering radio signals. These factors allow the radio channel frequencies to be reused in more closely spaced cells than with earlier narrowband analog systems. In real application, the GSM radio channel can tolerate an interfering signal up to 20% of (9 dB below) the desired signal. By comparison, analog signals can only tolerate interfering signals from 1.6% to 6.3 % of (18-12 dB below) the desired signal [2].

Slow frequency hopping and mobile assisted hand-over information (discussed later) provide interference averaging. These additional factors further increase GSM system frequency reuse.

Speech Coding

To allow several users to share a single 200 kHz wide radio carrier waveform, voice signals are digitally coded before transmission. Figure 6.8 illustrates GSM digital speech coding. From the microphone in the mobile telephone, the analog voice signal is sampled 8,000 times each second and digitized using a uniform binary code for each voltage sample, composed of 12 or 13 bits per sample. On the base side, the PSTN digitally codes speech using 8,000 samples per second, but the standard method of digital coding used in the PSTN involves use of a non-uniform binary code called Mu-law coding. In this system, the voltage steps between consecutive code values are larger at the high voltage end of the scale, and smaller near zero volts. This code value can be converted into a uniform 12 or 13 bit code as well. Every 20 msec time window of the digitally coded audio is supplied to a GSM speech coder. The GSM speech compression method is called Regular Pulse Excitation-Long Term Prediction (RPE-LTP) to encode the speech signal at a data rate of 13 kb/s. Because digital fading radio channels can experience significant numbers of bit errors, error protection bits are added to the most important information of the digitally coded audio. The error protection bits increase the bit rate to 22.8 kb/s. To help protect the data from short periods when the radio signal is poor (called fast fading), the 22.8 kb/s error protected data is interleaved (or distributed) over 8 adjacent slot periods. That is, a portion of the bits are transmitted in each of 8 consecutive radio transmit bursts or frames of time slots. At the receiving end of the radio link, the data from these 8 bursts are put back together to make up the digitally coded representation of the 20 msec time window of speech.

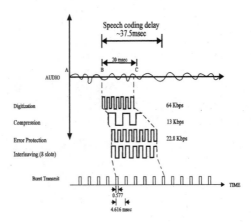

Figure 6.8, Speech Data Compression

The GSM speech coder can stop transmitting the digital voice signals when speech activity is low. When the speech coder senses no speech activity (that is, silence), it digitally encodes a 20 millisecond window of background noise. Then it shuts off the radio transmitter until some sounds are picked up again by the microphone. This may occur between phrases of speech, for example. This process is called discontinuous transmission (DTx), The speech coder uses voice activity detection (VAD) to determine if it is to transmit background noise only. When the VAD indicates low voice activity, the speech coder transmits occasional (once every 480 milliseconds) frames of digitally coded background noise only at an effective data rate of about 500 b/s. These occasional blocks, which contain the background noise characteristics, are called silence descriptor frames. When the silence descriptor frame blocks are received, they are expanded to re-create the background "comfort noise." Comfort noise prevents sudden disturbing changes in perceived sound characteristics when the caller stops talking.

One of the greatest concerns in any digital voice transmission system is the performance of the speech coder. If the speech coder does not accurately code and decode voice data, subscribers notice the inaccuracies as distorted speech, noise, or errors. Performance of the GSM speech coder in poor radio conditions was shown to be superior to that of analog cellular. The mathematical operation of the GSM speech coder is completely standardized in every detail, and it is therefore identical in each phone and system. This uniformity eliminates requirements for compatabilty testing of different manufacturers' equipment.

Dynamic Time Alignment

A non-zero amount of time is required for radio waves to propagate from the mobile telephone to the base transceiver station. Mobile telephones transmit in bursts, so the BTS must receive all bursts in proper time sequence without any time overlap. Without time alignment, a mobile telephone operating close to the BTS could overlap bursts with a mobile telephone operating far from the BTS. To compensate for this effect, commands sent from the base station are used when necessary to adjust mobile telephones' relative transmit time to align the times their signals arrive at the base station.

For dynamic time alignment to function properly, the BTS must determine how much offset time to use. Initially, when the mobile telephone must send a radio transmission to a base station over an unknown distance, it transmits a shortened burst (access burst) until the base transceiver station can calculate the required offset time (timing advance). The required timing offset is twice the path delay, combining the downlink delay (mobile telephone receive) and uplink delay (mobile telephone transmit). The Mobile telephone uses the received burst to determine when its burst transmission should start. The amount of dedicated guard time between adjacent time slots is 8.25 bits (30 μsec). This allows a distance of only 4.5 km before bursts overlap. When the distance from the cell site exceeds 4.5 km, timing must advance to ensure the transmit burst does not overlap with the adjacent time slot. In the 900 MHz GSM system, the timing can be advanced in 1/2 bit steps up to 237 μsec. This 237 μsec maximum without slot collisions limits the distance the GSM mobile telephone can operate from the cell site to about 40 km. In the DCS-1800 and

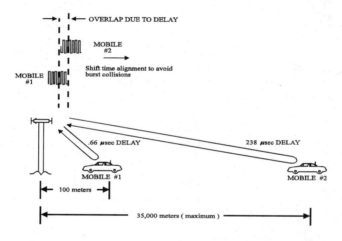

Figure 6.9, Dynamic Time Alignment

PCS-1900 system the same timing adjustment range applies, but because the mobile telephone transmit power is limited to a maximum of 2 watts instead of the 20 watts maximum for 900 MHz GSM sets, there are power limitations on the size of cells which come into play long before the timing limitation becomes dominant. Figure 6.9 illustrates the dynamic time alignment process.

Because the GSM system continuously transmits timing information on the control channels, mobile telephones can perform some time alignment independent of the system time alignment commands sent by the BTS. While operating on a traffic channel (dedicated mode), the mobile telephone can monitor other radio channels between receive and transmit burst periods to capture the system time from the FCH and SCCH channels. This information provides a relative time shift between other cell site radio channels, which the mobile telephone can temporarily store in memory. Time alignment to other radio channels allows the mobile telephone to self-synchronize to a new BTS radio channel before receiving a hand-over command, greatly assisting hand-over.

If the mobile telephone is farther than about 40 km from the cell site, bursts received at the BTS will begin to overlap with the adjacent time slot. If the adjacent time slot is not assigned to another user, or if it is assigned to another user at the same distance from the base station, it is possible for the BTS to receive the mobile telephone transmit burst in the adjacent time slot. This procedure increases the radius of the cell site to more than 75 km where mobile transmitter power is not already the limiting factor. If additional time slots are unused (delayed relative to the assigned time slot) and the BTS can

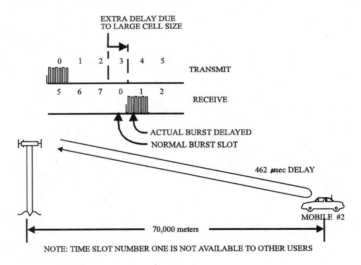

Figure 6.10, Extended Dynamic Time Alignment

receive the delayed signal, a mobile telephone could operate even farther from the cell site. Figure 6.10 illustrates the extended cell dynamic time alignment process. The mobile telephone is assigned to Time slot Number (TN) 0. Due to the excessive delay, even with the timing advance set to a maximum of 237 usec, the burst transmitted from the mobile telephone is received at the BTS in TN 1. Because TN 1 is not assigned to another subscriber, the BTS can receive the mobile telephone's transmitted burst partially in the adjacent slot (TN 1).

Discontinuous Reception (Sleep Mode)

To increase the time until battery recharge of a GSM mobile telephone, the system was designed to allow the mobile telephone to power off non-essential circuitry (sleep) during periods when paging messages will not be received. This is known as discontinuous reception (DRx). To provide for DRx capability, the paging channel is divided into paging sub-channel groups. The number of the paging sub-channel is determined by the last digits of the mobile telephone's phone number or another identification number called the International Mobile Service Identity (IMSI). The system parameter information sent on the BCCH identifies the grouping of paging sub-channels.

Figure 6.11 shows the DRx (sleep mode) process. Initially, the broadcast (BCCH) channel indicates to the mobile telephone which multiframes contain paging and access blocks, and which contain sub-paging classes. Mobile telephones can sleep during multiframes which are not part of its paging sub-channel. In the sleep mode, only a simple electronic timer is operating in the mobile telephone set, and the receiver and transmitter circuits are off and are not using power. Paging messages are contained in a 9-frame PAGCH block of the control channel multiframe. The PAGCH block contains either all page messages or channel assignment messages. The paging blocks can be divided into 9 sub-paging channel groups. Alternating between paging and access channel blocks, and dividing the transmitted information into paging sub-channels allows sleep intervals to range between 4 and 81 multiframes. These intervals translate to sleep periods ranging from 0.94 second to 19 seconds maximum.

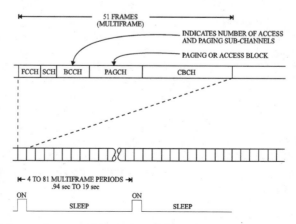

Figure 6.11, Discontinuous Reception (Sleep Mode)

Slow Frequency Hopping

Instantaneous radio signal fading and co-channel interference is often limited to particular narrowband radio channels at a specific time. GSM slow frequency hopping provides for the frequency diversity and interference averaging needed to improve the bit error rate.

Frequency diversity is created by using different radio carrier frequencies to transmit and receive different time slots in the same conversation. Because radio signal fades are usually limited to a narrow radio band of less than 1 MHz, radio frequencies separated by more than 1 MHz typically do not fade simultaneously [3]. Therefore, when the successive time slot bursts which make up a TDMA signal are transmitted on different frequencies, it is much less probable that a fade will occur on consecutive bursts. Frequency hopping, together with bit interleaving, normally causes a fade on only one frequency to have the effect of producing a few isolated bit errors after de-interleaving of the bits. The error correction and detection codes used in GSM each have a property called the constraint length. This is the maximum number of consecutive bits which can contain errors but which can still be corrected by the error protection code. For most of the GSM error protection codes, this is 5 consecutive bits. The result of not having long strings of consecutive bit errors, thanks to frequency hopping, is that the error correction codes used in GSM can completely correct the errors which do occur.

Cellular system frequency plans can be designed for a worst-case situation, placing nearby cells using the same radio channel frequencies at a distance that eliminates interference. The acceptable carrier to interference ratio (C/I) occurs when the desired signal is more than 5 times (7-9 dB) all interfering signals. Interference from nearby cells varies with mobile telephone location, activity, and power levels which are dynamically changing, so in many situations, there is very little interference from other cells. With interference averaging, frequency plans need not be designed for the worst case, and frequencies can be re-used in nearby cell sites more often than they could be without frequency hopping. The greater frequency re-use increases the system capacity.

On the 900 MHz band, there are 125 carrier frequencies presently allocated, and the European Union will soon allocate more. In the 1.9 GHz North American PCS band, a 30 MHz licensee has up to 75 carrier frequencies available. GSM's 125 radio channel possibilities (375 for the entire PCS-1900 band in all licensed segments) and 8 slots per channel, create a very large number of hopping possibilities. To create

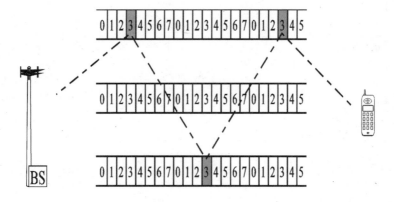

Figure 6.12, Slow Frequency Hopping Example (Symbolic)

the hopping sequence pattern, the BSS assigns two parameters: mobile allocation index offset (MAIO) and hopping sequence number (HSN). These two variables combine to form the hopping pattern. The result is a hopping rate of 217 hops per second for a specific mobile telephone, the same as the TDMA frame rate. Figure 6.12 illustrates a sample of frequency hopping using only 3 frequencies and time slot 3.

RF Power Control

Power control provides four important advantages: 1) it minimizes changes in base receiver RF signal strength from slot to slot; 2) it minimizes interference to nearby cell sites operating on the same radio channel 3) it increases the time interval until battery recharge for portable mobile telephones; and 4) it reduces out-of-band radiation [4]. Mobile telephones and base stations require only enough radio energy to maintain a quality radio link. Minimizing the transmitted energy allows the same radio frequencies to be re-used in nearby cell sites with less interference. Base transceivers and mobile telephones have several maximum power level classifications, and a minimum receive antenna power level of -104 dBm (-102 dBm acceptable for hand portables). Both the mobile telephone and the base station transmitter can adjust their transmit output power level in 2 dB steps. Commands received from the BTS can adjust the mobile telephone up to 15 steps below its maximum transmit power. Mobile telephones can only change power level 2 dB every 60 msec (13 bursts for full rate). Base station transmitters can adjust their output power level down to minimum of 13 dBm (20 mW).

In the mobile telephone, the RF transmitter typically consumes the most power, so transmitting at the lowest output power to maintain a quality radio link also increases time before battery recharge. Even if a higher transmitted power would not interfere with neighboring cells, its is advantageous to transmit at the lowest acceptable power. Doing this reduces the total electric power consumption and air-conditioning power consumption at the base station as well.

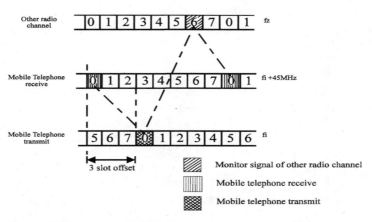

Figure 6.13, GSM Duplex Radio Channels

GSM radio frequencies must co-exist with other GSM frequencies from other cells, and from other radio systems (such as TACS and NMT) operating on nearby frequencies. Although the GSM transmitted signal uses GMSK modulation, which does not produce significant radio energy outside the designated bandwidth, the cycling (rapid changes or sudden turn-on and turn-off) of the RF energy between a high transmit level and an off period results in so-called spurious or out of band radio signals. The radio signal ramp up and ramp down specification (during the tail bits in figure 6.13) limits the amount of spurious emissions by gradually instead of suddenly increasing or decreasing the radio signal as the burst is transmitted.

Short Message Service (SMS)

The ability for a GSM mobile telephone to receive short messages allows it to operate like a pager. A GSM phone can receive short data messages of up to 160 characters from a short message service center (SC). Most of the SMS features are described in phase 2 of the specification.

Short messages can be sent while a phone is operating on a call or while it is idle. Short messages can be sent by a GSM device such as a mobile telephone, or by a short message entity (SME). An example of an outside message would be from a paging service company connected to the SC via a data link such as an X.25 network or the Internet e-mail network.

The SC's purpose is to store short messages, attempt to deliver them, and confirm their receipt. Short messages sent to a mobile telephone are typically stored in the SIM card (explained further below)allowing the user to receive and keep messages, and display them on any GSM mobile telephone with display capability.

Short messages can be broadcast (unacknowledged) or point-to-point (acknowledged). The mobile telephone may obtain status information from the SC about messages that it has already sent. When the mobile telephone's short message memory is full, messages are typically kept in the SC until the memory is cleared. The SC can delete messages already sent to a mobile telephone. The SC can indicate that the message is urgent, allowing the message to be displayed immediately on the mobile telephone display rather than stored in memory to be viewed later. The SC can request a reply from the user, and indicate a response to be automatically directed back to the sender. If the mobile telephone could not receive a message (for example, poor signal conditions or memory full), it can autonomously contact the SC and request its messages.

Subscriber Identity Module (SIM)

The subscriber identity module (SIM) is a complete miniature microprocessor and memory packaged in a removable card or computer chip carrier about the size of a thumbnail, containing the subscriber's identity and feature information. This includes subscriber identity codes, personal features such as short-code (speed) dialing and short messages, and a personal identity number (PIN). The PIN is used by the subscriber to restrict access to the SIM card to only people who know the code. Because SIM cards store the subscriber's unique information, the SIM card can be used in any phone that accepts a SIM card to place and receive calls. For example, a subscriber could use a GSM SIM card in a taxicab or a rented car which has a GSM mobile radio installed.

The full- sized smart card is the same size as a plastic credit card and typically slides into a slot on the bottom of the larger mobile phone. The small chip carrier is usually located under the back cover of portable mobile handset telephones. Because SIM cards are not radio technology specific, standards setting organizations are studying the possibility of using a SIM card with cellular phones of different radio access technologies (for example, TDMA and CDMA).

The SIM card has some memory which can store and retrieve information provided by the mobile telephone. This memory is sometimes used to store short messages. When messages are stored on the SIM card, the user can take the SIM card out of the mobile telephone and place it in an appropriate GSM card reader or another mobile telephone to display the messages at a later time.

Duplex Channels

Although the digital channel is frequency duplex (transmit on one frequency and receive on a different frequency), the mobile telephone receives and transmits at the different times. With this time separation, a simple radio switch can be used to connect the antenna to the the transmitter and receiver sections in hand portables. The radio frequency separation of forward (downlink) and reverse (uplink) frequencies on the 900 MHz band is 45 MHz (80 MHz for PCS-1900). The transmit band for the base station is 935-960 MHz on the 900 MHz band (1805-1880 MHz for UK DCS-1800). The transmit band for the mobile telephone is 890-915 MHz (1710-1785 MHz for DCS-1800). For the North American PCS-1900 system, the downlink base transmitter uses one of the 6 licensed sub-bands of the 1930-1990 MHz frequency range, and the uplink uses the corresponding sub-band in the frequency range from 1850-1910 MHz. GSM, PCS-1900 and DCS-1800 use the same 200 kHz carrier waveform bandwidth. Figure 6.13 shows the 3 slot time offset between the forward and reverse channel. The three slot offset allows for dynamic time alignment and for frequency synthesizer electronic oscillator carrier frequency tuning between the transmit and receive bursts.

The duplex radio channel structure allows the mobile telephone to tune to the radio channels of a neighboring base station, to synchronize, and to obtain channel quality information. Figure 6.14 shows that the mobile telephone can tune to another radio channel during idle mode (non transmit or receive mode) to measure signal strength, and report its own signal quality information (bit error rate — BER) and the signal strength of the channel it is operating on back to the BTS by means of a SACCH message.

RF Power Classification

Mobile telephones and base stations are both classified by power output. The RF power class specifications set the maximum allowable power transmitted during a burst period. Both the mobile telephone and the base station can decrease power from the maximum stepwise in 2 dB increments.

GSM mobile telephones have 5 different power classes. The class 1 mobile telephone has a maximum output power of 20 Watts (+13 dBW or +43 dBm) and a minimum output power level of -4 dBm. Transportable mobile telephones are class 2, with a maximum output power of 8 Watts. Hand portables are classified 3 through 5. Class 3 and class 4 hand portables' output power are 5 Watts and 2 Watts respectively. Class 5 hand portables are designated for micro cellular networks, with a maximum output power of 0.8 Watts. Because mobile telephones transmit in bursts, their output power is measured during the burst period. During full rate transmission, the average output power of the mobile telephone is approximately 1/8th the maximum burst transmit power. For a portable mobile telephone transmitting at 2 Watts, the average power is only 250 mW. For DCS-1800 and PCS-1900 there are only 3 mobile telephone power classes, and the maximum power class for these services is 2 W. This reflects the design of these two services which are intended to be used in small

Power Class	Mobile telephone Power (Watts)	Base Transceiver Power (Watts)
1	20	320
2	8	160
3	5	80
4	2	40
5	0.8	20
6	not applicable	10
7	not applicable	5
8	not applicable	2.5

Table 6.1, GSM RF Power Classifications

cells and hand-held mobile telephones only. Table 6.1 does *not* show the levels for these two low power systems.

While the mobile telephone transmits in bursts, the base station typically transmits continuously. However, the BTS can reduce power between bursts. The mobile telephone output power can be adjusted by commands from the BTS. Each RF power adjustment step is 2 dB with a total range of approximately 20-30 dB depending upon power class. The BSC is responsible for controlling power level settings of both the BTS and mobile telephone. Table 6.1 lists the GSM mobile telephone RF Power classifications.

Basic Operation

When a GSM mobile telephone is first powered on, it initializes by scanning for a beacon frequency and after initially finding more than one, it tunes to the strongest one it finds. To facilitate the mobile telephone finding a beacon frequency, the beacon frequency transmits a so-called frequency correction (FCCH) burst about one out of each 10 consecutive frames on time slot zero. This particular type of burst contains an absolutely constant frequency during the entire time slot, which is very easy for the mobile receiver to distinguish from all other GSM time slot transmissions. Since the same carrier frequency and time slot (slot zero) are used at other times during the 51-frame multiframe transmission schedule for the synchronizing burst and the broadcast channel (BCCH), the mobile receiver then stays tuned to this same frequency to see and receive all the needed initialization information. During initialization, it acquires all of the system information needed to monitor for paging messages and information about how to access the system. After initialization, the mobile telephone enters idle (sleep and wake cycle) mode and waits either to be paged for an incoming call or for the user to place a call (access). When a call is to be received or placed, the mobile telephone enters system access mode to try to access the system via a control channel. When access is granted, the control channel (or to be more precise, during the time slots which are scheduled to transmit the so-called access granting channel (AGCH) messages) commands the mobile telephone to tune to a stand alone dedicated control channel (SDCCH), where further call setup messages are transmitted back and forth, and then ultimately a digital traffic channel. The mobile telephone tunes to the designated channel and time slot and enters conversation mode. As the mobile telephone moves out of range of one cell site radio coverage area, it is handed over to a radio traffic channel at another nearby cell site.

After the mobile telephone has found the FCCH, it locks onto the synchronization channel (SCH) which provides the mobile telephone with most of the bits from the current value of the system hyperframe counter, which is needed to synchronize the mobile set to the base and ultimately to access the cellular system. The synchronization channel uses the S burst containing a long fixed training sequence (41 bits) which allows the mobile telephone to synchronize to burst messages on the traffic channel. Normal bursts, such as those which occur on the traffic channel, contain 26 bits. In both long and normal training bursts, the adaptive equalizer in the receiver can be adjusted to compensate for multipath radio propagation effects and improve the accuracy of the digital bits which are received.

After the mobile telephone has synchronized to the system, it monitors the broadcast control channel (BCCH) and the Paging and access grant channel (PAGCH). The BCCH continually sends system identification and access control information to all mobile telephones. This information includes control channel location and configuration, initial mobile telephone access power level, the neighboring cell site radio channel frequency list, and cell identity. The PAGCH is composed of two channels; the paging channel (PCH) and access grant channel (AGCH). The PCH sends pages to the mobile telephone to indicate that a call is incoming. The AGCH transfers messages that assign the mobile telephone to a traffic channel or to a type of channel called a standalone dedicated control channel, used in some systems.

The mobile telephone may monitor the cell broadcast channel (CBCH) which is an optional control channel that transfers short messages to mobile telephones. A single CBCH channel can transfer about one 80 character message every 2 seconds. Because the CBCH shares the same time slot as the BCCH, the CBCH messages can be received without missing any BCCH messages.

Access

A mobile telephone attempts to gain service from the cellular system by transmitting a request on the random access channel (RACH). This name applies to the uplink use of time slot zero on the beacon frequency. If the system is not busy, the mobile telephone attempts access by transmitting an access burst on the RACH channel. The access burst contains a 5 bit random number which temporarily identifies the mobile telephone attempting the access. The access burst also contains a 3 bit code which identifies the type of access requested, such as page response, call origination, or reconnection of an accidentally disconnected call (due to poor quality radio signals). If the system successfully receives the access request message, it sends the random number in the immediate assignment message on the AGCH channel, directing the mobile telephone to tune to a specific radio fre-

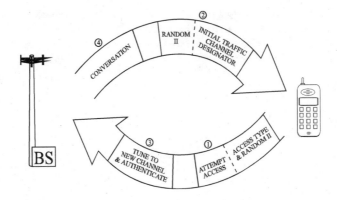

Figure 6.14, System Access

quency and time slot. After the mobile telephone tunes to its assigned channel, the system typically requests authentication. If the system authorizes service, a conversation can be set up. Figure 6.14 illustrates the system access process.

Because one or more mobile telephones may request access at the same time, a collision may occur. If the signal from one of the two contending mobile telephones is much stronger, it gets the cellular system's attention (as a result of the so-called capture effect) and receives a channel assignment. The other mobile telephone will wait a random amount of time prior to access the system again. If the signals of both mobile telephones interfere with each other, no channel will be assigned, and both will wait a random and normally each one a different amount of time before their next access attempts.

When accessing the system for the first time, the radio propagation time delay is not known. A normal burst received from the subscriber could overlap an adjacent burst period. To prevent this on initial access to the system, the mobile telephone transmits a shortened access burst with a longer training sequence. The access burst is short (it contains only 87 bits) which provides extra guard time to protect against overlapping the adjacent burst period. The long training sequence helps the base station to initially decode the first burst.

Occasionally, a mobile telephone and the system need to establish a connection for signaling purposes only. A special signaling channel called the traffic and access channel, one-eighth rate (TACH/8) is dedicated for this purpose. Its signaling rate is very low (one eighth of the data rate of a normal TCH), and no user data is sent on the TACH/8 except for short messages. This signaling channel can be used to manage off air call set up (OACSU). OACSU is a process of assigning a mobile telephone to a low data rate channel which maintains a connection (via TACH/8) while the call is being set up (ringing). When the call is answered, the mobile telephone is assigned to a regular traffic channel.

Paging

Paging is the process of sending a page message to the mobile telephone to indicate that a call is to be received. Page messages are sent on the paging and access grant channel (PAGCH). To increase the number of paging messages that a control channel can deliver (and also to preserve the privacy of the subscribers), a mobile telephone is assigned a temporary mobile subscriber identity (TMSI) when it registers in a system. The TMSI is shorter than the International Mobile Subscriber Identity (IMSI), which uniquely identifies the subscriber. If a mobile telephone has not been assigned a TMSI, the IMSI can be sent on the paging channel.

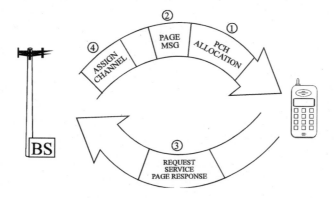

Figure 6.15, System Paging

Figure 6.15 illustrates the paging process. Initially (step 1), the mobile telephone synchronizes itself to the frame sequence in the multiframe, superframe and hyperframe. In the 51-frame multiframe sequence, the broadcast channel (BCCH) occurs after the FCH and SCH, near the beginning of the 51-frame cycle, and is then followed by the the paging channel time slots. Following the BCCH signals, paging messages will be sent on the PAGCH at a scheduled times. The mobile telephone then monitors the PAGCH for its page (step 2). In general, the mobile telephone will go through a sleep and wake cycle while doing this. After the mobile telephone is paged, it requests service from the cellular system (step 3), indicating that it is responding to a page message. The cellular system then assigns a SDCCH, and then a traffic channel (step 4) where, respectively, the subscriber will be authenticated, and may begin.

Handover

The terms handoff and handover are synonymous, but the GSM documentation uses only the word handover, while most North American documents and standards use the word handoff. Mobile assisted handover is a feature that helps the system determine when a hand-over to another radio channel is needed. Existing analog systems rely entirely upon receivers in the base stations to measure the signal strength of mobiles and determine when a hand-over was needed. To make the handover process faster and reduce the amount of equipment needed at the base stations, and to reduce the amount of data communication which must be sent between the base stations and the MSC, GSM mobile telephones continually return radio channel quality information to the BTS during conversation (dedicated mode). This information describes the signal strength and bit error rate (BER) on the downlink from the current base station and the base stations in surrounding cells as well.

Figure 6.16 illustrates mobile assisted hand-over. The mobile telephone initially receives a radio channel list of nearby cell sites from the BCCH channel (step 1). During idle periods (after the transmit burst and before the receive burst), the mobile telephone monitors up to 12 neighboring base station control channels (although only 6 are

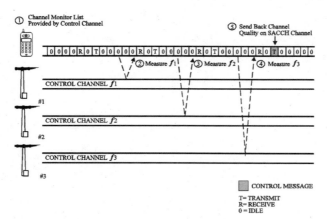

Figure 6.16, Mobile Assisted Hand-over

used in most situations) for signal strength on all channels and BER (steps 2-4). The BER is estimated by use of the error protection coding which is used on the BCCH messages. The Mobile telephone returns the channel quality information to the BTS on the SACCH channel (step 5). The system uses the channel quality measurements provided by the mobile telephone and other signal quality information provided by the BTS to determine when the system will initiate a hand-over, and the preferred target cell for the handover.

The transfer of a radio communication from one radio channel to another is called handover. A radio communication is handed over when the BSC determines that channel quality has fallen below a desired level, and another better radio channel is available. The BSC continuously receives radio channel quality information from the BTS and the mobile telephone.

Figure 6.17 illustrates the hand-over process. First, the mobile telephone monitors nearby cell sites' radio channels. The list of these channels is provided (step 1) via the broadcast control channel (BCCH) or via the SACCH if the mobile telephone is on a traffic channel in cell site #1. During idle periods between bursts, the mobile telephone measures the signal strength and channel quality of the radio channels on the list received from the base station serving it, and returns one of these measurements per second (approximately) to the BTS via the SACCH channel (time 2). When the BSC determines that the mobile telephone can be better served by another cell site, it sends a hand-over message commanding the unit to tune to a new radio channel (time 3). When the mobile telephone receives the message in

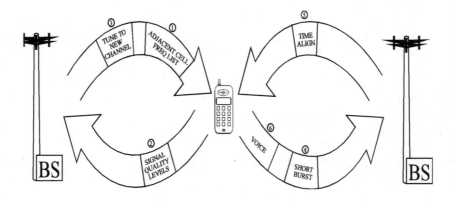

Figure 6.17, GSM System Channel Handover

a case where the distance to the new target base station is not initially known, it mutes the audio and tunes to the new radio channel. It then begins sending short access bursts (time 4) to avoid potential collisions with bursts sent from other mobile telephones. If the mobile telephone has pre-synchronized with the other cell site by decoding the FCCH and SCH channels, it immediately begins to transmit normal bursts and never mutes the audio. This latter pre-synchronized case is called a "seamless" handover, since there is no interruption in the digitally coded speech communication. If the mobile telephone has not pre-synchronized, the BTS determines the necessary time adjustment and commands the mobile telephone to time align (time 5). The mobile telephone can then un-mute the audio and begin voice communications again (time 6). If adjacent cells in the system are similar in size, differing by less than about 1 km, a seamless hanover is done in which no shortened access burst is required, and the mobile telephone sends normal bursts immediately.

References:

1. Balston, D.M., "Cellular Radio Systems," Artech House, MA, 1993, p. 188.
2. Ibid, p. 126.
3. Mouly, Michel, "The GSM System for Mobile Communications," M. Mouly et Marie-B Pautet, Palaiseau, France, 1992, pp. 218-221.
4. Balston, D.M., "Cellular Radio Systems," Artech House, MA, 1993, p. 181.

Chapter 7
Wireless Office Telephone Systems

Wireless Office Telephone Systems

There are several different types of wireless office telephone systems (WOTS) including Digital Enhanced (or European) Cordless Telephone (DECT), Personal Wireless Telecommunication Enhanced (PWT(E)), BusinessLink™, SpectraLink, and other wireless office technologies. This chapter provides a brief overview to the general features of WOTS systems and describes several commercial wireless office systems.

Wireless office telephone systems provide communication to office personnel and production workers who are highly mobile. These people are frequently away from their desk or other fixed telephone station set location. This group typically involves such people as guards, maintenance personnel, and shop floor supervisors. In a hospital, this group includes nurses and other medical staff members who are on call for emergency situations. The nature of their work prevents them from remaining at a fixed desk or other location, and at the same time makes instantaneous communication with these people very important, so that they can be involved in rapidly developing situations which require their presence and attention. Studies by several vendors indicate that 10% to 15% of the personnel in a typical industrial or manufacturing location, and a somewhat smaller percentage in a service location, fall into this highly mobile category.

It is interesting to note that in-office wireless systems are of less value to high-level executives in the office setting than to line and staff personnel of the type described above. This situation is contrary to the initial usual situation where a new and expensive "office toy" such as a new telephone with special features is first placed on the desk of the highest ranking executives. Certain executives are *intentionally* not reachable by telephone call during certain activities, such as meetings. At such times, their calls are taken by a secretary or administrative assistant, who then can judge if the matter is sufficiently important to disturb the executive at that time. If so, the customary method is via a hand-carried note on paper! Although there are situations in which the executive may wish to carry an in-office wireless telephone, one of the necessary features is clearly to *restrict those who are able to call the executive*, by using an unpublished directory number for the wireless set, or at least to identify the caller. In this way, only calls which have been screened by the secretary or administrative assistant reach the executive. A related need is to divert unanswered calls to voice mail or a similar service, so that the user of the WOTS handset is *not obligated to answer every call immediately* when their surroundings are not suitable for a private call.

A WOTS radio system allows for voice or data communications on either an analog (typically FM) or digital radio channel. The radio channel typically allows multiple mobile telephones to communicate on the same frequency at the same time by special coding of their radio signals. WOTS systems include many of the same basic subsystems as other cellular systems, including a switching system, base stations (BS), and wireless telephones.

Figure 7. 1 shows an overview of a WOTS radio system. A WOTS system typically has a switching system which is located at the company although other options such as a combined private branch exchange (PBX) are possible. The switch connects to several indoor and/or outdoor base stations via cables. These base stations communicate with wireless office telephones which can communicate with the system. A control terminal is used to configure and update the WPBX with information about the wireless office telephones and how they can be connected to the PSTN.

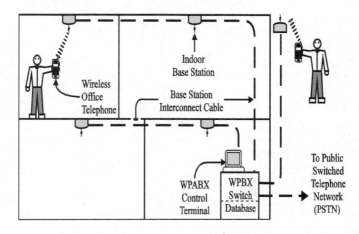

Figure 7.1, Overview of WOTS Radio System

In the past, and to a large extent today, these already identified highly mobile office people are served by a combination of independent short range radio systems, radio paging systems, and intercoms. To reach mobile workers, short range radio systems can be used which are not integrated with the telephone system. In these systems, a single person can be assigned the job of radio dispatcher which involves relaying messages to the mobile staff member who may be carrying a two-way portable radio.

Radio-paging systems are used to send one-way messages to mobile workers who carry pagers. In some cases, radio paging systems can be integrated with the private branch exchange (PBX) wired telephone system which is used in the office. Incoming callers may be given the option (typically by an audio message) to have the worker paged. The PBX system then will send a page message to either a private paging system or route it to a commercial (public common carrier) radio paging system. Like many other radio paging systems, these systems allow the sending of numeric message to the pager. The numeric message can come from the caller who enters in a telephone directory number from a touch tone telephone, typically representing a call back number. In some cases, the caller may enter a pre-arranged message code or in more sophisticated systems, an alphanumeric (text) message can be sent which conveys the actual message information.

Intercom and audio loudspeaker (public address - PA) systems or other audible systems such as chimes and other devices can be used to communicate messages to mobile workers. Like the short-range radio systems described above, the public address system is typically controlled by one person appointed to act as dispatcher. In some systems, the PA line is integrated into the internal telephone system as a pre-designated dialed number (on a PBX) or a reserved line on a key system. This allows anyone in the office to directly speak over the PA system. These lines are usually intentionally designed so that they cannot be accessed from outside the office telephone system for Centrex or similar systems in which permit direct inward dialing (DID). In certain situations a PA system is perceived as too "industrial" and a less disturbing method is desired, so a chime or other pre-arranged sound is activated by either a designated dispatch person or by any internal person calling a pre-designated number. The use of chimes was very popular for years in large department stores and other high class retail establishments. This type of system required each person to have a pre-arranged audible signal (such as 1, 2, 3 or more chimes in a burst) and a pre-designated call-back number. A similar audible signal can be used in an office for evening and weekend use when an attendant who usually handles incoming calls is not on duty. In this case, all incoming calls activate an audible signal which is heard throughout the office area. Any internal user can dial a "call pick-up" code or answer the line indicated by a flashing visible indicator on a key system. If necessary, the call may be transferred internally to the proper person.

All of these systems still have their applications, although they have several drawbacks. Systems requiring a dedicated dispatcher are expensive unless the amount of time the dispatcher devotes to this task and the salary of the dispatcher are clearly beneficial and cost effective to the office organization. When the volume of work permits the same person to handle both a related task, such as receptionist and/or internal telephone system attendant, and also the dispatcher function for an internal radio or PA/chime system, without falling behind schedule, such a method can be acceptable. Once the traffic for these special internal mobile users exceeds the capabilities of a shared dispatcher, yet is not sufficient to use the entire time of a single dedicated person, most business people will take a hard look at the cost of a full-time dedicated dispatcher before continuing to use such a system.

For intentionally reachable office personnel, such as maintenance or foreman staff, a mobile communication system which is not well-integrated into the telephone system can consume the time of those *other* office persons who are trying to reach the mobile staff member. Poorly designed or implemented integration makes the caller dial different numbers for different means of reaching the desired target person, typically only after a long ring-no-answer wait on previously dialed numbers.

Systems with a simple audible or call-back signal introduce delay before the called person can get the full message and respond. In many situations, particularly medical emergencies and other life-threatening situations, this is perceived as unacceptable. Although a more comprehensive PA or alphanumeric pager message addresses this problem to a degree, a direct and immediate voice telephone connection with the mobile staff member may be a much better solution in many cases. Hospitals prefer to not have visitors hear repeated details of a medical emergency, and even an emotionally neutral message, such as the name of a doctor called over a PA system, can disturb the quiet ambiance needed for the rest of hospital patients, particularly during evening hours.

Several technologies have been developed to provide a combination of privacy, better integration with the office telephone system, and immediate voice communication. Some of these are adaptations of technologies also used in the public cellular or PCS domain of so-called high-tier or low-tier systems, and others are technologies limited specifically to internal office areas within the walls of a building or the geographical limits of a campus.

Wireless Office Telephone

A wireless office telephone is a device which links the mobile worker to the wireless telephone system. Wireless office telephones are similar to cellular phones but are typically smaller and have features similar to a wired PBX phone. A wireless office telephone should be capable of giving the user as many features as possible which are available on an office PBX, in addition to mobility. This may include data and short messages as well as voice. Wireless office telephones are small and may have belt clips for wearing and vibrators instead of audible ringers to silently alert employees who may be sitting in a meeting and do not want to be disturbed. Because of the low power, antennas may be concealed within the plastic casing which allows a smaller size.

Figure 7. 2 shows a block diagram of a typical wireless office tele-phone. WOTS telephones have an RF section that is typically a low power multiple channel transceiver (transmitter and receiver) which is capable of sensing interference and changing radio channels. The logic section consists of a microprocessor and a moderate amount of memory. To identify individual WOTS handsets, a unique code is usu-ally stored in the handset and the WOTS switching system. This code is typically programmed by the system operator. An audio section routes the audio from the microphone and speaker to the modulator of the radio transmission section. The battery is typically charged from a desktop charging unit or portable cable such as a car's cigarette lighter adapter.

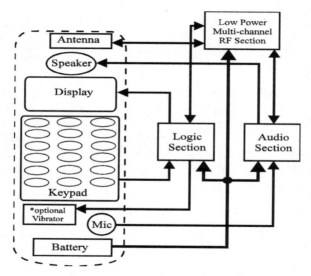

Figure 7.2, Wireless Office Telephone Block Diagram

Some wireless office telephones may only be used to transfer data. In systems such as the DECT system, the radio channel can be used as a high speed data link for computer terminals. In this case, the wire-less telephone may be a simple black box which has a cable that con-nects to the computer and an antenna to transmit to the system.

Wireless Office Base Station

The wireless office base station is the link between the radio transmissions sent to and received from the wireless office telephone and the WPBX switching system. Wireless office base stations are similar to cell sites as they can communicate directly with the WPBX switching system. Because these base stations are fairly close to the switching system, they are directly connected by cable. This allows power to be supplied by the wireless switching system and no battery backup power supply system is required.

Figure 7. 3 shows a block diagram of a wireless office base station. The cable which connects the base station to the switching system typically carries multiple voice and/or data channels. Both the power and data signal may be supplied over a single twisted pair or dedicated lines may be used for data and power. As the signals arrive at the base station, a communications controller divides the multiple channels, processes their signals, and routes them to the radio signal amplifier. Wireless office system base stations typically use a hybrid radio signal combiner and a single linear RF amplifier to amplify the low level RF signals for radio transmission. These base stations often have which allows for diversity reception. The received signals are demodulated and processed and routed back to the base station through the communications controller.

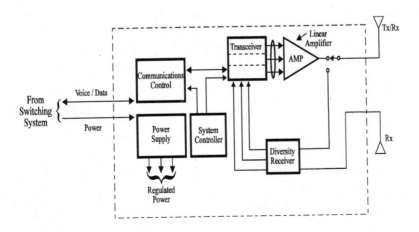

Figure 7.3, Wireless Office Base Station

The design objectives of a WOTS base station are similar to those of a general public PCS system, but there are several additional requirements.

WOTS base stations must be much simpler to install, relocate and service (diagnose or debug). Operations without skilled or highly trained staff are very desirable. Many WOTS base stations are almost "self configuring," implying that the system sets the frequencies of each base station automatically, to both optimize the overall frequency plan and to avoid interference with non-radio RF sources which may be present. Although many WOTS installations today are preceded and verified by a thorough set of accurate RF field strength measurements, the design objectives for most systems is to achieve and maintain proper WOTS operation without this costly and time-consuming step.

WOTS base stations generally do not have a fixed carrier frequency, in distinction to most wide coverage area cellular and PCS base station technologies. Most WOTS base stations have the capability to tune to any of the carrier channel frequencies. Several TDMA or TDMA/TDD WOTS technologies allow for the base station to change its carrier frequency during each distinct time slot. This rapid change in carrier frequency is described by the words "frequency agile." Since most WOTS base units use a hybrid frequency combiner which allows any frequency to be combined with others to the antenna without any frequency tuning, there is no restriction on changing the transmit carrier frequency as desired.

WOTS base stations generally have smaller size and traffic capacity and lower RF transmit power per station than any of the public PCS systems. A typical cell size in a WOTS installation is less than 600m (1900 ft) radius, and base transmit power is often less than 1 watt. The antenna assembly is typically factory installed in the base station and it is usually not possible to connect additional base units to an existing WOTS antenna. There are some exceptions to this, such as the use of a "leaky co-ax distributed antenna" or when directional antennas can be attached.

WOTS base antennas are almost always horizontally omnidirectional (360 degrees). Sectored base antennas are seldom used, despite their various well-known advantages. There has been some discussion of using so-called "smart" adaptive phased array which can focus the energy in a particular direction in WOTS equipment, but no commercial product using such technology has appeared at the time of publication. Smart antennas would reduce the interference to other base stations, increase the range, and increase the total channel capacity of the system by allowing increased frequency reuse.

Wireless Office Switching System

The wireless office switching system, which is also sometimes called a wireless private branch exchange (WPBX), coordinates the operation of all the base stations and WOTS handsets in the system. The switching hardware and software for the WPBX may be incorporated into the main office telephone system (integrated), may reside in a separate switching and/or control module (external), or be completely separate from any wired system (independent). Integrated systems allow one switch to serve all the base stations and wired telephones connected to the system. An external system is used when a radio system is added to an existing system or the older system cannot be directly upgraded to support handoff switching inside the main switch. Independent systems may be used when there is no wired system installed. An independent system may only consist of WOTS handsets that can access a public cellular system for office use at a reduced billing rate.

Figure 7. 4 shows a simplified block diagram of a WPBX. The core of the switching system is the time slot interconnection (TSI) switching unit. The switching unit connects base stations with the PSTN or with a wired PBX. The majority of PBX manufacturers use a proprietary electrical interface between their base stations and their switches. Whether the electrical interface is proprietary or based on ISDN, the wiring used for the connection is typically one or more wire pairs (2-pair "quad" or 3-pair cable) consisting of standard office telephone wire (typically 24 gauge copper). If the WPBX is an external system, there is an interface from the WPBX to the existing wired PBX. This is typically a proprietary interface. The WPBX may also directly connect to the PSTN or allow connection to other wired telephones. The WPBX contains a user database which contains the identification and authorized features for each WOTS handset that is permitted access to the system. A separate authentication area may be used to store secret codes to validate the identity and encrypt the voice to prevent eavesdropping.

In a complete WOTS installation with handoff, there is a continual process of signaling which occurs between all the handsets which are powered up but idle and the nearest base station(s). This signaling is analogous to the autonomous registration or attachment processes in public cellular and PCS systems, and it makes call delivery to the

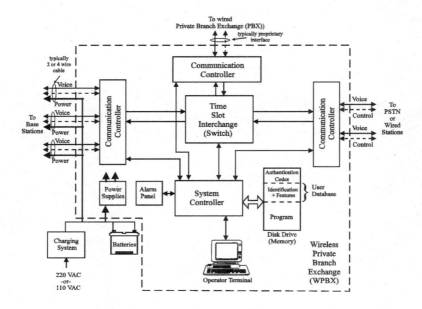

Figure 7.4, Wireless Office Switching System

handset feasible regardless of its location in the office or campus at the time a call is placed. This type of continual signaling for the purpose of maintaining a data base giving the present cell/sector location is completely unlike the signaling used in a wired office telephone system, where there is virtually no signaling between the PBX and the wired extension until a call occurs.

Systems Overview

A complete WOTS system typically includes a wireless private branch exchange (WPBX) switching system, base stations, wireless telephones, and connections to telephone lines. In many situations, wired telephones or wired PBX's may be connected to the WPBX system. There are four basic system configurations in WOTS installations: integrated, external, enhanced external, and independent wireless systems.

Integrated system designs use base stations which are connected directly to the central switch. This type is usually supplied entirely by the single PBX system vendor. Figure 7. 5 shows a block diagram of an integrated WOTS system. The integrated WOTS system allows wired feature telephones and wireless base stations to be connected to the same system.

External WOTS systems use a separate module which performs the wireless related interface and handoff switching operations. This type may be supplied by a third party vendor, and typically uses a PSTN type of extension interface, thus omitting some of the proprietary PBX features for the wireless handsets. Coordination between the wireless

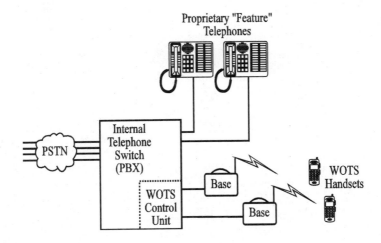

Figure 7.5, Integrated Wireless Office System

handsets and wired desk telephones of the users is handled by a PBX feature such as ring-multiple-stations-first-line-answer-seizes-call. Figure 7. 6 shows a block diagram of an external (adjunct) WOTS system. In this type of system, the WOTS control unit (typically called a "base station controller") uses standard telephone lines (type 2500) supplied by the wired PBX to connect to the PSTN when required. The WOTS system may also allow some wired telephones to be connected directly to its control unit. Because of the limited signaling capability available on the standard telephone line, advanced features such as station ID cannot be used by the WOTS control unit.

Enhanced external WOTS systems integrate the advanced features available to the standard PBX with the wireless transmission system. Figure 7. 7 that the proprietary extension lines from the wired PBX are connected directly to the WOTS control unit. This allows the WOTS control unit to interpret advanced messaging commands such

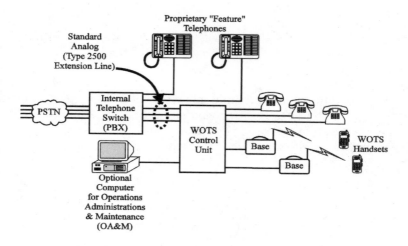

Figure 7.6, External Wireless Office System

as caller or station identification which states the extension which is calling and passes this information on to the wireless office telephone.

Independent wireless systems use a wireless system that is not interconnected to the wired telephone system in the office. In some cases, a wired system may not exist or the user may not own access to any telephone lines. Figure 7. 8 shows a wireless system where a cellular or PCS company has installed microcells in a building. The Cellular or

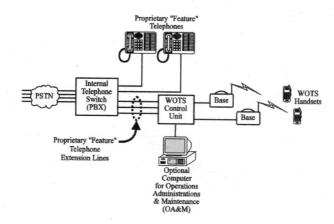

Figure 7. 7, Enhanced External Wireless Office System

PCS company can allow either low cost use or restricted use of wireless telephones when operating in the local coverage area. This is typically set up so the billing rate is a fixed fee for operation in one or more cell site areas for unlimited usage of local calls.

Frequency Reuse

Wireless office systems typically use very low power that allows frequencies to be reused in nearby cells. The frequency reuse plan of each

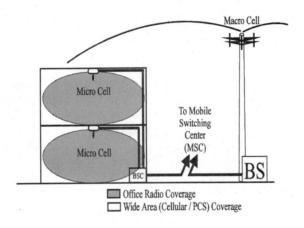

Figure 7. 8, Independent Wireless Office System

technology is dependent on the carrier to interference (C/I) ratio which is required for substantially error-free operation (this is covered in chapter 2). Depending on the radio technology used, the C/I ratio may be a value from approximately 9 to 23 dB, so that very significant differences exist between various technologies.

The amount of interference from nearby base stations which are operating on the same frequency will be determined by the radio signal attenuation (path loss) characteristics of the radio path. The path loss also tends to be very irregular inside buildings, depending upon the site specific characteristics of their construction and form. For most short range radio systems with no obstacles in the radio path (optical as well as radio line of sight between the base and mobile set), the overall theoretical design is usually based on the best case assumption of 1/r2 path loss, corresponding to just the spreading loss in free space without obstructions. This mathematical formula is also described as a path loss of 20 dB per decade of distance. However, in an office with

195

walls having metal construction studs or steel reinforcing rods, the loss with such obstacles in the RF path may be very severe, and is sometimes approximated by a formula such as 1/r6 corresponding to a path loss of 60 dB per decade of distance. Some authorities [1] suggest using an indoor path loss model with a loss rate which starts from 20 dB/decade close in to the base station and then increases in magnitude at greater distances. This particular case of very little path loss for nearby wireless telephones and much greater loss rate per unit of increased distance for more distant handsets is actually beneficial This results in the strong signals available to nearby mobile stations which are intended to communicate with the base station, and greatly attenuated interference from nearby base stations which are operating on the same frequency.

Figure 7. 9 shows a sample wireless office system that uses frequencies in different rooms and on different floors. Notice that the radio coverage areas are not strictly limited to specific rooms. Walls, pipes, furniture, windows, and doors can significantly affect the radio coverage patterns. The coverage patterns will change as people and equipment move throughout the building. Because most wireless office systems use dynamic radio channel allocation, when radio interference occurs as these radio patterns change, the system automatically selects new frequencies that have less interference.

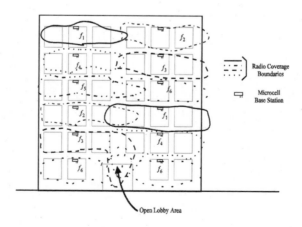

Figure 7. 9, Wireless Office System Frequency Reuse

Aside from placing the base stations sufficiently far from each other at installation time, the system designer and installer(s) of a WOTS installation usually do not need to do frequency planning for each cell along the lines of the detailed choice of the set of carrier frequencies used in each cell of a large public cellular system. Most WOTS technologies can automatically change their radio channel frequency (called frequency agility), and therefore the carrier frequency at each base station may change from one call to another. The explicit objective in the design of most WOTS systems is to provide radio coverage to all areas which telephone calls are expected to be completed.

When one observes the actual frequencies which are dynamically assigned to various conversations under high usage (traffic) conditions, the resulting geographical assignment of frequencies does, in fact, resemble frequency planning clusters used in cellular and PCS systems like the classic n=3, n=7 or (in a case of base stations which are inadvertently located too close to each other) n=12 frequency plan. When too many WOTS base units are installed too close to each other, the WOTS installation with automatic frequency allocation may not permit operation in one cell when mutually interfering radio frequencies in all the surrounding nearby cells are operating first. This is observed as an irregular blockage of calls, leading to occasional and non-consistent inability to begin a call in that region or dropped calls due to the lack of a usable channel for handoff. Indoor systems used in buildings pose a unique frequency re-use dilemma: multiple stories. The three dimensional characteristics of WOTS means a doubling or even tripling of the re-use plans from n=12 to as high as n=36! This can be a serious challenge to WOTS at 1920-1930 MHz where only 10 MHz of bandwidth is available.

Irregular call blockage symptoms can be identified with even more certainty when diagnostic software supplied with several WOTS systems displays the reasons for the lack of a channel as excessive inter-cell interference. It is important to distinguish the first problem of blocked calls due to offered traffic in excess of the system capacity (with all installed traffic channels already in use) from blocked calls due to the problem of high levels of interference from nearby cells. This is the result of the inability to use channels which were installed but which are never exercised due to excessive co-carrier interference between nearby base stations. The first problem situation is to be expected as the number of issued handsets and the offered traffic of the system increases. It can be corrected by installing additional base stations and/or base channels which are *properly* geographically separated from each other. If the call blockage problem is incorrectly identified as a lack of available radio channels, the blockage problem can actually be made even worse by installing additional base stations too close together!

When one discovers that a poor choice of base radio placement has been made in a WOTS installation, as described in the previous paragraph, the proper action is to relocate the base units further apart or reduce their transmitted power level. When done properly, this will ultimately allow all of the voice or traffic channels in all of the base stations to operate simultaneously when the offered traffic requires it. Placing too many base units too close together is wasteful of money because the customer is paying for radio base channel equipment which cannot be fully utilized. There is no backup or redundancy provided in most WOTS systems by placing base units too close together. Figure 7. 10 shows a test device which tests for the signal strength and interference levels so the correct placement of WOTS radio base stations can be accomplished.

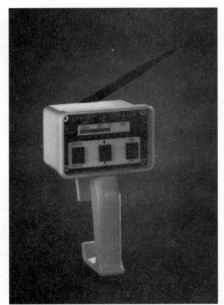

Figure 7.10, Wireless Office System Installation Equipment
source: Berkeley Varitronix

Capacity Expansion

The amount of simultaneous telephone calls (capacity) of a WOTS system can be expanded by either adding radio channels, adding base stations, or by allowing more users to share each radio channel (increasing the efficiency of the radio channel). It may be possible to install additional radio channels in each base station if all the available radio channels have not already been installed and if the nearby base stations are not already using most of the available radio channel frequencies. Figure 7. 11 shows two base stations that originally had two radio channels in operation. To increase capacity, a new radio channel is installed in each base station. In a system which has all the available channels in use by adjacent base stations, adding more radio channels will only increase interference levels and not necessarily increase capacity.

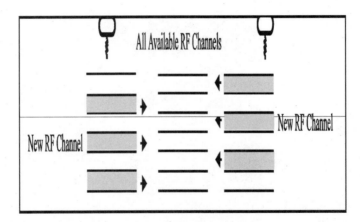

Figure 7. 11, Wireless Office Radio Channel Addition

When new base stations are installed, they should of course be geographically separated sufficiently so that the level of co-carrier interference which is expected from other base stations in the system under heavy usage will not cause high levels of interference. Excessive interference will cause the automatic frequency allocation algorithm in the WOTS system to block calls because no carrier channels are available at that location. Manufacturers give guidelines regarding the recommended minimum distance between base station locations. One should both observe these guides and also use whatever system diagnostic software is available to determine that the problem of under-utilization of installed channels does not occur, as described above. If it does occur, the base stations affected should be relocated further apart until blocking due to this particular cause is eliminated, as described in the previous paragraphs.

Figure 7. 12 shows how an original base station coverage area can be divided and an additional base station added. In 7. 12 (a), a building area is covered by one base station. In 7. 12 (b), two base stations are installed with reduced coverage areas (lower output power level of the base stations) are reduced so the interference they cause to other base stations are also reduced. Because these base stations are adjacent to each other, the new base station (in fixed frequency systems) should not use frequencies which are not the same as the original base station. In most WOTS systems, the frequency of the adjacent base stations will automatically change as needed to reduce interference.

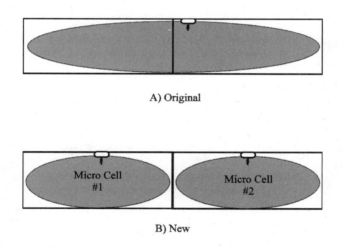

A) Original

B) New

Figure 7. 12, Wireless Office Base Station Addition

Only horizontally omnidirectional base antennas are supplied and used in most WOTS installations. Sectored antennas would be beneficial but require the proper location and orientation, and more technically qualified persons to install them. Improper orientation of a directional antenna could increase co-carrier interference and possibly produce a "hole" with no radio coverage. Proper installation of directional antennas also requires the use of test equipment and display of field strength on a geographical map of the service area to ensure that optimum RF coverage has been achieved. Because of the care and the amount of test equipment required, directional antennas and sectored cells are not widely used for WOTS.

Another way to increase system capacity is to allow more users to share a single radio channel. This can be accomplished by reducing the average number of data bits transmitted to each user. If each user could use only fi the amount of data bits, the other half could be available for new users. A reduction in the average number of data bits can be accomplished by using speech data compression technology similar to those used in digital cellular systems and a reduction in the error protection bits. Unfortunately, while speech data compression does reduce the average bit rate, the result typically involves reduced voice quality and a considerably more complex and power hungry digital speech coder. Most of the wireless office systems use simple 32 kb/s ADPCM speech compression which provides good quality audio signals, short transmission delay, and low power consumption. The average number of data bits for each user is also dependent on how many additional data bits are used for error protection. Because most wireless office systems are used at close range, the signal quality is often very good so most wireless office systems use few (if any) error protection bits.

System Control

Like cellular and PCS systems, a main switching system coordinates the connection of calls throughout a WOTS system. Most WOTS systems have added intelligence in the base stations and handsets to allow the dynamic selection of different radio frequency channels. During normal operation of most WOTS equipment, no particular human control of the equipment is needed. In fact, the success of an installation is often judged by how little human attention is needed to keep it functioning properly. However, human intervention is almost always needed in WOTS equipment for the following purposes:

Installation of new base stations or mobile sets requires some data information to be entered into the WOTS system. The installation of new mobile sets requires programming of identification data into the set, and usually the entry of matching information into the base system controller. This process is sometimes called "datafill." After the system is installed, testing is performed to ensure that all modules in the system (particularly all base stations) are functioning and are able to handle calls. Additional diagnosis and analysis of problem situations is also required.

Several manufacturers of WOTS equipment provide computer aided diagnostic and test software which is usually operated on a personal computer platform under a standard operating system such as DOS or Windows. The personal computer used for this purpose is usually connected to the WOTS control unit by means of a serial data connection cable. This computer need not be connected continuously for the WOTS equipment to operate, but it can and should be connected during installation, initial testing, and whenever diagnosis of some problem or potential problem is needed.

These diagnostic software programs can display important information which is relevant to the operation of the system. For example, when it is not possible to originate a call from certain locations in the WOTS service area, the diagnostic system can help by disclosing which channels are used, showing if a call attempt is detected by the base system, and the system's internal cause or reason for lack of service or unintentional disconnection. If certain channels are never used, the base equipment may be faulty and needs to be replaced, or that channel suffers from excessive interference and the system will not allocate it. If a call attempt is not perceived by the system, the mobile set may be in a location which has inadequate radio coverage or the setup channel is experiencing interference problems, which may be corrected by reconfiguration of the system. If calls are dropped, the distinct causes may be poor quality signal (possibly due to bad radio coverage or to excessive interference), incorrect mobile station identification (due, for example, to wrong supervisory identification tone) may indicate excessive co-carrier interference with another cell in the same system, and a call dropped because there are no target handoff channels available indicates insufficient system traffic capacity. Each of these causes is distinct and the solution is distinct and different in each case. Without the aid of the system diagnostic display, all the symptoms appear to be superficially similar and a proper diagnosis and solution is far from obvious.

Radio Interference

The problem situation which the automatic frequency assignment method cannot overcome is the case in which *all* nominally available frequencies suffer from strong interference in the same area or cell. With only an automatic frequency allocation method, the system will not allow *any* traffic in such a high-interference area. Users who walk into such an area while already engaged in a conversation will not be able to handoff their call and they will be unintentionally disconnected. They will not be able to start a new or replacement call either. This does not occur often, but it must be discovered and corrected in the few cases where it occurs for the system to operate successfully. In such cases, the interference is normally due to one or more of these causes, and the suggested diagnosis and cure is also described.

Unintentional RF radiation from a non-radio device, such as a computer, microwave oven, or other electronic or electrical equipment (such as a faulty sparking electric motor, fluorescent lamp or neon sign). This is usually corrected by repairing the faulty electrical device and/or using appropriate frequency filters in the power wiring to prevent the unintentional RF signal from getting back into the building power wiring. This reduces or limits the amount of radiation from that source.

Spurious RF radiation from a radio transmitting device which should not fall into the band of the WOTS system. For example, spurious second harmonic radiation from a 450 MHz band land mobile radio transmitter could fall at the high end of the 800 MHz band and interfere with a cellular band WOTS system, or it could fall at the low end of the 900 MHz band and interfere with an industrial, scientific, and medical (ISM) band WOTS installation. This problem is corrected by finding the source of the interference by means of a spectrum analyzer and directive antenna(s), and then fixing the faulty condition within the source so that only the intended frequency is radiated. In some rare cases the harmonics are not produced in the actual radio transmitter itself, but are the result of secondary absorption and re-radiation of the signal by other conductors near the transmitting antenna which contain electrically non-linear devices, intentionally or unintentionally. For example, a corroded joint in a metal rain gutter down spout or a metal chain-link fence could cause this effect if the location is very close to a powerful RF transmitter.

In some cases, a very strong RF or other electromagnetic field is introduced into the "front end" (antenna and first RF amplifiers) of the WOTS receiver(s) and this has the effect of producing an electrical current in the front end which exceeds the permissible maximum dynamic range of the receiver. This could be caused by proper operation (with no spurious harmonics) of a powerful RF transmitter at a location which is too close to the area of WOTS operation. This strong RF signal causes a distortion or corruption of the desired RF signal even though the RF transmitter signal itself is free of spurious harmonics. The harmonics are actually generated within the WOTS receiver in this case. The solution to this problem is to reduce the undesired strong electromagnetic field by one or more of the following actions: move the powerful RF transmitter antenna to a more remote location, reduced the transmit power of the interference source (particularly when it is stronger than is needed for its own communication purposes), shield the RF transmit antenna from the WOTS service area by means of conductive sheet metal, wire screens or grids, or closely spaced parallel wires (oriented parallel to the direction of electric field polarization of the interfering signal). In some cases where the size and weight of the filter are acceptable, the WOTS receiver can have a frequency filter installed to attenuate the undesired strong RF signal. This is usually not feasible for a handset, but may be acceptable for a base radio unit.

Basic Wireless Office Operation

Similar to cellular and PCS systems, some WOTS systems use a continuous control channel to coordinate access to the system. WOTS systems typically use the same type of radio channel to serve multiple purposes. A portion of at least one of the radio channels from a base station is used to coordinate access to the system. This portion of the channel is referred to as a control channel. The control channels allow the wireless telephone to retrieve system control information and compete for access. After a wireless telephone has been recognized by the wireless office system, it is assigned to a new radio channel, time slot, or channel code on the existing channel to allow voice conversation. This is commonly called a voice channel or traffic channel. While voice and traffic channels are primarily used to transfer information being sent by the user, the system needs to send and receive some digital control messages. This control information is usually sent in parallel with the users information.

When a wireless office telephone is first powered on, it initializes by scanning for a "beacon" signal or pilot control channel and tuning to either the strongest channel or a predetermined (programmed) channel. In this initialization mode, it determines if the WOTS system is available. If the system is not available, some WOTS telephones have the capability of using other wireless systems (such as a public cellular system). After initialization, the wireless office telephone enters idle mode and waits to be paged for an incoming call or for the user to place a call (access). When a call is to be received or placed, the wireless office telephone enters system access mode to try to access the system via a control channel. When it gains access, the base station commands the wireless office telephone to tune to a voice/traffic channel. The wireless office telephone tunes to the designated channel (or new time slot) and enters conversation mode (voice or traffic channel).

Access

The initial access of the mobile set to the base equipment in various WOTS systems follows the method used in the "big brother" system designed for general or public use. All systems have a call setup channel (sometimes called a common control channel) which is reserved for all the wireless office handsets which are not currently engaged in a call. The control channel in an analog system is a particular carrier frequency (one per cell or sector) which is not used for voice but handles only digital messages and information related to call setup, registration of mobile sets, and similar background operations. In a TDMA system, one time slot on one carrier frequency in each cell or sector is designated for this purpose. When there is a large amount of data to be exchanged for call setup or other non-voice-call purposes (such as alphanumeric short messages), additional supplementary time slots may be allocated as well to increase the setup message capacity. In a CDMA system, one or more spreading codes are used to identify control channels.

In general all systems have some form of access contention resolution for the first attempt by a wireless office telephone to send a message to the base station on the setup channel. This first message may be due to the wireless office telephone attempting to begin a call, or it may be the attempt by the telephone to respond to a paging message, or it may only be attempting to register with the system so its presence will be known. Because there is no direct communication between different wireless office telephones, two or more may attempt to transmit a first access message simultaneously. Each technology has a means to prevent or minimize the occurrence of such collisions. In analog cellular systems, the base station periodically transmits a busy/idle (B/I) status bit interspersed among the other general base transmitted data. By examining the B/I bits, analog WOTS telephones can monitor if the system is busy prior to attempting access to the system. In general, most of the time the reverse or uplink frequency of the setup channel is idle, and access attempts occur only infrequently. The WOTS telephone is designed to observe the status of the base receiver by means of the value of the B/I bit, and only transmit when the base receiver is idle. Furthermore, the base receiver should become busy almost immediately after the mobile begins to transmit. The system design standards require that the mobile set must stop transmitting if the B/I bit value does not change from idle to busy within a brief prescribed time after the mobile set starts to transmit. This situation could be due to the WOTS telephone not being able to put a sufficiently strong RF signal on the base receiver antennas, or it could be due to mutual interference with another uncoordinated access attempt by another mobile set, which prevents the base receiver from clearly receiving either signal. In either case, the mobile set will wait a random time and then try again automatically and repeat this if necessary for a number of time specified previously by the base station (typically 3 to 5 times). The mobile set may also have all the indications that the base receiver has become busy as a result of receiving the access signal from the mobile set, based on the B/I bit, but it may in fact be due to another stronger signal which dominated or captured the base receiver. In that case, the mobile will eventually not see the next logically expected message response from the base station. This will also cause a prescribed number of automatic retry attempts.

Figure 7. 13 shows the basic access operation for several of the wireless office telephone system in use. Initially, the user requests service from the WOTS system, typically by pressing a TALK or SEND button and the WOTS handset transmits an access request message along with its identification information. The WOTS system validates that the user is in its database and is authorized to receive service (step 2). The WOTS system then connects a phone line (step 3) and assigns the WOTS telephone to a voice channel where the dial tone can be heard (step 5). In systems which use a design based on public analog cellular, the user dials before pressing the SEND button, and the dialed number is transmitted by the handset during step 1 instead of step 6. Then the user hears ringing tone during step 5 instead of hearing dial tone. In a proprietary design, following the dial tone of step 5, the user then dials the phone number and when the person dialed answers, conversation begins (step 6).

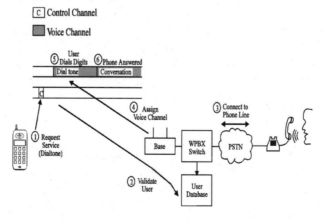

Figure 7. 13, Wireless Office System Access

As noted above, in systems designed to follow the methods of the public cellular network, pre-origination dialing is used. This allows the user to enter the dialed telephone number into the handset first. After all the digits are entered, the wireless office telephone transmits its request for service along with the dialed digits. When pre-origination dialing is used, no dialtone is received and the voice channel is assigned immediately so the user can hear the phone ring. GTE Tele-go, which is described in another chapter, actually uses pre-origination dialing, but hides this from the user by producing an internally generated dial tone in the handset and using some internal logic to automatically SEND the dialed digits after the proper 7 or 11 digit sequence has been dialed. A GTE Tele-go system is now available only in a single line base station configuration and is intended to be used on a PSTN (not WPBX) telephone line, and would therefore not normally be used in a WOTS installation.

Paging

Paging is the process used to alert a wireless office telephone that an incoming call is ready to be answered. To alert the wireless office telephone of an incoming call, the WOTS base station(s) send(s) a paging message. In general paging messages for the same telephone handset are transmitted on the setup channel in several cells in the system where the handset is likely to be. In a WOTS installation, in most cases the paging is done in all the cells, since there is not usually a problem with excessive paging activity. However, most systems have designed in ability to page more selectively as an option. This usually requires that the system be partitioned into several logical location areas (LAs), and that a distinct LA identification code be transmitted periodically in the setup channel by all the cell base transmitters which are in that LA. A different LA identification code is transmitted on the setup channel in cells which are part of a second LA, and so forth. The administrator of the system may arbitrarily assign any contiguous group of cells to each distinct LA, although the proper method for doing this is to make up groups of cells into LAs which have about equal amounts of call setups occurring per minute or per hour. (In a system in which this option of partitioning is not used, all the cells transmit the same default LA value). The wireless office telephone will transmit a registration or localizing identification message when it crosses a boundary between two different LAs. The telephone is aware that it has crossed a boundary because the LA identification code seen in a new cell setup channel is different from the LA identification code just seen in the previous cell setup channel. Wireless office telephones which move around from cell to cell are continually re-tuning to different setup channels when the signal from the previous setup channel becomes weak or contains detectable errors, in the normal course of operation. (The original autonomous registration method, based on a counter in each wireless telephone with increment values transmitted occasionally by the base stations, designed for public analog cellular systems, is not used in WOTS designs).

Paging is only the first step in a delivery of calls to wireless telephones. The telephone responds to a page message with an access type message which contains code values which signify that it is responding to a page. The access process for page response is similar to the access for a wireless telephone originated call, but the data element containing the dialed digits, which appears in the case of pre-origination dialing in a wireless telephone originated call setup access message, is not present. The third essential step in the call delivery process is a channel setup command from the base station to the telephone instructing it to re-tune to a distinct channel, different time slot, or channel code for use during the voice call. In some technologies there are a number of additional operations which occur during the setup of the call as well, such as authentication, setup of privacy encryption, and the like, which are done in different ways in different systems.

Short message service (SMS) is sometimes described as similar to alphanumeric paging because it places a short alphanumeric message on the visual display of a telephone handset. SMS is used in several cellular, PCS and WOTS technologies. SMS message delivery begins with a paging operation so that the desired wireless telephone will respond, but then rather than causing the wireless telephone to retune to a voice or traffic channel in the normal way, the telephone is automatically re-tuned to a special channel on which it receives the digitally coded short message. Most SMS systems can also deliver a short message to the telephone while it is already engaged in a voice conversation. In that case, there is no paging involved, since the base system already knows in which cell and on which channel to communicate with the desired telephone.

Figure 7. 14 shows the basic paging operation for several of the wireless office telephone system in use. A caller dials the WOTS telephone number which rings the incoming line to the WOTS switching system (step 1). The WOTS switching system verifies the location of the WOTS telephone in its database (step 2) and sends a paging message to the wireless telephone on the control channel (step 3). When the WOTS telephone detects it's page, it starts to ring (step 4). When the user answers the phone, typically by pressing TALK or SEND, the wireless telephone transmits a request for service indicating it is responding to a page message (step 5). The WOTS system then assigns the WOTS telephone to a voice channel (step 6) where conversation can begin (step 7). In a system design which follows the method used in public cellular systems, in step 4 the handset is first assigned to a special channel (which may be the same channel used for voice, in some designs) and then the ringing occurs.

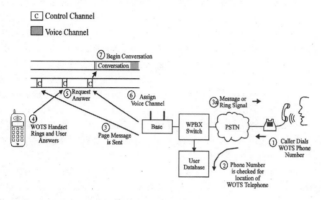

Figure 7. 14, Wireless Office System Paging

Handoff

Handoff permits the user to move from cell to cell without losing the call. The handoff process involves some method of measuring signal strength and/or quality (such as estimated bit error rate (BER) on a digital channel based on the results of error protection decoding) on the present voice or traffic channel, to determine when the quality has deteriorated to the extent that a handoff is advisable. Radio signal strength indication (RSSI) measurement is a normal byproduct of the action of the automatic gain control (AGC) in the receiver section of the base and wireless telephones. The threshold value for this handoff initiation may involve either or both of these parameters (RSSI and/or BER) and may be set distinctly in each cell by the system administrator, but in most WOTS installations there is a default value set by the manufacturer which is satisfactory in most cases and which allows operation of the system almost immediately after the various modules are put in place.

At the same time that the present voice channel quality is being measured, other measurements occur which are used to estimate the signal quality if a handoff occurs to any of the immediately adjacent cells. For analog WOTS technologies, such as IS-94, tunable locating receivers included with each base station measure the RSSI from mobile set transmitters in adjacent cells. The central control unit of the WOTS equipment coordinates these measurements by sending digital control messages to each locating receiver to indicate which frequencies it should tune to and measure. The central control in turn makes use of an adjacency data table which is one of the items of data fill which is entered manually by the installer, and which must be kept current as the cell configuration grows or changes. In an adjacency table each cell is identified by a cell number (*not* the same as the LA identification, which is shared by several cells). For each cell in the system, the adjacency table contains a list of entries giving all other cell numbers which are geographically adjacent to that cell, or which are otherwise sufficiently close that a handoff between them is feasible. Several WOTS systems have automatic or semi-automatic capabilities to initialize and/or update this adjacency table as a normal part of the system operation. Adjacency tables are used by all types of cellular, PCS and WOTS technology (not just analog IS-94) for various purposes. When a particular wireless telephone is operating in a specific cell, the central control unit sends a message to the locating receivers in all the cells identified by that group of entries in the table, requesting a RSSI measurement. In some WOTS designs, this locating receiver measurement occurs continuously. In other designs, this occurs only when the wireless telephone in a given cell passes a "initiate handoff" threshold condition on the RSSI or BER as noted above.

In most digital technologies used for WOTS, the mobile set receivers are used to provide information about the RSSI (and in some cases the BER) of the signals from adjacent cells. This process is called mobile assisted handoff (MAHO) or mobile reported interference (MRI). The wireless telephone's receiver is used for only part of the time to receive signals, during the time slot designated for the voice channel. During other time slots, particularly when the wireless telephone is not transmitting, the receiver can tune to another carrier frequency and measure the RSSI. The base station sends a message to the telephone as soon as it enters a cell using a voice channel. The message includes a list of the carrier frequencies which the mobile receiver should scan during these measurements.

Figure 7. 15 shows the basic handoff operation for several of the digital wireless office telephone systems in use. Digital WOTS telephones typically have the capability to monitor multiple channels while a call is in progress. The WOTS telephone can determine the channel quality of its one signal by the signal strength level and the bit error rate. The WOTS telephone monitors the control channel(s) of a nearby base station (step 2). When it determines that the channel quality of the other channel is significantly better than its present channel, it may request to be transferred to the new base station (step 3) or the base station or switching system may determine it requires transfer to a new channel. If the new base station has a radio channel available, the base station assigns the wireless telephone to the new channel (step 4). The telephone then tunes to the new channel (step 5) and the WOTS switch changes the voice path from the PSTN from base station #1 to base station #2 (step 6).

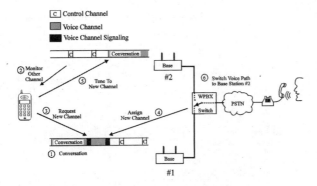

Figure 7.15, Wireless Office System Handoff

In most systems base stations also have the capability to measure the channel quality and request a handoff. This is important in analog systems where the WOTS telephone cannot measure the quality of other channels when a call is in progress. In this case the system must use the channel quality as reported by the base stations to determine when a handoff is required.

Digital Enhanced Cordless Telephones (DECT)

The DECT system is a dedicated wireless office system which is capable of serving wireless telephones and high bit-rate data devices (32 kb/s or more). The DECT system is a digital system which uses a relatively wide radio bandwidth on each base station, which allows up to 12 simultaneous wireless telephones to share each channel. The DECT system has been adapted to several different frequencies and data rates and is also being used for the US PCS system.

DECT and its North American version PWT(E) constitute a third generation standard air interface TDD/TDMA system which has certain technical similarities to the CT-2 and Ericsson CT-3 systems. DECT has the advantages of high base station channel capacity due to multiplexing 12 channels in one base station, which should in turn permit the design and construction of equipment with a low cost per channel. On the other hand, the complexity and cost of a 12 channel base station is somewhat greater than a single channel base station, even though the pro-rated cost per channel may be lower for the 12 channel base station. Because of this distinction, single channel base stations such as CT-2 or GTE Tele-go have been applied more frequently in residential or other low traffic applications, where one channel is all that is desired and the total system cost for that application is better served by means of a single channel base station. Multi-channel base stations and/or multiple base stations with handoff capabilities are more desirable in high usage applications. In these cases the higher total system cost is not objectionable, since the per-channel or per-user cost is normally lower. Figure 7.16 shows an Ericsson SuperCordless telephone handset which is used in the US PCS system that uses the PWT technology.

Figure 7. 16, SuperCordless handset
Source: Ericsson

History

DECT was originally developed by the ETSI technical standards committee in the late 1980s. It was first intended for wireless office use, but it is also adaptable to public low-tier PCS and home cordless use as well. It has strong historical similarities to the Ericsson CT-3 wireless office system, and some generic similarities to CT-2, but has more channels per carrier and other improvements over those system as well. DECT is a global standard and uniform frequency allocation has been made in the 1880 to 1900 MHz band for DECT in all European countries. Aside from basic call processing message standards on the air or radio link interface between the base and mobile handset parts, DECT also has a completely standardized set of so-called vertical features of the type typically available on a PBX, such as 3-way conference, including the human interface aspects of these features. Therefore, a user of a DECT handset should be able to use the learned features on any DECT system regardless of manufacturer.

The DECT standard was established by ETSI in 1992 and is published in the form of the ETS-300 series of documents. The level of standardization for DECT is very complete. The air interface is standardized regarding carrier frequencies (in all European countries), modulation, coding and the like. In addition, call processing aspects of the system operation have also been standardized to an extent which is not found for other PCS systems. This is due to the fact that DECT is primarily viewed by some vendors exclusively as a WOTS system where it will be used with feature-rich PBXs. It was therefore judged desirable to establish standard human interface procedures so that operations such as hold, transfer, 3-way conference, etc. , could be implemented in a user-friendly and standard way which would be easier to develop and also would be uniform regardless of the manufacturer of the DECT handset and WPBX used.

Personal Wireless Telecommunication (PWT) is an adaptation of DECT for the North American market. PWT is used for low power unlicensed operation (1920-1930 MHz) and uses TDD duplexing. PWT-E (enhanced) is used for the licensed operation (PCS blocks D,E, & F) and uses FDD duplexing along with higher power. A major technical difference between DECT and PWT is the modulation method, and consequently the radio channel bandwidth. DECT uses GMSK digital FM modulation with one bit per symbol, very similar to that of GSM. In contrast, PWT uses a (/4 DQPSK modulation method, which is nominally the same as the modulation used for the North American TDMA digital cellular system described by the IS-54 and IS-136 standards. Since there are 2 bits of data per phase change (symbol) of DQPSK, the symbol rate and consequently the radio signal bandwidth, of the PWT signal is half as broad as the DECT signal. The DQPSK modulation also has some instantaneous power variations as well, which requires linear RF power amplifiers in PWT, whereas DECT can use Class C RF power amplifiers, which are less complex and somewhat more power efficient.

The radio frequencies used for PWT in North America do not align with the frequencies used in Europe. First, there are twice as many PWT carrier frequencies allowed in a given bandwidth as corresponding DECT carrier frequencies, due to the difference in modulation and signal bandwidth. Second, PWT is usually deployed in either the unlicensed low power voice portion of the 1. 9 GHz band, or in one of the smaller licensed sections (section D or E or F). Because of the TDD method of duplexing the uplink and downlink, both directions of radio transmission occupy the same part of the spectrum in contrast to the need for an uplink and a downlink in frequency division duplex (FDD) technologies.

DECT and PWT equipment is made by a number of manufacturers, including Alcatel, Ericsson, Matra, Nokia, Philips, and Siemens. In various national markets, Ericsson uses the same trade name Freeset for both DECT (12 channels per RF carrier) and their proprietary CT-3 technology using 8 TDD-TDMA channels per carrier. Ericsson also uses the name SuperCordless for PWT-E.

System Overview

Figure 7. 17 shows a DECT telephone system. The DECT system contains wireless devices such as handsets and data terminals, base stations, a wireless PBX, and interfaces to the wired telephone network and sometimes to a computer/data network. The DECT system uses time division duplex (TDD) multiplexing which allows the same radio channel frequency to transmit and receive. The transmit and receiver

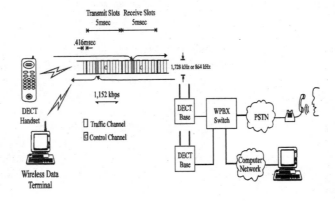

Figure 7. 17, DECT Telephone System

214

bursts are grouped together into frames. The gross channel data rate is 1152 kb/s or 1. 152 Mb/s and a single time slot can provide 32 kb/s data to a DECT handset. Multiple time slots can be grouped together to supply much higher data rates (up to 880 kb/s) for data products such as a computer network devices. While the DECT wireless PBX controls the overall operation of the base stations, DECT base stations and handsets have the capability to measure for interference and other available radio channels and request a handoff. DECT base stations can automatically change their frequencies to avoid interference.

DECT uses TDMA/TDD, which has several technological advantages: the handsets can be made very simple without need for complex signal processing thus eliminating the need for an adaptive equalizer. It is possible to put a "two-way" adaptive equalizer into the base station, since the uplink and downlink use exactly the same radio frequency. The benefits of a more complex adaptive equalizer and the use of diversity combining on the two base station antennas can also be used during base transmit RF bursts (rather than only on base receive bursts as is done with FDD) so that both the uplink and downlink benefit, and the cost per channel or per user is much smaller than if equalizers were installed in each handset. Also, the complexity, weight and, to a small extent, the power consumption of the handset is thus reduced. It must be emphasized that the quality of performance of TDD systems is dependent on the quality of the diversity and adaptive equalizer used in the design and implementation of the system. There is no inherent improvement arising only from using TDD itself, without an adaptive equalizer.

TDD combined with TDMA means that no frequency filter duplexer is needed in the mobile set, nor (in most designs) in the base station as well. This implies a reduction in bulk or size, and a slight reduction in power supply requirements. FDD systems with continuous transmission require a filter to prevent the continuous transmit signal from bleeding back into the receiver and causing undesired interference or intermodulation.

The digital signals used for encoding the speech and for the call processing signals are the same in both DECT and PWT. Speech coding is done according to the ITU G. 712 ADPCM standard at 32 kbit/s. Thus a firmware chip set designed for DECT can be used for PWT, and vice versa, with the minor difference of the symbolic carrier frequency numbers used for the RF carriers in the two systems.

Another advantage of using TDD is that a single radio channel can be used for two way (full duplex) communication. In FDD systems (such as most cellular and PCS systems), two frequencies are required for two way communications. In some cases where one must work around pre-existing radio transmissions (which is a particular problem in some parts of the 1. 9 GHz PCS band, for example) it is often more feasible to use one frequency channel than two separated frequency channels.

DECT/PWT is also easily adaptable to data communications, particularly those which are in the range of bit rates which are envisioned for Internet and wireless office LAN networks, such as 64, 128 and 256 kb/s. Although the net traffic channel throughput of a single DECT/PWT time slot is only 32 kb/s, it is possible to construct a special DECT/PWT terminal which uses more than 1 of the 12 time slots in a DECT/PWT time frame, and thus achieves any of the three bit rates noted above by using 2, 4 or 6 timeslots together in the same device. Standardization of multiple slot or multiple channel versions of DECT/PWT is now underway and this capability adds to the versatility of DECT/PWT technology in an office setting because it supports portable data terminal use about the office.

Figure 7.18, SuperCordless Base Station
Source: Ericsson

A DECT/PWT base station transmits and receives 12 TDMA time slots and can thus communicate with 12 separate handsets engaged in 12 separate conversations. DECT/PWT uses time division duplex (TDD) to separate the uplink and downlink signals. First the base station transmits 12 time slots of information, each slot being 0.416 ms in duration, to the 12 handsets (downlink direction) during a total time of 5 milliseconds. Then it receives for the next 5 milliseconds while the 12 handsets transmit, each one transmitting alone during one of the 0.416 ms time slots. This uplink transmission occurs on precisely the same frequency as the down link. Only the time of the two transmissions distinguishes the uplink and downlink, rather than the frequency. Figure 7.18 shows a PWT base station.

To extend the coverage area of DECT/PWT systems, some manufacturers furnish an optional free-standing radio repeater for DECT/PWT which does not require wires or a microwave link between itself and the central switch. It does require a power supply, which can be ordinary local electric outlet power at 110 or 220 volts, or lower voltage power derived from a step down transformer. The advantage of using a lower voltage for power is that a less expensive wiring installation will still meet electric wiring safety codes. For example, spare wire pairs in a telephone cable may be used to feed low voltage power to a repeater.

A DECT/PWT repeater is located so that it can extend the radio coverage from a DECT/PWT base station into an area which would not normally have adequate RF signal level due to range or shadowing by an obstacle. The repeater looks like a handset to its master base station, but it looks like a base station to the handset which is the "end of the line" for the radio voice path.

Basic Operation

A PWT base station and wireless telephone (handset) both transmit and receive on the same carrier frequencies. There are usually 8 frequency assignments for 1920-1930 MHz U-PCS band in the US. Even if only one 12-channel base station is installed in a system, it is designed with a so-called "frequency agile" radio oscillator, so that it can transmit or receive on a different carrier frequency in each time slot. Among other interesting results arising from this agility is that a handoff can be performed between two neighboring base stations by having the new or target base station take over the same time slot and frequency which was used by the previous base station, in addition to a more conventional (for other systems) handoff in which the handset changes frequency and/or time slot!

The basic operation of DECT/PWT involves TDD transmission of 12 RF bursts or time slots from the base station transmitter during a 5 millisecond time window, followed by a 5 ms time window for the base receiver to operate. Each time slot of the frame has a nominal bit interval of 480 bits during 0. 416 ms, at a physical bit rate of 1152 kb/s. The time average bit rate in each direction is just half of that, 576 kb/s, of which 48 kb/s is nominally allotted to each channel or time slot. Of this, 32 kb/s is used for the ADPCM speech coder data, corresponding to 320 bits per time slot, and the rest is used for synchronization, a separate low bit rate control channel, error protection bits, and 60 bits per time slot for guard time.

During each 0. 416 millisecond time slot, a DECT/PWT radio transmits 480 bits, of which a guard time interval equivalent to 64 of these 480 bits is reserved at the end of the time slot, and the transmitter is off, to prevent simultaneous reception (overlap) at the receiver from signals transmitted in different time slots by transmitters which are near or very far from the receiver. The total instantaneous bit rate is thus 480/0. 416 or 1152 kbit/second. 320 bits in each time slot are from the ADPCM voice coder. These bits can also be generated by a data terminal device such as a personal computer when the PWT system is used for data purposes. When a higher bit rate is desired, special data handsets can be used which operate on 2 or more of the time slots, and thus permit a data rate of 64 kbit/s or higher. The transmitted bit pattern in each time slot also includes 48 data bits used for call processing signals, and 32 data bits used for synchronization. The 48 call processing data bits are protected by an additional 16 error protection bits which are calculated using the CRC16 error detection code. The 32 voice or customer data bits have no additional error protection bits provided, but this is consistent with the usually excellent low bit error rate on these short range radio links.

The handset can, of course, scan all the allowed carrier frequencies for various purposes. It can scan other frequencies during the time slots when it is not receiving its own voice channel, so that it can provide the data on the best nearby base station for a mobile assisted hand off (MAHO). This is a feature which also is used in other TDMA systems. When a handset is turned on or first enters a service area, it scans all the frequencies and time slots, seeking a working base station. Even if a base station is not carrying any voice channel traffic at certain times, it will still transmit a "beacon" or pilot signal on one of its time slots, to make handsets in the vicinity aware of its presence. When all 12 channels are in use for conversations, the base station identification can be transmitted in a control data field of each base station transmission. The handset in communication with the base station does not need that base identification data, since the handset and the base are already in communication. This identification data is transmitted periodically just so that *other* non-conversing handsets which are operating in the vicinity can use this data to recognize the system identification.

The PWT base station normally has 2 antennas, and each manufacturer can implement a proprietary means of providing diversity and/or adaptive equalization using the signals related to the two antennas. The advantage of TDD is that the base station diversity or equalizer mechanism, when used for both transmit and receive signals, improves signal quality and produces a low bit rate in both directions, without use of diversity or an equalizer in the handset. This, of course, makes the handset less complex and lower in cost. The use of dual antennas reduces the likelihood that a radio signal will be in a deep fade on both antennas at the same time. The reduction in radio signal fading also makes the desired signal more resistant to interference and noise, so the range of coverage or effective cell size will increase somewhat.

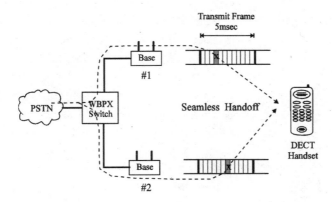

Figure 7. 19, DECT Seamless Handoff

One of the key advantages of the DECT system is its ability to perform a seamless handoff. Figure 7. 19 shows that two base stations can simultaneously communicate with the DECT handset at the same time. This process involves using different time slots on each of the base stations to communicate with the DECT handset during the preparation for handoff. When a DECT handset performs a handoff, it changes the voice traffic channel from the old base station during one frame interval to the new base station during the very next succeeding frame interval. There is no "lost" or missing digitized speech information, and no perceptable gap in the audible voice.

System Parameters

In Europe DECT carrier frequencies are reserved on the 1880-1900 GHz band, with 10 carriers starting at 1881. 792 MHz and ending at 1897. 344, spaced 1728 kHz apart. Because DECT and PWT are TDD systems, each of these carrier frequencies is used for both uplink and downlink. In North America, the bandwidth of PWT is only 864 kHz due to the use of the more radio spectrum efficient DQPSK modulation. The PWT carrier frequencies are typically located in the U-PCS band starting at 1920 MHz and ending at 1930 MHz. The DECT system proposed for PCS is called PWT-1900 and it uses the PWT radio channel structure. The main proponent of this system is Ericsson, who calls their product SuperCordless.

Figure 7. 20, IS-94 Cordless Telephone System
Source: Astronet

The maximum RF Power of PWT-E handsets is 200 milliwatts, and they are typically commanded by the base station to use a lower operational power level, down to as low as 5 milliwatts. Smaller cells only require smaller transmit power, and handsets in the inner part of a cell near the base station require less power as well. Greater base station power (1 watt or more) and greater antenna height can be used to increase cell coverage, but the general objective in a PWT-E system is to use small cells of less than 1 km (0. 6 mi) diameter. The cost of the base equipment is low, and the overall cost per customer is not excessive when these small cells are used. As traffic level grows, more modules can be installed at base locations to increase the traffic capacity.

IS-94

An IS-94 system is an integration of cellular network technology with wireless office or in-building systems. The interim standard 94 system uses the same type of cellular radio channels for cellular and wireless office telephony. Figure 7. 20 shows an IS-94 cordless telephone which is an IS-94 product which can access either a cellular system or a wireless office system.

History

Since 1985, a number of different systems for private base station use with 800 MHz band TIA-553 standard analog cellular handsets have been developed. In general, these systems involve the use of a private setup or control channel which permits the private base station to command the mobile set to use a voice channel supported by the private base station. These private setup channels are not one of the 21 standard setup channels reserved for that particular use in the public cellular network. Instead, some carrier frequency which is not in use by the public cellular network in that area or cell is used for private setup. Similarly, the voice channels used by the private system are carrier frequencies not used in that cell or area by the public cellular network. Private systems use the lowest RF transmit power levels and thus do not usually cause perceptible interference with the public network. In addition, most private systems which use the 800 MHz North American cellular band are sold or leased to the end user by the cellular licensee in that area, who is viewed by the customer as responsible for the technological steps required to ensure that there is no deleterious intersystem interference between the public and private systems.

An important feature of all of these systems is the ability of the handset to operate on both the private and the public cellular systems. However, there was initially no standard radio interface among many of these systems, which effectively gave the end customer no choice among vendors for additional handsets in a private system. Each private system had a distinct way of modifying the algorithms in the call processing software or firmware in the handset to make it seek the private setup channels when in the private system and seek the public setup channels when outside of its own private system radio coverage area. In addition, there were incompatibility problems which occasionally prevented some handsets from seeking the public cellular setup channels when they were in the radio coverage area of an incompatible private system. To correct these problems, the TR-45 standards committee of the TIA developed the IS-94 standard, later incorporating some of the same technology into IS-91-A as well. This describes a standard air interface for private/public cellular system operation which is now followed by several manufacturers.

The TIA standards IS-94, covers dual-system handsets which can be used on both the public 800 MHz cellular network and also on special home base stations. The later standards document, IS-91A, is a comprehensive revision of EIA-553 (which was, itself, an earlier comprehensive revision of IS-3-D), the general analog cellular 800 MHz North American standard. IS-91A incorporates the features of IS-94, so a handset made to the IS-91A standard may have data setup in its read only memory (ROM) to work with a home or WOTS base station as well as a public base station. IS-91A incorporates a number of special features and capabilities which were originally expounded in individual standards documents, such as IS-88 for NAMPS, and IS-94 for private and in-building services. For clarity, we will separate our discussion of systems which are compatible with IS-94 (wireless PBX) and IS-91A (wireless PBX and wireless residential extension).

Several manufacturers market IS-94 compatible systems for wireless office use in conjunction with a PBX or Centrex office telephone system, and their configuration in connection with that application is also described in the chapter on wireless office telephone systems. Astronet (Mitsubishi) uses the trade name OffiCell for its in-office system and ContiNet for their low-tier public cellular system. Panasonic-Matsushita uses the trade name BusinessLink. Other manufacturers such as Motorola and Oki manufacture IS-91A equipment which is compatible with IS-94 systems.

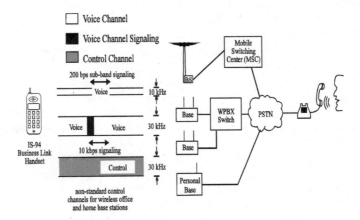

Figure 7. 21, IS-94 Telephone System

System Overview

An IS-94 system is composite of cellular and wireless office systems. Each of these independent systems can work with the same single wireless handset. There are one or more base stations. Each base station has a set of base antennas which covers a cell with RF channels. In most current systems, the antennas are omni-directional (360 degrees) in horizontal azimuthal coverage, rather than sectored. Each cell contains a setup channel and one or more voice channels. There is also a scanning or locating receiver for use in the handoff process (except for single-cell installations, where this can be omitted). The setup channels are in general, not on the 21 reserved carrier frequencies used by the public cellular network. The base transmitters operate at *very* low RF transmit power levels, and the cells are consequently very small, typically less than 500 meters (1500 ft) in diameter. The wireless handsets (called mobile stations), when used with these private systems, are likewise commanded by means of messages from the base stations, to operate at minimum RF transmit power levels. Figure 7. 21 shows a IS-94 telephone system.

The mobile stations have a more complicated sequence of operations which they follow to lock onto a suitable setup channel when the mobile set power is turned on, compared to a mobile set used only on the public network. To explain the distinction, first recall the setup channel acquisition method of a public network mobile set. Public network mobile sets first scan only the 21 setup channel frequencies which are designated for their home carrier in the public cellular system. They measure and save in memory the signal strength indication of each such carrier frequency. They then go back and examine the strongest carrier frequency, by receiving its continuous digital information transmission in a testing mode. The mobile set can determine, by means of error detection codes used with the base station transmitted data, whether it is receiving the signal substantially free of errors or not. If not, the mobile set will try the second strongest carrier frequency with a similar test. If it cannot get a good signal on either frequency, it will then try to start over using the 21 assigned setup channel frequencies of the other public cellular operating company. If it cannot get a good error free signal on either the A or the B public cellular operator system, it will display some visible indication such as "service not available." Although it continues to seek a good setup channel continuously when its power is on, it merely repeats the processes described in this paragraph indefinitely until it finds a good signal. This could occur, for example, if a customer first turns on the mobile set in a country area with no cellular service and then moves or drives into an area of good RF signal coverage.

In contrast, an IS-94 set is pre-programmed with several items of information which are only used in the private system(s) for which it is intended. This information is in addition to the standard information which permits the set to work on the public cellular network. This private information includes:

An internal list of channel numbers or channel ranges which it must scan when seeking a private setup channel. There can be a "permanent" list and also a temporary list which is generated and modified by the wireless telephone from time to time as it finds that certain frequency channels are more often used than others in the areas where it is used. Use of a temporary search list of this type increases the probability that it will find a proper setup channel quickly. There can also be multiple permanent and temporary lists for a mobile station which is intentionally used on several private systems.

A (or several) private system identification number(s) which is/are used by base stations in its own private system. This or these private system identification numbers are used in a manner analogous to the system identification code (SID) in a public system. The private system base stations transmit their system identification numbers periodically. One or more private telephone number (MIN) values, which may be different from the public MIN of the set. For example, the private MIN may be a 3, 4 or 5 digit extension number which is used to make calls to this wireless telephone when used in a private system.

The dialing procedure in both public and private modes of operation is pre-origination dialing. The user enters all digits of the telephone number and then press the SEND button. The user can dial in the same way they are accustomed to with an ordinary cellular telephone.

IS-94 equipment is marketed in each city by the local cellular operating company whose frequency band it uses. It is generally rented or sold with the intention that its installation will remain in the service area of that cellular operator. Systems are available with a single base station for home use, but the more popular and cost effective package is a set of base stations for an in-building mini-cellular system, usually used in conjunction with a PBX, Centrex, or other small business telephone system. The multi-cell system can handoff calls within its own cells and do many of the other things a big public cellular system can do. IS-94 is not designed to handoff a call between the private office system and the public cellular system. The features and limitations are described in more detail in the chapter on wireless office telephone systems.

Basic Operation

In the private system, the processes of call setup, handoff, and call release are substantially the same as the corresponding processes in a public cellular system. The substantial differences include the use of non-standard control channels and interference detection for the private systems. Some private IS-94 cellular systems also have only one cell and base station, and in that case handoff is not possible to a second cell.

Each base station in an IS-94 system performs like its "big brother" base station in a standard public cellular system. The handsets contain data tables which specify that the handset, when turned on, should first search the designated special frequencies assigned to that IS-94 installation. When it finds a private setup channel, the handset will stay tuned to that channel unless the received bit error rate deteriorates, and then it will search for another private setup channel. If it cannot find a satisfactory private setup channel, it will then search the normal public system setup channel frequencies to determine if it should go into "public" cellular operation. In addition, even if it is receiving an adequate public setup channel, it will periodically rescan the private assigned frequencies to determine if it has re-entered the private domain. In short, the private domain is preferred when adequate setup channel signals are available from both a private and public base station.

Paging, page response, mobile originated call setup and other system operations take place on the private setup channel much as they do in the public setup channels. Most manufacturers of IS-94 sets use the narrow band NAMPS 10 kHz analog FM mode for voice channel transmission. This gives more channel capacity in a system where there is no frequency reuse, particularly when the system has no actual internal frequency reuse because there are only a very few cells. This requires that the co-channel interference from the public cellular system is 23 dB lower than the NAMPS signal at the receivers inside the operational area, which is important to check because there are occasionally strong public system signals due to changes in public system frequency plans made after the initial installation of the IS-94 system. In many IS-94 systems, the local public cellular operator can dial into the IS-94 system using a modem and can securely make system modifications to track any changes in the public cellular system which would otherwise affect the IS-94 system operation. This includes the frequencies assigned for use in each cell.

Because IS-94 private systems share the same radio spectrum with the cellular systems and potentially with other nearby IS-94 users, the private systems typically check to determine if the radio channel is in use before assigning a voice channel. The public cellular system operator may or may not reserve certain channels for a particular IS-94 system operation, but in any case, the system should be configured to use the carrier frequencies which have the minimum radio background noise level. When a channel is not absolutely reserved for IS-94 use in a certain area, the automatic scanning capabilities of the IS-94 system will prevent that system from using it when there is already an RF signal there from the public cellular system, so its use will automatically vary from moment to moment. The public cellular operator can dial into the IS-94 system and, using a secure modem connection, make needed changes in the preferred frequency assignments to minimize RF carrier interference with the public PCS system, particularly when the frequency plan of the public system is modified. In the eyes of the FCC, the public cellular system operator is responsible for proper configuration of both the public cellular system and the private IS-94 system so that there is adequate traffic capacity for the public users and no harmful interference.

System Parameters

The frequency bands used for the IS-94 system is 824 to 849 MHz for the reverse link (telephone to base) and 869 to 894 MHz for the forward link (base to telephone) with the exclusion of the standard AMPS control channel frequencies. IS-94 base stations can transmit their 30 kHz wide control channel on any non-AMPS dedicated control channel frequencies. The voice channel bandwidth can be either 30 kHz (standard AMPS) or 10 kHz (NAMPS). Voice modulation is FM and signaling is 10 kb/s blank and burst for AMPS channels or 200 bit/second sub-band signaling for 10 kHz NAMPS channels. Threshold values for handoff initiation (minimum RSSI and/or maximum BER) may be partially under user control, and partially automatically controlled by the system. These values are, in general, different in each cell.

Other Wireless Office Systems

There are several other wireless office systems including SpectraLink and Companion.

SpectraLink

The SpectraLink system is a wireless office telephone system which is produced by SpectraLink Corporation. SpectraLink equipment has a unique radio interface in the WOTS product lineup, and to an extent in the entire PCS radio frequency band. SpectraLink uses a TDMA Frequency Hopping (FH) radio interface in the 902-928 MHz band. SpectraLink's design uses a low power RF transmitter level which falls within the limits of Part 15, section 247 of the FCC rules. This is the part of the US radio technical regulations which permit low-power unlicensed radio transmission. This frequency band is intended for equipment such as certain types of radio devices, medical equipment such as diathermy devices to create an artificial fever or for thermal therapy, or certain types of radar-like distance measuring equipment. The FCC permits the use of devices such as the SpectraLink equipment and certain radio data communications systems (such as radio-based data LANs) when they meet the same limitations on radio power stated in Part 15. Such devices are often generically described as "Part 15" equipment. The radio coverage range is limited by the small power transmitted (100 mW in the case of SpectraLink), but this is normally acceptable in a properly configured cellular WOTS installation. Figure 7. 22 shows a SpectraLink wireless office telephone.

Figure 7. 22, SpectraLink handset
Source: SpectraLink

The FH radio link signal is a feature which is not used in most other PCS systems. FH is a technique of spread spectrum which has been used in some military systems for avoiding intentional interference by an enemy "jammer." The mobile and base radio use both TDMA and frequency hopping, jumping to a different carrier frequency for each time slot transmission by the base and the mobile units. In the SpectraLink system the sequence of frequencies used for frequency hopping is apparently random (called "pseudo-random"), but is actually predictable and known to both the base radio and the handset so the two can remain in communication. FH may lead to an occasional "collision" between two different handsets which attempt to use the same frequency during the same time slot, but FH also gives the radios the advantage of avoiding a local radio channel fade which may only affect a limited number of radio frequencies. By properly programming the mobile units when they are installed in the system, they can be set to use only a group of carrier frequencies which is known, by means of a prior RF site interference survey, to have no local sources of interference (for example, from a wireless LAN device). Using a permissive, though proprietary technique, the SpectraLink system is synchronized and thus "collissions" do not occur. The only other PCS-related radio system which is in commercial use with pseudo-random FH is the Geotek ESMR system, which is not a WOTS system. Figure 7. 23 shows a SpectraLink base station. SpectraLink also makes a separate and distinct base controller box for use with the Nortel Meridian™, Lucent Definity PBXs and Norstar™, Lucent Merlin, Legend, and Comdial key telephone systems. SpectraLink has developed the necessary proprietary interfaces as described in figure 7. 7, and the handsets emulate the native wired stations of the host switching system. Both of these widely used wired office systems use proprietary digital station set loops employing so-called time compression multiplexing (TCM) or "ping-pong" loop signals. The technology of TCM loops is described briefly in the next section.

Figure 7. 23, SpectraLink base station
Source: SpectraLink

Nortel CompanionTM

Companion™ is the Nortel version of the cordless CT-2 standard, upgraded for North American use with wireless office features. The Companion™ product was introduced in Hong Kong in 1992, in Canada in 1993, and in the United States in 1995. In 1996, The Companion™ Microcellular system was introduced which allows a single wireless telephone to access either the Companion™ system or a cellular system. With Companion™ Microcellular, the users cellular telephone number can appear as a PBX extension and can be assigned to the same extension number as a wired PBX. This allows calls to go simultaneously to both the wireless handset and wired PBX telephone extension.

Companion™ is usually described as CT-2+ because of the improvements over the original CT-2 common air interface standard developed in England. These improvements mainly relate to the use of handoff, which is not present in the CT-2 standard. The Companion™ system uses a controller which is adaptable to all the small business telephone systems made by Nortel, including Meridian™ PBXs and Norstar™ key telephone systems. In addition, addition, it can also work with any PBX or Centrex system having a standard analog extension telephone line. Figure 7. 24 shows a Nortel Companion™ handset. This Companion™ C3050 portable phone includes two-line display, calling party name display, call line identification, visual message waiting indicator, call transfer, a 50 item directory, and a silent alerter.

The Companion™ system operates in a variety of spectrum frequencies depending upon the frequency allocated for a particular country. In Canada, the Companion™ system operates on the 948-952 MHz band which has been dedicated for that product. In the United States, the Companion™ system operates on the the 1. 9 GHz unlicensed PCS band. The general commercial product is normally furnished using the first two bands noted in this paragraph. Neither of these are the same as the CT-2 frequencies used in Europe and Asia where there are CT-2 systems in public operation in France, Germany, Singapore, Hong Kong, etc. Companion™ WOTS equipment is particularly appropriate for an installation in which there is a very large geographic area to be served, including the capability to perform handoff between calls.

Nortel Meridian™ and Norstar™ office telephone systems each use a proprietary method for transmitting digital signals on a two-wire extension line between the proprietary wired telephone extension and

Figure 7. 24, Companion™ System

the PBX or central key service unit (KSU). The proprietary telephone sets used in these two Nortel systems have the analog-digital convertor (or coder-decoder — CODEC) built into the proprietary telephone set. This CODEC converts between the analog speech waveforms associated with the handset microphone and earphone, and industry standard Mu-law 64 kbit/second digital coding. This digital signal is also called pulse code modulation (PCM). It is the standard method for digital transmission of speech in the North American PSTN. Because these Nortel systems perform the conversion into digital form at the telephone set, they are less susceptible to electrical interference which could cause audible noise or degradation in an analog system. The installation can be made with a direct digital link from the PBX or electronic key system to the PSTN via a T-1 carrier installation. A digital link can also be established using ISDN Basic Rate Interface (BRI). Either of these two digital connections to the public telephone system will preserve the high quality of the digitally coded speech, and also support high bit-rate digital data communications if desired by the customer. Of course, a traditional analog subscriber line can also be used to connect these Nortel systems to the PSTN as well.

The TCM signals on the extension loop wiring between the proprietary wired telephone extension set and the central PBX or KSU permit the use of only two wires (one pair) while transmitting digital signals in both directions apparently simultaneously. Other two-way digital transmission method require four wires (two pairs). A TCM signal consists of alternate bursts of digital pulses in opposite directions on the wires. During a brief time interval of 125 microseconds, the central PBX or KSU generates a burst of bits for half of the time interval, and then the proprietary telephone set generates a similar burst during the second half of the interval. The burst consists of 8 bits of digital information from the speech coder, and several other bits used for synchronization and call processing messages. Because the bursts of electrical pulses each travel in opposite directions on the wire pair, each burst occurring while the other burst is inactive, it is simple to separate the received burst of pulses from the transmitted burst by means of a fast synchronized electronic switch. If we transmitted the bursts of pulses in both directions continuously and simultaneously, it is still feasible to separate the relatively weak received signal from the strong transmitted signal at one end, but the method is much more complicated and expensive. In order to send the required number of bits per second, the actual pulse rate in each TCM burst must be twice as fast as would be used for continuous transmission. Because of this, the group of bits which comprise the burst are transmitted in half the time that would have been needed for continuous transmission. This is the origin for the term "time compression" since the total time to transmit the bits in the burst is "compressed" to half of the time used for continuous transmission.

Companion™ WOTS equipment is particularly appropriate for an installation in which there is minimal traffic (typically no more than one conversation per cell) but a very large geographic area must be served, including the capability to perform handoff between cells. In addition, a relatively inexpensive home base unit for home cordless use is available for the Companion™, thus allowing the same handset to be used in both the work and home environments. In Canada, some public installations of CT-2+ are operating in some shopping centers and airports. The public installations do not, at this time, support mobile destination calls. When the frequency band of the public, home and office CT-2+ systems are the same, the Companion™ handset becomes a viable telecommunications device in all three environments.

References:

1. Rappaport, T. S., "Wireless Communications," Prentice Hall and IEEE Press, 1996, pp.123-133.

Chapter 8
Cordless Telephone Technology

Cordless Telephone Technology

Cordless telephone technology allows customers to use a wireless telephone within approximately 50 to 100 meters of their local telephone line. Cordless telephone technology has evolved from the first generation of home Cordless Technology generation (CT-0), through various generations of public cordless or low-tier systems such as CT-2, Tele-Go, and unlicensed spread spectrum cordless telephones. There is also a certain degree of overlap in both technology and scope of application between some the systems described here and those described in Chapter 7 on wireless office telephone systems.

Wireless systems can be divided into high tier and low tier categories. High tier systems provide wireless communications services through a wide coverage area such as a cellular system. Because of the base station interconnection and switching complexity, and large investment in providing radio coverage through a large geographic region, these systems are typically more costly to use. Low tier systems provide radio coverage to limited regions and use low power base stations and wireless telephones. Although the name "low tier" may sound inferior to some people, its intentionally reduced range of coverage and its relative technical simplicity results in a system which is economically very competitive with high tier systems, and in many cases even superior. Low tier systems are relatively less expensive to install and maintain than their high tier counterparts. In some cases, one wireless telephone may be capable of accessing both high tier and low tier systems.

History

Residential and several office cordless telephones have been a popular consumer product for over 20 years. Some estimates show that over 60% of all US homes have at least one cordless telephone [1]. AT&T Bell Laboratories demonstrated a prototype cordless telephone in the 1960s. The first commercially available cordless telephones used the 27 MHz citizen's band, mostly with amplitude modulation (AM). Although they performed the primary function of giving the consumer portability, they had several undesirable properties, which were mostly corrected by later improvements.

The quality of sound for early generations of cordless telephones was inferior to wired telephones, particularly near sources of radio interference. This was remedied in later developments by use of frequency modulation (FM) and a new group of frequencies in the 49 MHz band for cordless sets, which was not subject to the interference from citizen's band radios. Conversations on AM (and some FM) cordless phones were subject to eavesdropping by anyone who had a suitable radio receiver or scanner. In the 1990's, cordless telephones began to use digital radio transmission and produced even better speech quality. Use of digitally coded speech and/or scrambled analog speech in later designs helped to improve the privacy of cordless conversations. Some cordless telephones use wide bandwidth radio signals in the 900 MHz frequency range which allows a larger geographical operating range.

Unauthorized users with a compatible cordless handset could access the base station of a residential cordless handset, and place calls even from outside the home of the victim. This is a form of theft of service, since the victim generally had to pay for any such calls. Scrambling or encryption of the signals from the handset, together with other methods which require authentication of the handset, help to prevent this.

The earliest generation of home cordless telephones are sometimes described by the jargon term CT-0 (cordless telephone, generation zero). The improved versions with FM or digitally coded and/or scrambled voice and signaling is sometimes described as CT-1, implying a baseline for further developments which officially or unofficially have the designations "2," "3," and so forth. Both CT-0 and CT-1 cordless telephones are only usable in a small area around a single base station. That base station is normally connected to the telephone line of the owner (either residential or a single office telephone line) and is *not* intended to serve the general public. CT-2 cordless telephones were created to allow cordless telephones to be used in public locations. CT-2 telephones could be used in the home and in areas which were served by public CT-2 base stations.

Although cordless systems were not originally designed for large cell coverage and radio signals which could experience rapidly signal fades when used in fast-moving vehicles, future improvements in antenna and equalizer technology could greatly extend the range. In the US, a form of DCT with general public PCS coverage is being promoted by Ericsson under the name DCT-1900 or Supercordless. This system is planned to have substantially complete multi-cell public radio coverage, at much lower cost than high-tier PCS or cellular systems. In the early 1990's, wireless telephones started to have multi-mode capability which allowed them to operate with low tier home cordless and high tier cellular systems. The evolution of cordless telephone technology is shown in figure 8.1.

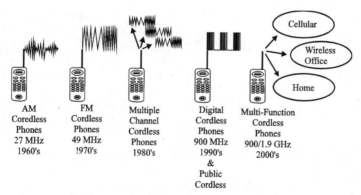

Figure 8.1, Evolution of Cordless Telephone Systems

Residential Cordless Telephones

The cordless telephone connects the customer to the telephone network. A residential cordless telephone is usually similar in shape, size and weight to the handset portion of a corded telephone. Many of these comments also apply to public cordless or low-tier PCS systems as well. In many cases, the cordless handset can be clipped to a belt or carried in a pocket, folded down to a smaller size. Cordless telephones normally have rechargeable batteries and they are fitted so that the batteries receive recharge power when the handset rests in a cradle which is part of the base station. The handset normally has a microphone and earphone disposed for normal use when held like the handset of a corded phone, next to the head.

The controls on a cordless handset are often no more complex than those on a wired handset: namely a push button dial keypad and some type of button to press which ultimately requests a dial tone from the cordless base station. This button can be pressed briefly for a so-called "flash" (less than 2 seconds interruption of dial tone or during a call in progress) which is used with some wired telephone features such as conference calling. It can also be pressed and held to disconnect and get a fresh dial tone. In some designs, there is a separate button for the "flash" which automatically produces the correct interrupt timing regardless of how long the user holds the button down. Some cordless sets have features which include speed dialing buttons which allows the rapid dialing of several pre-stored numbers, intercom conversations between the handset and base, and other features.

Because of the intentional short range of residential cordless and low-tier public systems, the transmit power of the handset is as small as practical, sometimes below 10 milliwatts. This, in turn, permits long talk time using a battery with a low energy storage capacity, and thus low bulk and weight, particularly when compared to the average cellular handset. Figure 8.2 shows a typical block diagram of a cordless telephone.

Residential cordless telephones have evolved from single channel AM transmission to multiple radio channel digital transmission. The RF section is typically a low power multiple channel transceiver (transmitter and receiver) which is capable of sensing interference and changing radio channels. The logic section consists of a microprocessor and a small amount of memory. To reduce the susceptibility of the cordless base unit to access by another unauthorized handset, a unique code is usually stored in the handset and the base station. Some of the code can be changed by the customer through a keypad setting or several slide switches. An audio section routes the audio from the microphone and speaker to the modulator of the radio transmission section. The battery is charged from a desktop or wall mounted base station which supplies power to each of the assemblies in the handsets.

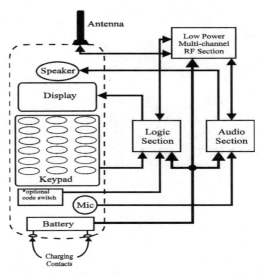

Figure 8.2, Cordless Telephone Block Diagram

There are various features that may be included on cordless telephones which include a digital display, private signaling, and signal strength monitoring. The digital display can be used to show the dialed phone number or in some models, caller identification number and name. An out of range warning indication such as an audible and/or visible warning to inform the user that they are moving the handset out of radio range of the base. This allows the user to stay in the range of adequate radio coverage, or reposition the base station to modify the range to cover your normal area of circulation. Private paging and intercom features may be used to call to the user from the base even when no external call is in progress. This permits the base and handset to be used as a home radio intercom. It also helps you to find the handset if, for example, it is under the couch cushions, by causing it to ring or produce another loud sound due to a command at the base.

Cordless Base Station

The cordless base station is the link between the cordless handset and the public switched telephone network (PSTN). In its simplest form, the cordless base station has no external controls for the user. It has a cord for electric power and a cord for connection to the telephone line (typically using a standard RJ-11 connector). Most cordless base stations provide a cradle where the handset can be placed which allows the bottom of the handset to make electric contact with a recharging power source for the handset's rechargeable batteries. The electric power may be fed via two metal spring-loaded contacts on the surface of the handset, or via a magnetic coupling coil which is not visible from the outside.

Many base stations have other amenities, up to and including the full functionality of a telephone at the base station independent of the cordless handset. Some base stations also are combinable with or have other built in telephone accessories such as an answering machine (so-called "voice mail"), calling number identification, etc. The cordless base station always contains the following minimum technical functions:

1. **A connection to a telephone line.** (also called a telephone loop) Senses ringing so calls can be answered, and which can close the electric circuit between the two wires of the telephone loop to begin a call or answer a ringing incoming call. The operating condition in which direct current flows through the loop interface while the two loop wires are in electric contact is called "off hook," while the opposite standby condition with no direct current flow, is called "on hook."

2. **A two wire to four wire converter.** The voice circuit is handled by a directional coupler of the so-called "hybrid coil" or "induction coil" type. This is similar to a device in an ordinary wired telephone. Its purpose is to separate the incoming and outgoing voice signals, so that each one exists in a separate pair of wires. This allows the two voice signals to be handled separately via the radio link to the cordless handset. In most cases, a separate radio frequency is used for the two voice signals. The cordless microphone signal is transmitted to the base station on an "uplink" or "reverse" radio carrier frequency, and the earphone signal is transmitted from the base station to the cordless handset on a distinct "downlink" or "forward" radio carrier frequency.

3. **Radio transmitter and receiver, and associated antennas.** The base station radio transmitter does not transmit until it is requested to do so either by an incoming call or when the customer uses the cordless telephone to request a call to be originated. If the base station has multiple radio channel capability, it regularly scans its channel list to determine channels which are not in use.

4. **A radio signaling system.** A process of signaling between the base station and the cordless telephone which allows the dialing (call origination) and paging or ringing (incoming call alerting and answer-

ing) by the customer. This signaling system is typically proprietary to the manufacturer. Because these systems are typically unique for each manufacturer, a cordless telephone from one manufacturer usually cannot be used with a base station from another manufacturer.

Either the cordless telephone or base station must contain a ringing device or speaker so that it will produce audible ringing (or other alerting sound such as beeping). The system must provide a method for the users to indicate that they wish to begin using the cordless set via the uplink radio channel. This may be done by turning on the handset transmitter when it was previously off, or it may be done by transmitting a special signal via a handset transmitter which is continually on. This signal will cause the telephone line (loop interface) to close the electric circuit path, so that a call can be dialed (originated) or a ringing call can be answered. Figure 8.3 shows a diagram for a cordless base station.

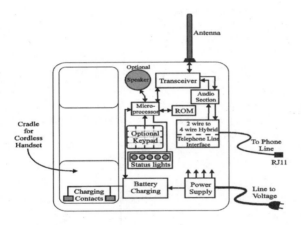

Figure 8.3, Cordless Base Station Block Diagram

PSTN Connection

A cordless base station connection to the wired telephone network through a telephone line. The base station must be capable of interpreting signaling messages or changes in state of the telephone line connection. There are three widely used telephone line connections used with cordless base stations: standard telephone line, ISDN, and primary rate digital interface:

1. **A standard analog telephone line.** This is the standard telephone line connection used in most residential homes and is sometime referred to as plain old telephone service (POTS) or a 2500 set compatible line (referring to the number type designation for the standard touch-tone dial telephone). Because of the lack of more sophisticated signaling features, this type of connection cannot efficiently support multi-cell public cellular systems with handoff. However, it is adequate for non-handoff systems or non-handoff installations of more sophisticated systems.

2. **An ISDN BRI digital interface.** This interface supports two normal voice grade channels in digital 64 kbit/s form, as well as a 16 kbit/s channel which can, in theory, be used for multi-channel base stations and/or more sophisticated signaling between a multi-cell public cordless or low-tier PCS base station and the switch. All of these digital signals are multiplexed onto a single pair of ordinary telephone wires. At this time, the availability of switch software which can effectively use this interface to support a multi-cell public cordless system is limited, but a number of manufacturers have announced that they intend to make this available.

3. **A primary rate digital interface.** The primary rate digital interface is used for a multi-channel base station, using a digital channel for signaling. In North America, the main example of this type is a T-1 (or DS-1) digital multiplexed signal with a digital bit rate of 1.544 Mbit/s, conveyed on two pairs of telephone wires.

In addition to these basic types of PSTN connection, there are a number of proprietary interfaces which are designed for use with particular types of private branch exchange (PBX) equipment and thus belong in another chapter on wireless office systems. Most of these proprietary interfaces use one, two or three ordinary telephone wire pairs.

For residential cordless base stations, the telephone line (loop) interface to the PSTN is usually a standard single line telephone line. The cordless base station is typically designed so that it draws no power (and no electric current) from the telephone line until the user begins the connection phase of the call. During a ringing signal, the cordless base station, like several other types of electronic telephone devices, need not draw any power (or direct current) from the ringing signal on the telephone line in order to provide power for the alerting sound. The cordless base will only start to draw direct current after the user has answered the call (the cordless base goes "off hook"). This implies that it has a ringer equivalence rating, according to the designation used in FCC rules (Part 68), of zero, and many of these zero equivalent telephone devices (such as an answering machine) can be connected to a single telephone line without drawing excessive ac ringing current.

In neither the public nor the proprietary systems are there yet standard call processing message protocols for call processing to be used on these interfaces. The need for this is recognized, but the large variety of different public cordless and low-tier PCS system technologies has delayed the development of a standard protocol here. Figure 8.4 shows how a cordless base station typically connects to the telephone line and shares the telephone line with other telephone devices.

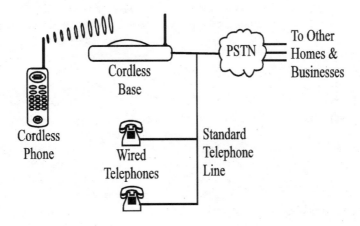

Figure 8.4, Cordless Base PSTN Connection

System Overview

Residential cordless telephones are not typically part of a coordinated wireless system. Each cordless telephone is acting unaware of any other cordless handsets installed in the vicinity, other than the radio channel scanning which some cordless sets perform to find a channel with minimum radio interference. When an excessive number of simple cordless sets are installed too close to each other in the same vicinity, there is a limit on the amount of radio traffic which these uncoordinated sets may readily carry. Under those conditions, the problem of excessive offered or attempted radio traffic will be manifested in one of two ways. For sets which do not first scan for interference or which do not inhibit transmission in the presence of interference, there will be unacceptable interference as more and more users join the pileup on the radio channels. For handsets which inhibit their transmission unless there is a channel which is sufficiently free of prior radio interference, the initial users *will* get assigned a radio channel, likewise for the second user and so on..., but once all available channels are assigned, all later attempts will be blocked. This second situation is somewhat more desirable than total interference anarchy, since at least some of the users will have an interference-free conversation. But neither situation is good, if these users actually have a need for more traffic than the system can handle.

In a multi-cell system with a coordinated overall design, the size and the number of voice channels in each cell can be set by design. Higher traffic density can be achieved in high traffic areas by installing a larger number of smaller cells. Some carrier frequencies may be used in the high traffic areas and radio interference on these frequencies can be prevented by setting up the system so that these same frequencies will not be simultaneously used in other peripheral cells with lower traffic.

For a public cordless or low-tier PCS system, the same design and development tools for proper radio coverage and choice of cell base station location can be used as are used for high-tier public cordless and PCS systems. The main difference is quantitative rather than qualitative. Since the size of low-tier cells is definitely smaller, a much larger number of cells will be needed to cover the same service area compared to a high-tier system, perhaps 8 to 20 times as many cells. It is interesting to note that with an equivalent number of total voice channels in both systems, the low-tier system may often be able to satisfy very high traffic demands (in Erlangs per square km or per square mile) that the high-tier system cannot handle. This is a direct result of using more cells and lower radio transmit power. In addition, the lower design target cost of low-tier public PCS base equipment can make the two systems comparable in price, despite the large difference in the number of cells.

The overall structure of a simple public cordless system with no handoff and originate-only service (such as the original British CT-2) can theoretically be optimized with regard to overall capacity, by choosing optimal locations for the base stations and setting their choice(s) of operating frequency and RF power to minimize cell to cell interference. However, this would either require manual settings and adjustments to existing base stations as new base stations are installed, or some remote control data network capability. Therefore it is not normally part of the objective of these simple public cordless installations.

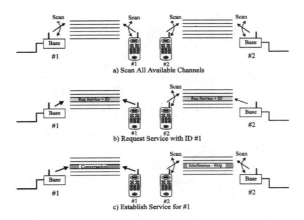

Figure 8.5, Cordless System Control

Figure 8.5 shows how most cordless phones are controlled as a system. Because cordless telephone systems rarely have a dedicated control channel to provide information, both the cordless handset and base station must continuously scan all of the available channels (typically 25 channels) regularly to see if communications is required (figure 8.5 (a)). When service is required, one of the units will choose an unused channel and transmit a pilot tone or digital code on a radio channel with its own unique code to request service (figure 8.5 (b)). The base station and nearby base stations will receive this request for service and lock onto the radio channel. When the handsets home base station has locked on to its request for service, it will authorize service and conversation can begin (figure 8.5 (c)). When a nearby base station detects the request for service, it will determine that the message is not intended for it and will not process the call and scanning will continue.

Frequency Reuse

When two radio transmitters operate close to each other on the same frequency, interference occurs which can cause distortion or cross-talk. Radio channel frequencies can be reused when sufficient distance exists between two transmitters which are using the same frequency. While cellular and PCS systems can use frequency planning to coordinate the distance between transmitters, cordless telephone systems are not frequency coordinated.

The ability to reuse frequencies is dependent upon the maximum allowable level of interference which is called carrier-to-interference (C/I) rating of the signal. The question of whether or not a particular system installation will exceed the C/I rating is directly related to the radio signal attenuation of nearby transmitters which are operating on the same frequency. This is called the path loss of the radio signal propagation. The C/I ratio required by most of the cordless systems is in the range of 14 to 18 dB. This means the combined power of nearby signals which are operating on the same frequency should not be above 1.5 to 4.0 percent of the desired signal which is being received from its own base station.

For most cordless systems, the very low power (typically 10 milliWatts) used restricts radio coverage to small cell sizes (less than 50 to 100 meters) in which there are a few obstacles such as building walls to radio wave propagation. Therefore, it is reasonably accurate to describe the path loss of radio wave propagation by the empirical formula $P=F/rk$, where the exponent k is in the range of -2 (ideal) to -4 (partially obstructed). This is different from the model used for large cell cellular and PCS systems, where the value of k is more accurately in the range of -3 to -4. This same information is often stated in another way, by describing how much attenuation occurs per 10 times distance change, which is also called a "decade" of distance. For example, we compare the signal power received at 50 meters distance compared to 500 meters. An unobstructed path loss (k = -2) has 100 times less energy at 10 times the distance and an moderately obstructed path loss (k=-4) has 10,000 times less energy at 10 times the distance. When the power ratio is measured using the logarithmic ratio unit called the decibel (abbreviate dB), we would say that k=2 corresponds to a signal reduction of 20 dB per decade of distance; and k=4 corresponds to a signal reduction of 40 dB per decade of distance.

For UHF frequencies, such as 800 or 900 MHz, or 1800 or 1900 MHz bands, other base radios which are sufficiently far distant and which have low enough antenna height so that they are effectively beyond the horizon (hidden by the curvature of the earth), there is no significant interference. However, for lower frequency units such as cordless phones which operate on the 27 MHz or 49 MHz radio bands, there is the possibility of interference from distant transmitters. Low frequency radio transmitted signals have the property that they effectively flow over the horizon and follow the "curvature" of the earth. When high power transmitters (such as a trucker's citizens band radio) are located beyond the horizon, the signal may bend in this way, or "skip," to a distant location. "Skip" is a name for the multiple reflection of HF radio waves which are guided along a belt between the somewhat electrically conductive surface of the earth and the electrically conductive ionized layers in the atmosphere.

The phenomenon of radio skip and the distance of interference sources is more pronounced at twilight or during the dark hours in the vicinity of the base station, since the ionized layers in the atmosphere form at a higher altitude during those hours of the evening, night and early morning. The higher altitude provides a longer geometric path for the skip refection points. This effect causes the radio signal to be detected at various distances from its point of origin, at each skip reflection point on the earth. Skip is well known on the 27 MHz citizen's band. It is not legally permitted to utilize skip to communicate with a distant radio on this frequency band, but when the interference from other distant radios occurs it is not useful for communication, but does have the undesirable effect of making some of the carrier frequencies unusable and thus reducing the total conversation traffic which can be handled. Figure 8.6 shows how low frequency signals are reflected which can be a problem for 27 MHz cordless telephones.

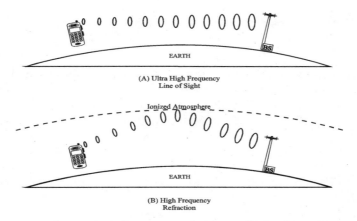

(A) Ultra High Frequency
Line of Sight

(B) High Frequency
Refraction

Figure 8.6, UHF -vs- High Frequency Cordless Radio Signals

The frequency plan for cordless telephones changes dynamically. Figure 8.7 shows how frequencies for cordless phones can be may be assigned. In figure 8.7 (a), cordless telephone users in each of the three houses initiated calls with their cordless telephone. As each user initiated their call, the cordless handset selected a radio frequency that was not in use. After a period of time, the callers in houses numbers 2 and 3 finished their conversations and the cordless telephone user in house number 3 initiated a new call. Because frequency number 2 became available when the caller in house number 2 finished their call, the cordless system in house number 3 was able to use frequency number 2.

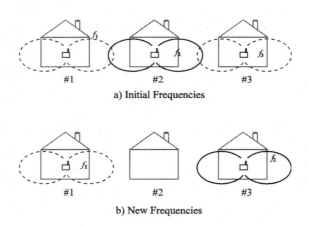

a) Initial Frequencies

b) New Frequencies

Figure 8.7, Cordless Radio Frequency Channel Assignment

When more base stations are installed in a situation in which all the legally permitted carrier frequencies are already in use in a region, the new base station will ultimately reuse one of the 50 frequencies. However, it will naturally chose a frequency which is already used by the most distant previously installed base station, since its own internal processor will chose the frequency with the lowest level of pre-existing radio signal strength in order to minimize interference. As more additional base stations are installed, still more of the frequencies will be reused.

Capacity Expansion

Some of the same means of expanding capacity are usable for cordless telephone systems as are usable for large-cell cellular and PCS systems. This includes installing more base stations, decreasing the channel bandwidth, adding more radio channels, or using digital speech compression. Some of these mechanisms are less feasible in cordless systems in which there is no coordination between the owners of each base station. The capacity of a cordless telephone system is only an issue in dense urban areas where there are many potentially interfering cordless telephones.

The system capacity in a geographic region can be increased by adding more base stations with very small radio coverage areas. Because cordless telephones and base stations already transmit at very low power levels, many can be located in a small geographic area without significant interference. If each base station has the capability to serve a specific number of calls (typically one at a time) and is not interfered with by nearby base stations, each base station that is added will increase capacity.

By reducing the radio channel bandwidth, more channels can be fit in the same amount of radio spectrum. This is done in FM analog systems by reducing the amount of modulation for the audio signal (modulation index). This reduction in modulation index increases the susceptibility to co-channel interference. A typical cordless telephone channel bandwidth is 25 kHz, and the IS-94A radio channel bandwidth is only 10 kHz. Digital systems can use a different type modulation (more efficient) to reduce the channel bandwidth. This makes it more susceptible to amplitude and phase distortions from nearby interference. The end result is that if there is little co-channel interference, then decreasing the bandwidth can increase the total number of available channels.

The capability to serve more users is also possible by adding radio channels in new radio spectrum. This was the expansion given when the FCC authorized the 49 MHz band for cordless telephone usage. The total amount of available channels is determined by the amount of bandwidth made available by the FCC. It is unlikely new radio channels will be allocated for cordless telephone operation. However, cordless telephones have recently started using a new band of frequencies in the 900 MHz region which allows for hundreds of radio channels. Unfortunately, this radio channel spectrum is unregulated and there are many potential sources of interference.

Increasing capacity can also be the result of reducing the average number of bits allocated to each user for a digital communications system. This will either reduce the channel bandwidth or allow the channel to be further divided into more time slots or new spread spectrum codes. Digital speech compression describes a group of methods for digitally encoding speech at a low digital bit rate, which allows the voice to be represented with a very small number of bits per second. Unfortunately, compressed voice typically reduces the audio signal quality as the exact reproduction is not possible. As a result of reduced voice quality, none of the major residential cordless telephone systems use a significant amount of voice compression. Figure 8.8 shows the various ways to increase capacity in a cordless telephone system.

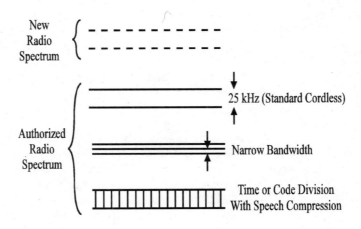

Figure 8.8, Cordless Radio System Capacity Expansion

System Control

A network of cordless telephones is controlled by the independent operation of multiple cordless handsets and base stations. Individual control of the handset and base station pair is determined by industry standard signaling or the proprietary signaling processes developed by each manufacturer. In some cases, wireless telephones may be capable of accessing more than one base station. If the wireless telephone is capable of accessing a different base station, the signaling process may be different when the wireless telephone is accessing a home base station (private) or a public base station. The system control for low tier and cordless telephone systems involves:

1. Communication between the wireless telephone and the base station.
2. Communication between the base station and the telephone network.
3. Communication between base stations.
4. Independent interaction between radio equipment and other radio equipment.
5. Authorization of service.

The communication between the wireless telephone and the base station involves radio signaling protocols (message transfer processes). Wireless telephones can request service from the base station or can be commanded by the base station. Generally, wireless telephones scan radio channels to check for busy channels before originating messages. They also continually scan radio channels to look for messages that are addressed to them. These messages and how radio equipment reacts to them are defined in a radio communications specification. Most cordless telephones use proprietary signaling specifications developed by the manufacturer of the cordless telephone.

Communication between the base station and the telephone network involves call delivery (receiving a call) and call origination (initiating a call). Call delivery involves sensing that an incoming call is present (a ring signal on the telephone line) and alerting the base station or cordless telephone (ringing at the wireless telephone or base station). Before a message is sent, an alerting tone or code is sent first. As a wireless telephone scans, it looks for this pilot tone or code to determine if a message is incoming. After the pilot tone or code is detected, a more extensive message follows. In public or semipublic cordless systems, billing records are created as calls are completed.The billing records are stored at the base station, and generally transferred to the service provider at a later time.

In some cordless systems, base stations may be either connected together through a switch or through the telephone network. Most simple cordless telephone systems do not have the capability to hand-off to another cordless base station. This greatly simplifies the issue of system control, mainly by doing away with most of it! The base unit still exchanges call processing messages with the handset to set up and intentionally release the connection.

Because cordless telephones independently operate among other cordless radio equipment, interference must be detected and avoided. This can be categorized as both interference avoidance prior to initiating a call and avoidance after a call is in progress. Prior to establishing a call, the cordless telephone can sample the radio signal strength of each radio channel to see if any activity is present. After a cordless telephone has been assigned a voice channel, it must detect interference though monitoring of the channel quality of its received signal. For digital signals which incorporate error protection coding bits, this may be done by monitoring of how many bits are received in error in a predetermined period of time. In an analog system, this may be accomplished by checking for the presence or absence of a supervisory tone or some other measurable signal component. In either case, most cordless phones have the ability to request a new radio channel either manually or automatically while a call is in progress.

Figure 8.9 shows a diagram describing how most cordless telephones initiate a call. In step 1, the customer starts the call typically by pressing a TALK or LINE button. The cordless telephone then scans the available channels to see which channels are not busy or have the least radio interference. When it finds an idle channel, it initiates a request for service (step 2) by placing a tone or digital signal on an unused channel. The base station also scans all of its channels and determines that the cordless telephone is requesting a call. When the base station accepts the request, it sends out a dial tone back to the cordless telephone (step 3) as a result of connection to the telephone network (step 3A). The customer can then dial the digits (step 4) which are sent to the telephone network (step 5).

Authentication of a wireless telephone ensures that only the correct telephone can access the base station for service. After a wireless telephone requests service, a code sequence is usually checked to verify the identity of the wireless telephone. This security process may range from a simple check of a code switch which is set by the cordless telephone user, or a complex process of checking various databases with a encrypted key. For public cordless base stations, a table of authorized users may be stored in the base stations memory which is checked prior to allowing access. This table may be loaded into memory each day by a central control system automatically calling into the public cordless base station via the telephone line.

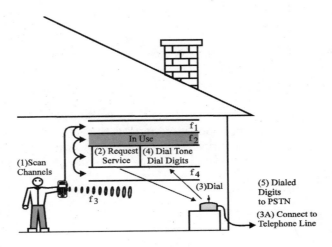

Figure 8.9, Basic Cordless System Operation

Radio Interference

Low tier PCS and cordless telephones suffer from the same causes of radio interference as other cellular and PCS systems; both co-channel or so-called "in band" interference (from other handsets) and other sources of radio interference which radiate on other carrier frequencies, including adjacent and alternate carrier frequency transmission from other radios which produce excessive undesirable or "spurious" radiation on the wrong frequencies, and devices which should not generate radiation but do so as a byproduct of their main operation, such as computers, microwave ovens, etc. In addition, all radios are subject to random noise caused by the random thermal motion of electrons in the antenna and other wiring.

In band interference which comes from other handsets was a significant challenge in early cordless handsets which only were capable of operating on one radio channel. If another cordless phone were operating nearby, it was often possible to hear their conversation quite loudly and the user was unable to complete a call undisturbed. In band interference has been virtually eliminated by the dynamic radio channel selection process used by most cordless telephones. When interference is detected, the cordless telephone coordinates with its base station to change radio frequencies.

Figure 8.10 shows the dynamic channel interference avoidance process. In step 1, the cordless handset detects interference on the radio channel. It sends a message to the cordless base station requesting to change to a new frequency (step 2). Next (step 3), the cordless base station scans other channels to find an available channel with less interference. The cordless base station then sends a command to the cordless telephone to change to the new channel (step 4) where conversation continues. This entire process occurs in a fraction of a second and is often imperceptible to the user.

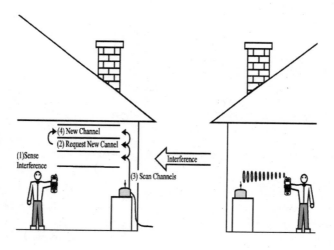

Figure 8.10, Dynamic Radio Channel Interference Avoidance

Random or non-co-channel radio interference and noise comes from improperly operating neon or fluorescent lights, series motors (of the type used in electric drills or elevators) with sparking commutator brushes, etc. Electronic equipment such as computers and electronic wristwatches also produce low level RF noise signals, although the amount of RF noise they are allowed to emit is limited by FCC Part 15 rules. Electronic devices including cordless handsets are often shielded to limit the amount of RF noise they produce. However, if a low-tier handset is used very close to a device which produces such interference, there may be a benefit by relocating the source of the interference, adding additional shielding (sometimes in the form of household aluminum foil!), or using a power cord assembly which contains an integral RF filter.

The 900 MHz ISM band systems can suffer interference from other services such as microwave ovens, RF equipment used for industrial electronics systems, etc. The Spectralink wireless office system (described in another chapter) changed their original developmental CDMA design to a FHMA design to avoid this. To optimize performance, they use frequencies in the frequency hopping plan which are chosen distinctly at each installation site to avoid pre-existing local sources of 900 MHz band interference.

Residential Cordless Handsets

Residential cordless telephone handsets can be categorized primarily by the radio frequency band on which they operate. In most cases, a distinct modulation technology is used on each band, as well as other features. In general, residential cordless equipment is not designed to link multiple cells, and as such does not support handoff. The equipment uses low power, is limited to about 200 meters range at most, and qualifies for unlicensed operation.

27 MHz "Citizen's Band"

This was historically the original band used for residential cordless telephones as well as other registered but nominally un-licensed citizen's radio equipment. The modulation method was amplitude modulation (AM) to meet the channel bandwidth requirements of the citizen's band (CB). Thus the equipment was quite sensitive to RF interference, particularly from other CB radios with higher transmit power. In general the audio noise level was and is noticeably worse than wired service. These early cordless systems were equally susceptible to eavesdropping as well. Despite all these technical shortcomings, cordless sets were popular from the beginning.

In general, the radio technology of these systems was to use a separate CB radio channel for uplink and downlink, so that the user would not need to push a button to change from talk to listen. In early units, the sensing of a RF signal (called a carrier) played the same role in supervision as the dc loop current in a wired telephone. A standard rotary dial in the handset was wired to interrupt the carrier signal from the handset during each dial pulse. This worked reasonably well except in the presence of strong carrier frequency interference. A touch-tone dial could also be provided in the handset. For ringing, a suitable audio alerting tone or tone pair was generated at the base station in response to the normal high voltage ac ringing current on the wired loop. This audio tone or pair was transmitted via the down-link radio channel to the handset, which had an audio amplifier which produce a loud earphone signal from this tone until the user pressed a button or some other switch to answer. This button turned on the uplink carrier signal (symbolically analogous to drawing loop current in a wired telephone) and this was sensed by the base receiver, which turned off the alerting tone(s). Operating the "answer" or "talk" button or switch also turned down the loud audio amplifier, so the handset earphone could be held to the ear without an overly loud audio level for further conversation.

There were some hazards from use of the earphone as a loud alert device. If users held the earphone to their ear and the loud alerting tone was generated just at that time, damage to hearing could result. This could also happen if the normal "answer" switch mechanism did not function correctly when the handset was pressed to the ear. A number of modifications were developed to reduce the hazard to hearing, and many of them are still used in other cordless sets as well. One method was to avoid using the earphone as a loudspeaker altogether. A separate loudspeaker, or use of an electromagnetic microphone as the alerting loudspeaker, reduced the hazard for most users since the earphone never produced a dangerously loud audio level. There is still some danger that a child may hold the microphone to the ear due to lack of experience with a telephone handset, but either approach greatly reduces the chance of hearing damage.

Although they are simple and inexpensive to manufacture, 27 MHz cordless sets are not popular today because of their perceived typically poor audio quality, primarily due to CB radio interference, and the problem of lack of privacy and fraudulent use of one's own base station by another compatible handset. Frequency scanning base and handsets which seek the CB radio channel(s) with the minimum interference level, and use of automatic authorization signals to prevent unauthorized use have somewhat reduced these problems, but have not eliminated them.

49 MHz Cordless

In the 1970s, additional radio channels for residential cordless use were made available by the FCC in the 49 MHz band, and a few channels in the 43 MHz band, giving 25 channels in total. These channels had adequate bandwidth for use of FM, which produces a much better audio signal. Many of the sets produced for use on the 49 MHz band used automatic scanning, so that the "quietest" channel is chosen dynamically each time a call is made. The signal was much less susceptible to interference from other cordless sets as well, due to the so-called capture phenomenon of FM. However, when there are a sufficient number of cordless sets operating near to each other, they will generally produce some spurious clicks and pops in the audio due to FM interference.

Many of the 49 MHz cordless sets use some form of authorization code and analog speech scrambling to prevent unauthorized use of the base station by another handset and to improve the level of privacy. The term "privacy" here implies a moderate level of protection, since the technology of the authorization and the scrambling is susceptible to a fairly brief "exhaustive search" attack. Most such cordless sets have a manual adjustment using a rotary switch which allows the user to set one of a small number of codes in that handset and base unit. After such a matching setting, the handset and base will work together but not with other units having different code settings. The same code setting controls both the transmission of authorization signals and the details of the voice scrambling process. The voice scrambling process may use several techniques originally developed in the 1930s and 1940s, such as audio sub band inversion, or adding an audio signal which resembles random audio frequency noise to the transmitted signal and then subtracting a synchronized replica of this apparently random signal at the receiver. These technical methods are adequate to prevent casual or inadvertent eavesdropping with a radio scanner, but are not good protection against a determined malicious neighbor. Because the choice of codes is limited, a determined eavesdropper, who knows the make and model of cordless set used by the victim, can try all the codes on a similar make and model in the vicinity of the victim's home base, and will with sufficient patience, eventually be able to eavesdrop and make unauthorized use of the base station.

900 MHz ISM Band

The newest band for residential cordless use is the industrial scientific and medical (ISM) band at 902-928 MHz. Cordless equipment is made for this band using either FM or direct spread spectrum (DSS) code division multiple access (CDMA) or frequency hopping spread spectrum (FHSS). FM units may have a selection of up to 100 channels, although many use only 25 channels. Radio transmissions below an effective radiated power level of approximately 100 milliwatts on this band are permitted under FCC Rules, Part 15, for devices which are primarily intended for another purpose than radio communication. For example, a computer, a television receiver, a microwave oven or an industrial RF heating system for drying the glue in plywood manufacturing. Of course, the RF power level on the *inside* of a microwave oven or inside the stack of plywood is much higher than 100 mW, but the radiated power outside the equipment is permitted if it is limited to the level specified by law. This low transmitter power level permits these cordless phones to typically operate for up to 8 hours talk time between recharges. The FCC also allows devices which are designed and intended for radio communication use on this band as well, if they meet the same radiated power requirements as the non-communication devices. There is no licensing and no legal protection against RF interference on that band. If you discover that there is intolerable interference from some other equipment on that band acting on your own receiver, you cannot get the FCC to do anything to reduce or prevent RF transmission from such other devices if they meet the rule regarding legal power level.

All narrow band FM 900 MHz cordless sets have RF scanning mechanism to find a portion of the ISM RF spectrum which is locally free from serious interference. Due to the large number of channels available, if the area of use is also free of other interference from non-radio devices, a large number of simultaneous conversations should be able to exist simultaneously without interference.

Digital speech coding is used in many 900 MHz cordless telephones. The typical methods are delta modulation using some form of continuously variable slope delta modulation (CVS-DM) or adaptive differential pulse code modulation (ADPCM). Digital coding of the speech provides some inherent protection against eavesdropping since an ordinary scanner cannot "decode" the digitally coded speech. Furthermore, digital coding makes the audio even more resistant to radio interference than FM alone.

Some 900 MHz cordless sets with either type of coding or modulation have a range of up to 300 meters (about 900 feet). This greater range is due primarily to the fact that the antenna is a bigger fraction of the wavelength of the 900 MHz radio signal, and therefore the antenna absorbs a greater amount of power from the radio wave. The wavelength of a 900 MHz wave is about 30 cm (one foot), while the wavelengths of 27 and 49 MHz waves are 11 meters (about 33 feet) and 6 meters (about 18 feet) respectively).

CT-2 (Public Cordless)

The CT-2 cordless telephone system allows the same cordless telephones to be used at home and in public places such as shopping malls, hotels, and train stations. The popularity of home cordless telephones, together with a serious problem with vandalism and inadequate repair resources for public coin telephones in Great Britain in the 1980s motivated the concept of public cordless telephone service. British Telecom, then the only public telephone operating company in most of the UK, had great difficulty for a number of years to maintain service in public telephones. As a result, they were receptive to the concept of a public cordless telephone service, officially designated CT-2.

History

Initially, in Britain, three competitive firms began to offer service around 1988. The subsequent business history of CT-2 service in Great Britain was a tragedy or farce worthy of the great dramatists of the English language, such as Shakespeare or perhaps Gilbert and Sullivan! From plans made public at the outset, it was clear that the management of all three CT-2 service competitors recognized clearly that public CT-2 service must be very cost-competitive with both cellular and home cordless sets. They also recognized the importance of a single common air interface (CAI) to permit technological compatibility between the three competitive systems. At first, they could not agree on a single CAI and made only desultory efforts at technical conferences to resolve the issue. Although the basic technology used in the three types of CT-2 sets were very similar, the details of the signaling were sufficiently different that the three were incompatible. This implied that a CT-2 customer could use a CT-2 handset only in the vicinity of a public base unit of a compatible type. The other two were incompatible.

Many observers accused the three British competitors of dragging their feet on standardization, each one delaying meaningful negotiation while pushing their own incompatible product onto the market in the hope that their own signaling protocol would become the *de facto* standard merely because they were able to get more equipment installed faster than the competition. Ultimately, as the total Great Britain market for CT-2 equipment later declined, there were more vigorous efforts to write a CAI standard, and it finally appeared in 1990, but it did not save the CT-2 industry in Great Britain. All three original British competitors are now no longer in the CT-2 service business, and a "resurrecting" investment, after its initial business demise, in one of the systems, called by the trade name "Rabbit," has also ceased to do business. The joke among CT-2 detractors after that was the slogan, "The Rabbit Died," which was also a catch phrase years ago for the so-called "rabbit test" for pregnancy.

However, CT-2 equipment is somewhat successful in Hong Kong, and is still used to a very limited extent in France and Germany. An improved version of CT-2, known as CT-2+, is used to a limited degree for public installations in Canada. Furthermore, CT-2 was the ancestor of several other improved versions which have improved technological features, such as DECT, DCT, WACS/PACS, etc. These new public cordless or low tier PCS systems appear to be meeting with initial business success all over the globe.

Post Mortem: After the business fiasco of CT-2 in Britain, several pundits appraised the reasons for the problems in retrospect. The British business problems were partially unique to that time and place, but some are perhaps universal and should not be repeated in other cordless systems.

The cost of a CT-2 home kit (home fixed part – base unit — and portable part or handset) was comparable to the purchase price of a cellular handset. The CT-2 portable part was smaller, simpler and less costly to manufacture than the cellular handset, but the purchase price of the actually higher-cost cellular handset was partially subsidized by the cellular service provider, so there was no price advantage to the CT-2 equipment.

The pricing of air time for many of the CT-2 services was comparable, and in some cases even higher, than similar air time charges for E-TACS public cellular service in the UK. For example, air time charges of 50 P (fifty new pence — half of a British pound sterling) per minute was the price in the so-called "City of London" financial district (comparable to Wall Street in New York), while cellular service air time there was only 25 P per minute.

The technical quality of much of the early generation CT-2 equipment was inferior to other wireless technologies. It was apparently rushed to market without sufficient time for careful design and product unit testing, and the operating distance range between fixed and portable part and the sound quality both suffered. Although later production was improved in quality, a bad reputation and bad consumer perceptions lingered despite the reality of product improvement.

The originate-only operation of CT-2 was not suitable nor satisfactory for certain users who mainly need to be reached by others. Most people who tried a radio pager and CT-2 combination to overcome this did not find it fully satisfactory.

Customers were continually frustrated by the incompatibility of the three services. Each service posted signs in public areas having radio coverage, but customers complained that the signs were most frequently for the "other two" incompatible CT-2 operators. Seeing 2 out of 3 signs for the other two systems is logical, but very irritating to the customer.

After an initial spurt of sales to a few brave consumers at the end of 1989 sales never increased significantly in Great Britain, with the sad result noted above.

System Overview

The basic parts of the CT-2 system are dual function wireless telephone which can operate with a home base station or public cordless base station. The CT-2 system uses time division duplex (TDD) operation. In the TDD system, the base station transmits for approximately 1 msec and the cordless handset replies with a 1 msec burst. Each 100 kHz single CT-2 radio channel can service 1conversation at a time.

In a CT-2 system, the handset is called the *cordless portable part* (CPP) and the base station is the *cordless fixed part* (CFP). There are two types of CFPs; a *home* fixed part, which is connected to a residential or office telephone line and a *public* fixed part which allows other CPPs to use the base station when in a public place. When operating with a *home* CFP, the CPP can *originate* and *answer* calls. When operating with the public CFP other, the CPP can only *originate* calls. The inability for CPPs to receive calls in public places was overcome in later CT-2 designs such as the CT-2+ system deployed in Canada. These advanced CT-2 specifications have also allowed other advanced features such as handoff.

Calls originated in the public network are ultimately billed to the customer. This is accomplished with public CFP equipment by storing billing related data on a hard disk mounted inside the CFP. This billing information consists of the identification code of the handset which placed the call, the dialed number and the date, time of day, and the duration of the connection. The handset provides a digital identification code over the air interface when a call is placed. Periodically (perhaps once each day or less frequently) a central billing computer with a suitable modem and telephone line connection dials the line connected to each public fixed part, and after exchanging suitable identification via the modems, dumps the billing data from the individual public CFP to the central billing computer. From there, the billing information is sorted and separated according to the identity code of each originating caller, so that it can later be merged with the other calls made by the same customer and printed on a single bill.

Because the original CT-2 systems did not support handover, the customer needed to remain in the radio coverage region of the fixed part for the entire duration of the call. In most cases, the radio coverage area is as large as perhaps 300 meters radius. This is more than adequate for home use, and permits service in many high traffic public areas such as an airport, train station, or shopping mall.

In an installation with non-global radio coverage and no handover capability, customers are made aware of service availability by means of signs posted in the radio coverage area. In England and continental Europe, a number of trade names were used on these signs for various CT-2 public systems, including PhonePoint, TelePoint, PhoneZone, Rabbit, and (in France) Système BiBop. In some cases, a visible indicator on the handset also shows when it is in an area with radio coverage from a fixed part. The actual public fixed part equipment is normally installed out of sight behind a wall, which also deters vandalism as well.

Because of the extremely low radio transmit power (maximum 10 milliwatts), CT-2 system is only capable of operating at a short range of a few hundred meters. The digital transmission has no error protection so rapid movement producing fast fading can result in a poor quality audio signal. The use of CT-2 CPPs in fast moving vehicles typically suffer from a problem called fast fading which causes digital bit errors which are beyond the design limits of most low tier systems.

Speech quality is usually good because of a good radio channel signal and high bit rate digital transmission signal. The radio signal is typically good because of the typical close distance of a CPP to the CFP.

The 32 kb/s ADPCM speech coder produces audio quality which is imperceptably different than the standard public telephone network, although it is not suitable for high bit-rate modems or fax machines. The CT-2 system also provides digital encryption which provides privacy to the user.

The transmit timing is primarily controlled by the CFP except during the random access from the CPP. This is because the CPP only transmits after it receives a transmit burst from the CFP. The FSK modulation method chosen allows a channel data rate of 72 kb/s. Because the same radio channel is used for communication in both directions, 36 kb/s is available in the forward (base to cordless) and 36 kb/s is available in the reverse direction (cordless to base). Four kb/s of the 36 kb/s data is used for control signaling and guard (no transmission)

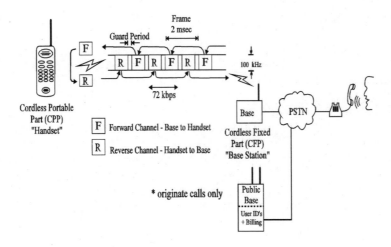

Figure 8.11, CT-2 System Overview

time to prevent overlapping signals which leaves 32 kb/s available for the speech or data signal. Figure 8.11 shows a CT-2 radio system.

Basic Operation

In the CT-2 system, there is no continuous "broadcast" setup channel so the CFP and the CPP both continually scan all CT-2 radio channels to check for messages. If either the CFP or CPP requires to send signaling, such as when the CPP wants to originate a call, then the radio must first scan to find a clear (unused) channel to send a message.

The CPP and CFP use matching handshake codes to ensure signaling messages are received correctly. When a user wants to initiate a call, the CPP scans all available radio channels and chooses an unused (free) radio channel and begins to transmit a request for service. A free channel is defined as any radio channel that has a minimal signal strength RF signal when compared to all other CT-2 radio channels. When a CFP detects that the CPP wants to originate a call on a radio channel, it responds to the request on that radio channel which includes the handshake code. When the CPP detects that the CFP is available and has responded with the correct handshake code, a voice connection with the telephone network is made. This provides a dial tone to the CPP to allow the user to dial a telephone number. A pre-connect dialing method is also supported by CT-2 as well.

In CT-2+ incoming calls are also supported. When an incoming call is present to a CFP, the CFP scans all available radio channels and chooses an unused (free) radio channel and begins to transmit a page message. The CPP should respond to this page message within a short period of time. If the CFP does not receive any response from the CPP, it may select another unused channel to transmit the page message because the CPP may have skipped over the previous radio channel due to the detection of an interfering signal. When a CPP detects that an incoming call is to be received (matching telephone number in the page message), the CPP begins to ring. After the user answers the call, the CPP sends a message to the CFP that the call can be completed along with the correct handshake code and a voice connection with the telephone network is made.

The CT-2 system does provide the capability for the CPP to adjust its RF transmit power level in one 16 dB step (2.5% of normal power). Lower power is used when the CPP is very close to the CFP. This power control can reduce the interference to other nearby CT-2 cordless telephones.

System Parameters

The original frequency band used for CT-2 radio service is 864 MHz to 868 MHz. With a radio bandwidth of 100 kHz, this provides for 40 radio channels. The CT-2 standard has been modified into several other industry standards which operate on different frequencies including 900 MHz ISM band and the 1900 MHz PCS band. Each radio channel has a data transfer rate 72 kb/s. Because of the TDD operation, this provides 36 kb/s data transfer in each direction. 32 kb/s of this data is available for speech or data communications. The modulation method is frequency shift keying (FSK) of the GMSK type. The normal power for a CT-2 handset is 10 milliwatts and it may be commanded to reduce it's power level by one step which is 16 dB lower than normal power.

GTE Tele-Go™

Tele-Go™ is a dual-service analog 800 MHz cellular telephone which can communicate with the cellular system or a home cordless base station. The Tele-Go™ system allows the user to make and receive calls via a standard telephone line (local telephone system) when the handset is located within radio range of its home cordless base station. When the user carries the handset outside of the range of it's cordless base, it functions like a standard cellular telephone, and allows the user to make and originate calls via the public cellular system.

The Tele-Go™ system requires the home cordless base station to communicate via the PSTN with the cellular switching system. To allow Tele-Go™ to operate, special signaling and new call processing software must be installed in the public cellular switching system which communicates with the Tele-Go™ cordless base stations.

While Tele-Go™ does allow handoff between different cells in the public system, it does not handoff between the public cellular system and the cordless base station during the same call. The Tele-Go™ handset is assigned a telephone (directory) number (cellular MIN) which is stored in the cellular system, and this is the only number which callers need to use to reach the Tele-Go™ handset in either the cellular or home service area. When the user is in the public cellular system, a call to the directory number of the handset will be connected via normal cellular radio setup channel paging and page response.

When the user is in the radio coverage area of the cordless base and its receiver is not locked onto the public cellular system setup channel, the cellular system will call-forward the call to the wired home

Figure 8.12, Tele-Go™ Handset and Enhanced Cordless Base Station (ECB)
Source: GTE

telephone line via the PSTN. The home telephone number is entered into the cellular system's data base for this purpose, and does not need to be changed to support Tele-Go™ service.

An individual Tele-Go™ handset can be set up to work with multiple cordless base stations, and an individual cordless base station can be set up to work with multiple handsets. In that way several members of the family can each have their own Tele-Go™ handset, or a user can plug in a separate cordless base station at the home and the office, and use the same handset at both locations. Figure 8.12 shows a Tele-Go™ handset and cordless base station.

History

Tele-Go™ was developed by Robert Zicker of GTE Mobilnet, Atlanta, as an approach to providing more marketable services from the cellular bands that GTE operates in several cities. There was a recognition that while many people had a cordless telephone in the house, they had only occasional need for a public cellular call, perhaps only a few each month. The size and weight of many public cellular handsets was comparable to the size and weight of many residential cordless telephone handsets. There was a significant opportunity to provide an "all in one" service with one handset which was a cordless phone in and around the home, with no air time charges, yet which can also go with the customer into the public domain and still make and receive calls with the same telephone directory number, and which operated in precisely the same manner (using "dial tone first" dialing procedures) regardless of which base system was serving it.

For many customers Tele-Go™ strikes just the right balance between cost and convenience, and has the personal safety and other advantages of carrying a cellular phone in case of emergency on the road or in the crowd. At the same time, the very same set is valuable around the home, without airtime charges. Tele-Go™ systems are only rented to the customers. The equipment is marketed under several additional trade names by various license holders in addition to GTE Tele-Go. For example, US Cellular uses the name CarryPhone, and Telstra Australia uses the name Personal Phone.

System Overview

The basic parts of the Tele-Go™ system are dual function wireless telephone and an enhanced cordless base station (ECB). The wireless handset has the capability to operate with the ECB and the public cellular system. The ECB can also communicate with the cellular system.

Tele-Go™ service should not be described as "low-tier." In the public cellular system, the Tele-Go™ handset can do anything and everything that an ordinary cellular handset can do. If the customer contracts for roaming service to support operation in other cities, that is technologically feasible as well.

Tele-Go™ handsets can both originate and answer calls in both the public cellular system and private cordless systems.

The dialing procedure in both modes of operation is always "dial tone first." The user can dial in the same way they are accustomed to with an ordinary wired home telephone.

The Tele-Go™ system design is a proprietary design owned by GTE, and is covered by a number of patents, some of which are issued and others are pending. It operates in a manner which is quite distinct from the superficially similar IS-94A TIA standard. It is marketed in each city by a licensee or by GTE itself, by agreement with the local cellular operating company whose frequency band it uses. The handsets are made under license by several manufacturers, and the ECB is now in its second generation of product development for cost reduction.

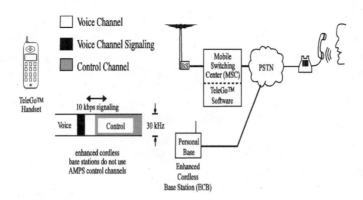

Figure 8.13, Tele-Go System

Figure 8.13 shows the basic parts of a Tele-Go™ system. The Tele-Go™ cordless handsets are dual mode which allow them to operate with both the cellular network and a Tele-Go™ enhanced cordless base station (ECB). The Tele-Go™ system requires no change to the cellular switching system. The cellular system is instructed to route telephone calls to the ECB when the ECB has detected the Tele-Go™ handset is within range of the ECB. This is performed through the process of call forwarding. When the handset detects the ECB, the handset will call the cellular system through the ECB and automatically update the call forwarding feature so that calls to the Tele-Go™ handset telephone number will be forwarded to the home telephone number. While the handset is near the ECB, it will briefly communicate with the ECB about every 3 minutes to ensure it is within radio coverage range. If the handset cannot communicate with the ECB, it will automatically contact the cellular system and deactivate the call forwarding feature.

A unique feature of the Tele-Go™ system is its process of programming ECBs and handsets. The Tele-Go™ system uses a master mobile unit which is mounted on cellular towers. This master unit can program ECBs. ECBs can also program Tele-Go™ handsets. This allows the cellular service provider to deactivate the home cordless base station if the customer does not pay their bill.

Basic Operation

In the Tele-Go™ system, the radio voice channel used with the ECB is 30 kHz analog cellular and there is no continuous "broadcast" setup channel. The ECB and the Tele-Go™ handset are both continually scanning a pre-determined set (normally 5) of the 416 800 MHz cellular carrier frequencies assigned to the cellular operator. None of these five frequencies are the public cellular system's setup channels. There are three situations where the ECB will begin to broadcast a setup channel. One situation is a periodic short reply transmission just to "reassure" the Tele-Go™ handset that it is still in radio range of the ECB. If the Tele-Go™ handset does not get this periodic signal from the ECB, it will start a normal public cellular setup channel search. Another situation is triggered by an incoming call on the home wired telephone line. The third situation is when the user dials a number on the Tele-Go™ handset first. This dialing will cause the Tele-Go™ handset to transmit a signal to the ECB even if the ECB is not currently transmitting anything. Once the base sends a setup channel signal, the Tele-Go™ handset exchanges messages with it which lead to this same carrier frequency changing its role from a setup channel to a voice channel. Neither the ECB nor the Tele-Go™ handset re-tune to a different frequency to change from call setup to conversation. This is completely different from standard cellular and IS-94A, in which one frequency is reserved only for use as a setup channel in each base station, and continuously transmits identification information. It is also in contrast to the normal cellular setup situation in which the handset is commanded by the setup channel to change to a different channel for voice communication, since the public cellular setup channel(s) are legally prohibited from voice channel use.

No particular channel among the 5 is reserved for any particular Tele-Go™ installation, so if there are several in the same neighborhood within radio range of each other, the first one to make a call while the others are idle will chose the carrier channel with the minimum RF background noise and interference level. If one is already in use, the second Tele-Go™ system to make a call will chose another carrier channel, and so on. Although it is theoretically possible to place more than 5 Tele-Go™ installations so close that they will cause blockage to each other, the cellular operator who rents the equipment can control this situation by assigning different or additional carrier frequencies to that area if the number of installed sets exceeds the number of existing channels in the area.

The nature of the voice modulation and the digital call processing control signals (which are transmitted via FSK modulation) are basically the same as standard public cellular radio. The modifications to the design of a cellular handset to make it compatible with the Tele-Go™ system are completely within the call processing software or firmware, and do not affect the so-called "hardware" aspects of the handset. The differences relate to three aspects of the set's operation: First, the sequence of channel scanning to find a private ECB, or, failing to find that, a public cellular setup channel; Second, the special call setup messages and sequence of operations to set up a call in conjunction with the ECB; Third, the special "dial tone first" dialing control software, which allows the user to dial calls exactly the same on wired telephone sets, and on the Tele-Go™ set in both the home and public environments. When the user presses the OFF HOOK button to dial (originate) a call, the earphone produces a dial tone. This dial tone is generated inside the handset by software, and is exactly like a North American precise tone plan dial tone. Then, as the originating digits are dialed, the dial tone is silenced. Special software/firmware in the Tele-Go™ set recognizes special dial codes in the North American numbering plan (NANP) and dialing plan. There is also a time out which will abort the dialing process if the originator waits too long without entering a digit and has not complete a valid dialed number sequence. This software recognizes, for example, that the three digits 911 (and certain others like 611 in some cities) form a complete dialing sequence, and then proceeds to transmit the proper dialing message, including the case of the proper call setup message when the handset is operating in the public cellular system. It also recognizes that a sequence beginning with a 1 should have 11 digits in total, and that other local dialed numbers must be 7 digits total. The set can be programmed and/or reprogrammed to recognize local special features such as the mixture of 7 and 10 digit dialing without a 1 prefix, which is used for local calls in some cities. The user answers a ringing incoming call by pressing the OFF HOOK button as for normal cellular.

System Parameters

The frequency and RF power properties of the voice and signaling of a Tele-Go^{TM} system are physically the same as the corresponding ordinary analog (AMPS) cellular telephone. The local public cellular system operator allocates a set of typically 5 frequencies which are preferred for Tele-Go^{TM} equipment in each part of the city, to minimize mutual interference between the ECB and the public system. The Tele-Go^{TM} handset creates its own dialtone so the perceived operation is very similar to a standard telephone. Other than this, there are no outstanding distinctions between the system aspects of Tele-Go^{TM} and ordinary cellular technology.

PACS

Personal Access Communication System (PACS) is a short range cordless telephone system which is highly integrated with the wired telephone network. PACS was developed with influence from Wireless Access Communication System (WACS), a Bellcore[1] standard and on the Japanese Personal Handy Phone (PHP) which was defined in a standards document issued by the Japanese research center for radio (RCR) standards group. While PACS is a short range cordless telephone system, there are methods which can be used that allows the range of a PACS system to be extended to make them competitive with wide area (high-tier) systems such as cellular telephone systems.

PACS is typically used in conjunction with a telephone central office switch together with ordinary wired telephones. To allow PACS to operate, special signaling and new call processing software must be installed in the switching system which communicates with the PACS radio base stations. This combination of wired and wireless service provided by a traditional wired local exchange carrier is now permitted under the new 1996 Telecommunications Act, which permits local exchange carriers to enter other businesses and also permits other firms to provide local telephone service in competition with established local telephone operating companies. The technical ability to

1. In 1997, Bellcore was the research and development organization of the 7 US regional Bell operating companies.

support wired telephones, cordless telephones, and fixed wireless devices while using the same infrastructure platforms (such as switches, transmission links, etc.) is a way for telephone companies to compete with wireless service providers. This is an important capability for both the traditional established local telephone system operators who are trying to defend their own market share, and as a tool for a new competitor to quickly penetrate the market.

History

Several wireline telephone operating companies are exploring the installation of PACS low-tier PCS systems in densely populated parts of the city, such as airports shopping centers, and the like, and also along heavily traveled highways.

The PACS specification SP-3418 was issued in June, 1995 by the Technical Ad Hoc Group 3 of the T1/TIA Joint Technical Committee (JTC).

System Overview

Figure 8.14 shows the basic parts of a PACS system. The cordless handsets are called subscriber units (SUs) and the fixed equivalent of a subscriber unit which provides dial tone services is called a wireless access fixed unit (WAFU). WAFUs convert the PACS radio channel

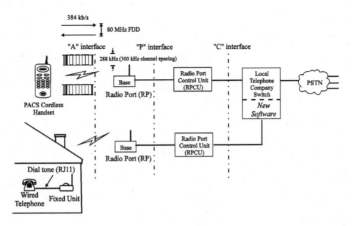

Figure 8.14, PACS Cordless Telephone System

into a dial tone signal which can be connected to standard telephone devices such as wired telephones, fax machines, and answering machines. Base stations are called radio port (RP) and the base station controllers which may be separate from the RP are called radio port control unit (RPCU). RPCUs are typically connected directly to the wired telephone network.

The PACS system defines how each of these parts are connected together. The A interface defines the radio signaling between the subscriber units and the radio ports. The P interface specifies how the RPs communicate with their controllers. The C interface is the communication standard between the RPCU and the switching system which may be the local telephone switch or a private branch exchange (PBX) switch.

PACS uses high quality 32 kb/s ADPCM speech coding to provide quality voice communications, although it is not intended for high bit rate modems or fax machines. When the end customer requires a high speed data connection, two of the time slots can be combined to provide 64 kb/s data transfer capability. This *will* support high bit rate modems and fax machines. The PACS radio system also has the capability of dividing up time slots for low data rate operation. This allows the PACS system to use speech compression to increase capacity or to allow low data rate devices to efficiently transmit their information.

Access manager (AM) is a telephone communications gateway between the wired telephone network and the PACS network. An AM capable switch provides the voice path and signaling translation between the PACS network and the PSTN.

Transmit and receive diversity in the subscriber unit and RP are required. Diversity transmission and reception uses two antennas which either combine two radio signals or select the strongest of the two radio signals. This improves the radio link quality of the system by reducing the effects of radio signal fades.

The PACS system provides for many different types of voice, messaging and data services. The primary voice service is 32 kb/s ADPCM speech coding with 16 kb/s and 8 kb/s anticipated for the future. PACS allows messaging services such as email, fax, video, and other forms of messages. The messages can be sent in fast mode where they interrupt speech service, or as slow messages which will wait until voice conversations are finished before transfer. PACS allows for both circuit switched (continuous) and packet switched (momentary) data transfer. It is also possible to allow simultaneous voice (primary) and data (secondary) channels by using more than one time slot for each channel. Advanced test modes are also included to allow the service operator to check the operational status of radio and telephone equipment while the PACS equipment is in operation.

Basic Operation

The PACS radio channel assignment system is called quasi-static autonomous frequency assignment (QSAFA). The QSAFA system uses channel quality measurements to allow SUs and RPs to independently coordinate their radio channel frequency assignment without the need for system help. If the SU or RP checks for interference either prior to attempting a call or during a call and finds it, they will change frequencies. This simplifies the network interconnection and eliminates frequency planning requirements. Figure 8.15 shows a PACS Radio Port.

Each RP continuously broadcasts (transmits) a control channel time slot. This control channel provides system information and marks if a

Figure 8.15, PACS Radio Port
Source: Hughes Network Systems Co.
Photography: John Keith Photography

time slot is available for communication. This same time slot may be used for voice communication after the SU has requested service although another frequency channel or a different time slot may be assigned. The SUs scan the available radio channels a lock onto to the control channel of the nearest base station.

During conversation, SUs and RPs continuously measure the radio channel quality through the use of word error indication (WEI), received signal strength indication (RSSI), and quality indication (QI) measurements. Either the SU or RP can request a new radio channel or time slot. New radio channels are assigned by a automatic link transfer (ALT) command and time slot changes are the result of a time slot transfer (TST) command. The PACS system is capable of handoff between RPs. To allow handoff, RPs communicate with each other through the radio port control unit (RPCU).

The PACS system does use automatic power control (APC) in the subscriber unit only. The output power of the SU can be adjusted in 1 dB steps over a 30 dB dynamic range. The output power of the RPs are adjusted only when they are installed. There is no dynamic power control of the base stations.

System Parameters

The PACS system's frequency bands are uplink (SU to RP) 1850 MHz to 1910 MHz and for the downlink (RP to SU) 1930 MHz to 1990 MHz. PACS uses frequencies in both the normally uplink and downlink bands as bi-directional frequencies. The duplex radio channel separation is fixed at 80 MHz. The PACS system uses 8 slot TDMA operation with bi-directional TDD (non overlapping time slots between transmit and receive) to simplify radio design. The bandwidth for a single radio channel is 288 kHz although the channel spacing is at 300 kHz. The modulation method is offset DQPSK with a channel data rate of 384 kb/s. Speech coding is 32 kb/s ADPCM with the provision to allow higher data rates for user data (for example, computer Internet connection) and lower data rates which allow more users to share the system through the use of speech compression. The maximum power for a subscriber unit is 200 milliWatts with an adjustable range of 30 dB in 1 dB steps (0.2 milliWatts minimum). The maximum power for a radio port is 800 milliWatts which is typically adjusted only when the RP is installed.

IS-91A

The IS-91A system is an integration of cellular network, wireless office, and home cordless telephony. The interim standard 91 revision A (IS-91A) system uses the same type of cellular radio channels for cellular and home cordless telephony. IS-91A is an industry standard for wireless residential equipment (WRE) that allows cellular telephones to operate with a personal base (PB) and wireless office base stations. IS-91A incorporates the features of the IS-94 wireless PBX standard.

History

The IS-91A specification was created to combine IS-94 (wireless office) and allow to add compatibility of a cellular telephone to communicate with a personal base station. The IS-91A specification was issued on November 6, 1995 by the Telecommunications Industry Association (TIA). While one of the original objectives of the IS-91A specification was to include the business operation of the IS-94 specification, proponents of the IS-94 specification started to expand their scope to also include cordless operation in the IS-94 specification.

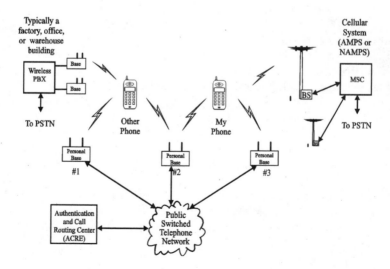

Figure 8.16, IS-91A Telephone System

System Overview

Figure 8.16 shows the basic parts of a IS-91A system. The cordless handsets are called mobile stations (MS) and the home cordless part is called a personal base (PB). The MS is capable of communicating with either the cellular system, wireless office system, or various PBs. A PB converts the IS-91A radio channel into a dial tone signal which can be connected to standard telephone devices such as wired telephones, fax machines, and answering machines. PBs communicate with a authorization and call routing equipment (ACRE) center the wired telephone network. The ACRE contains information about the PBs and can direct calls to or from PBs. The ACRE can provision maximum RF power levels, maintain its own phone number, can set a maximum call duration (usually set to several days), and authorize and set re-authorization intervals. The communication between the PB and ACRE is specified in IS-680, the Authorization and Call Routing Equipment Compatibility Standard.

IS-91A uses the same FM modulation as the AMPS or NAMPS system. The IS-91A radio system also has the capability to coordinate between PBs and the cellular system through the ACRE. The ACRE is a telephone communications gateway between the cellular telephone network and parts of the IS-91A network and is typically owned by a cellular service provider. The ACRE switch and database provides control information to the PBs and can complete a voice path or signaling translation between the telephone network and the PBs.

Basic Operation

The IS-91A radio channel assignment uses a interference sensing system to select authorized channels that were assigned by the ACRE. The interference avoidance system allows mobile telephones and PBs to independently coordinate their radio channel frequency assignment without the need for real time system help. If the MS or PB checks for interference either prior to attempting a call or during a call, they will change frequencies. This simplifies the network interconnection and eliminates frequency planning requirements.

Neither the PB or MS continuously transmit a pilot or control channel. When service is required, the PB or MS will transmit a request for service. Both the PB and MS scan the available radio channels to find a request for service.

System Parameters

The frequency bands used for the IS-91A system is 824 to 849 MHz for the reverse link (telephone to base) and 869 to 894 MHz for the forward link (base to telephone) with the exclusion of the standard AMPS control channel frequencies. When IS-91A base stations transmit, they can use any non-AMPS dedicated control channel frequencies. The voice channel bandwidth can be either 30 kHz (standard AMPS) or 10 kHz (NAMPS). Voice modulation is FM and signaling is 10 kb/s blank and burst for AMPS channels or 200 b/s sub-band signaling for 10 kHz NAMPS channels.

References:

1. SuperPhone conference, Institute for International Research, New York NY, January 22-24, 1997.

Chapter 9

Wireless Telephones

Wireless Telephones

Wireless telephones may be mobile radios mounted in motor vehicles, transportable radios (mobile radios configured with batteries for out-of-the-car use), or self-contained portable units. Whether mobiles, transportables, or portables, their functions are almost identical. Because all wireless telephone equipment provides benefits to the subscriber of the service, we refer to all types of wireless telephones by the same name, as wireless telephones or mobile stations (MSs). Multi-mode wireless telephones are capable of operating in more than one mode (usually an analog mode and a digital mode).

Wireless telephones can be divided into the following sections: user interface, radio section, signal processing section, power supply, and accessory. The user interface, sometimes called a man machine interface (MMI), allows the user to originate and respond to calls and messages. The radio frequency (RF) assembly converts the baseband signal (analog or digitally coded voice) into RF signals for transfer between the base station and the wireless telephone. The signal processor section conditions the voice (audio and digital compression) and controls the internal operations of the wireless telephone (logic). The power supply provides the energy to operate the mobile telephone.

In addition to the key assemblies contained in a wireless telephone, accessories are often connected to adapt the wireless telephone to perform an optional feature such as hands free operation. The user interface, radio section, signal processing, power supply, and any attached accessories must work together as a system. For example, when a portable wireless telephone is connected to a hands free accessory, the wireless telephone must sense the accessory is connected, disable its microphone and speaker, and route the hands free accessory microphone and speaker to the signal processing section.

User Interface

The customer controls and observes the status of the wireless telephone through a user interface. The user interface consists of audio interface, display, keypad, accessory connection assemblies, and software to coordinate their operation.

Audio Interface

The audio interface assembly consists of an earphone and/or speaker and a microphone which allows the user to talk and listen on the wireless telephone. These assemblies are located in a handset although they can be replaced by units in the hands free assembly. For portable telephones, the entire unit acts as the handset with the speaker and microphone being incorporated into the single unit. Because of their

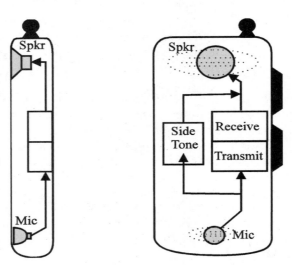

Figure 9.1, Audio Interface

small size, portable units have microphone systems with a substantial amount of gain. This allows the microphone to still pick up normal conversation even when it is not placed directly in front of the speakers mouth. This is especially important for the smaller portable units that have a length smaller that the distance between a person's ear and mouth. As in the land line system, a sidetone is generated to allow users the capability to hear what they are saying into the microphone. Figure 9.1 shows an audio interface block diagram.

Display

The display assembly allows the user to see a status of the telephone. Cordless telephones typically have a limited amount of display information such as a ringer light. Cellular and PCS telephones use pre-origination dialing which requires a dialed digit display. This display allows the customer to change the dialed digits before the call is initiated. An IN USE symbol or TRANSMIT list is displayed when a call is initiated or received, indicating that RF power is being transmitted. The display may also indicate other available features such as a received signal strength indicator (RSSI), call timer, or other services. In recent years the display has been used to implement advanced service features which include calling number identification (caller ID or calling line ID — CLID), name and number storage, and selecting preferences for the Mobile telephone operation.

There are two basic types of displays used in wireless telephones; Liquid Crystal Display (LCD) and Light Emitting Diode (LED) display. LCD displays require a minimal amount of power to operate and can be masked to create custom icons. However, LCD displays do not operate as well as LEDs in the cold, and require a backlight for use in dimly lit situations. Unfortunately, LED displays have a moderate amount of power consumption and thus requires the LED displays to be turned off when they are not in use.

Keypad

The keypad allows the user to enter information directly to control the phone. The information entered would include dialed numbers, commands to receive and originate calls, along with selecting feature options. A voice activation unit could be an option in replacing the keypad.

Layout and design of keypads vary greatly between different wireless technologies because of the various ways the customer controls the telephone. Cordless telephones usually allow the customer to dial a phone number without confirming the dialed digits. Cellular and PCS telephones require the customer to initiate the call by pressing a SEND button and disconnect the call by pressing the END button. A typical wireless keypad will contain keys for the numbers 0 to 9, the * and # keys (used to activate many subscriber services in the network), volume keys and a few keys to control the user functions.

Because portable telephones can be very small, this limits the number of available keys and some keys may have more than one function. For example, to turn the power on, the customer may press and hold the END button. Figure 9.2 shows how a keypad varies for a cordless phone and a cellular or PCS telephone. Wireless office telephones may have a combination of the two keypads.

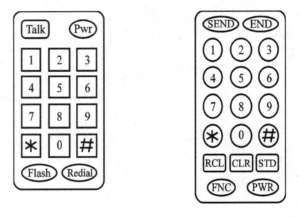

Figure 9.2, Typical Keypads

Accessory Interface

Accessories for mobile telephones can be attached via an optional electrical interface connector (plug). The accessory interface connector typically provides control lines (for dialing and display information), audio lines (in and out), antenna connection, and power lines (in and out) to connect to and from accessory devices. No standard accessory interface connection exists for wireless telephones. Each manufacturer, and often each model, will have a unique accessory interface. The accessory connector is normally on the bottom or end of the wireless telephone.

The types of accessories vary from active devices such as computer modems to passive devices like external antennas. A hands-free kit includes an external microphone and speaker to allow the subscriber to talk to the telephone without using the handset, usually in a vehicle. An external power supply (such as a car battery) may be used to charge the battery. The wireless telephone may provide power from the telephone's battery to external devices such as computer modems. An antenna connection allows the use of high gain external antennas which may be mounted on a car. When a data device such as a modem is used, this typically requires an audio connection (for the modem data) and a control connection (to dial and automatically set the wireless telephone's features). Other smart accessories (such as a voice dialer) may require audio and control line connection. Figure 9.3 shows a sample accessory connection diagram.

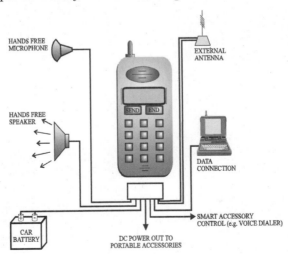

Figure 9.3, Typical Accessory Connection

Radio Frequency Section

The radio frequency section consists of transmitter, receiver, and antenna sections. The transmitter converts low level audio signals to proportional shifts in the RF carrier frequency or phase. The receiver amplifies and demodulates low level RF signals into their original audio form. The antenna section converts RF energy to and from electromagnetic waves.

Most wireless telephone designs today use a microprocessor and digital signal processors (DSPs) to initialize and control the RF section of the mobile telephone. RF amplifiers vary in their type and conversion efficiency. Analog mobile telephones and some digital units use nonlinear (Class C) amplifiers and most digital mobile telephones use linear (class A or AB) amplifiers. The efficiency of the RF amplifier is the rating of energy conversion (typically from a battery) to RF energy. Because the RF amplifier typically is the largest power consuming section during transmit, the higher the conversion efficiency, the longer the battery recharge life during conversation (transmission). While Class C amplifiers add some distortion to the radio signal and may be above 50% efficient, linear amplifiers add very little distortion and are sometimes only 30-40% efficient. The terms Class A, B, AB, and C refer to categories of amplifier design widely used in the electronics industry. Class A, B and AB amplifiers draw battery current through all of the time of each cycle of the RF waveform. A Class B or AB amplifier uses two complementary sections which each amplify a part of the waveform (one section only amplifies the positive voltage half-cycles of the waveform, while the other section amplifies the negative half-cycle). A Class C amplifier draws only a single short pulse of current from the battery during each cycle, which is why it is more efficient.

Transmitter

The transmitter section contains a modulator, a frequency synthesizer, and an RF amplifier. The modulator converts audio signals to low-power level radio frequency modulated radio signals on an intermediate frequency (IF). A frequency up-convertor, in conjunction with the frequency synthesizer, produces radio signal output on the assigned channel. The RF amplifier boosts the signal to a level necessary to be transmitted and received by the Base Station.

The transmitter is capable of adjusting the power levels to transmit only the necessary power to be received by the base station. To conserve battery life in portables and transportables, the RF amplifier may turn off its power during periods when the mobile operator is not talking, which is called Discontinuous Transmission (DTx).

Figure 9.4 displays a typical transmitter section block diagram. The RF section receives its low level input signal from the modulator. The RF amplifier is capable of adjusting its output level as a result of commands sent by the Base Station. A safety circuit is used to inhibit the transmitter in the event of an equipment malfunction. If the software is stuck in a loop with no exit or some other malfunction, it cannot reset the timer, which consequently runs out and stops the unit from operating until it is reset or repaired. This type of timer is used in many real-time computer controlled systems in addition to cellular and PCS systems.

The final amplification stage of the RF amplifier provides RF energy in excess of the energy which is transmitted by the antenna because of the losses of filtering and isolation. A portion of the radio energy is sampled by a power detector which is used to control the transmitted energy to a precise level. Some wireless telephones (usually high power mobile telephones) use an isolator which prevents radio energy from being reflected back into the transmitter in the event the antenna is disconnected. The high power radio signal is then passed through a duplex filter (duplexer) or transmit/receive (T/R) switch

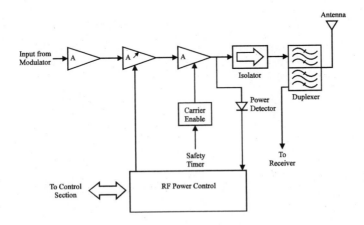

Figure 9.4, Transmitter Section Block Diagram

which prevents the transmitter energy from entering into the receiver assembly.

Receiver

The receiver section processes the low power level received RF signals into a low frequency signal that can be demodulated into the original audio signal. The receiver contains a receiver amplifier with automatic gain control (AGC), frequency down-converting RF mixers, intermediate frequency amplifying and frequency filtering sections, and a local oscillator signal source for the mixers.

Figure 9.5 shows that the low level RF signal is first passed to a filter (part of the duplexer in some cases) which removes unwanted radio signals outside the band of channels designated for the wireless telephone. Radio signals within the desired band of channels are amplified and converted to a lower frequency by the 1st mixer. A frequency synthesizer (frequency generator) tunes to the a frequency which converts the desired radio channel to a fixed Intermediate Frequency (typically 45 or 70 MHz). The output of the 1st IF amplifier is passed through a filter so only the desired channel within the allowable band of channels can be supplied to the 2nd frequency mixer. The second frequency mixer is used to convert the 1st IF frequency to a lower frequency which allows signal demodulation.

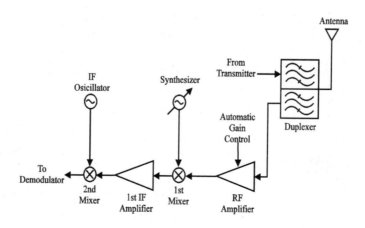

Figure 9.5, Receiver Section Block Diagram

Antenna Section

An antenna section focuses the transmission into a desired direction [1] and couples electromagnetic waves between the transceiver and free space. The antenna section consists of an antenna, cabling, duplexer, and possibly a coupling device for antenna connection through glass. The performance of the antenna system can enhance or seriously reduce the performance of the mobile telephone. The antenna may be an integral part of the transceiver section (such as a portable handset) or externally mounted (on the top of a car). Antennas can have a gain which describes the ability of the antenna to to focus or concentrate the radiated energy into a particular beamwidth area, rather than radiate it in all directions indiscriminately. This focused energy gives the ability to communicate over greater distances, but as the angle of the antenna changes, the direction of the beam also changes, reducing performance. For example, car-mounted antennas that have been tilted to match the style lines of the automobile often result in extremely poor performance.

Cabling that connects the transmitter to the antenna adds losses which reduce the performance of the antenna assembly. This loss ranges from approximately .01 to .1 dB per foot of cable. High gain antennas may be used to overcome these losses and possibly allow for a lower power output [2].

In early systems, separate antennas were used for transmitters and receivers to prevent the high power transmitter from overpowering the receiver. A duplexer or a transmit-receive (TR) switch allows a single antenna to serve both the transmitter and receiver. A duplexer consists of two RF filters; one for transmission and one for reception. A TR switch connects either the transmitter or the receiver to the antenna, but never at the same time, and is used only for certain digital systems.

Figure 9.6 shows a mobile telephone antenna system. The mobile radio transceiver is typically mounted in the trunk of the car or under a seat. An antenna cable is routed to the trunk and up to the rear window where it can be connected to the coupling box of a glass mounted antenna. The coupling box of the glass mounted antenna operates like a capacitor which allows the signal to pass through the glass to the base of the antenna. The antenna then converts the electrical energy into radio waves which are transmitted to and from the cell site.

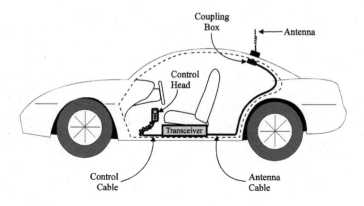

Figure 9.6, Mobile Telephone Antenna System

Signal Processing

Signal processing converts audio signals into a form suitable for radio transmission and converts received radio signals into their original audio form. Analog wireless telephones typically use FM modulation and digital wireless telephones use FM or phase modulation and digital signal processing.

Digital wireless telephones typically use DSPs or application specific integrated circuits (ASICs) to process all types of signals. If the wireless telephone has multi-mode capability (such as TDMA and AMPS), the same DSP can be used to process both the analog and digital signals. DSPs are high performance computing devices that are specifically designed to allow rapid signal processing. Advances in DSP technology allow increased signal processing ability of 40 to 60 million instructions per second (MIPS), lower operating battery voltage (3 volts -vs- 5 volts), and reduced cost. ASICs are custom designed integrated circuits and are typically created to combine several functional assemblies. For example, a single ASIC could contain all the control circuits necessary to connect the display and keypad to the DSP. By replacing several components with one ASIC, cost and size are reduced.

Digital cellular telephones require approximately 40 to 60 MIPS of processing power compared to 0.5 MIPS required by analog (AMPS) phones [3]. The digital signal processing requirements result from speech coding (1.5 MIPS to 8 MIPS), modulation and demodulation, radio channel coding and decoding, and radio signal equalization or rake reception. High performance DSPs which are required in digital Mobile telephones have only become available in the past few years.

Speech Coding

Digital wireless telephones have a speech coder that compresses (encodes) and expands (decodes) the digital signal information rate. On digital systems, the speech must be sent as digital data. When a subscriber speaks into the microphone it generates an analog signal, that is sampled (typically 8000 times per second) and then digitized and input to a speech coding section. This simply digitized voice is processed by a speech coding algorithm to create a lower bit-rate digital representation of the analog voice signal. When this compressed speech information is received, it is recreated to the original analog signal by decoding the speech data. Speech coders for digital systems are usually implemented in firmware that is executed by a DSP.

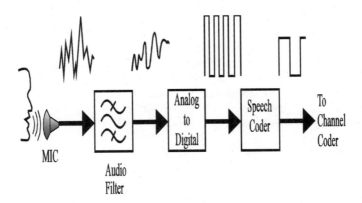

Figure 9.7, Speech Coding

The speech coding process (shown in Figure 9.7) converts acoustic signals (sound) into audio signals via a microphone. The filtered audio signal is converted to digital pulses (typically at 64 kb/s by using 8 bits to describe the voltage of each sample, and 8000 samples per second). The digital pulses are analyzed and compressed to a low bit rate digital signal (approximately 8 kb/s to 32 kb/s)s. The compressed digital pulses are sent to the channel coder which adds control signals and error protection processing.

Channel Coding

Channel coding involves adding error protection bits, adding error detection bits and multiplexing control signals with the transmitted information. Error protection and detection bits (they may share the same bits) are used to detect and correct errors which occur on the radio channel during transmission. The output of the speech coder is encoded with additional error protection and detection bits according to the channel coding rules for its particular specification. This extra information allows the receiver to determine if distortion from the radio transmission has caused errors in the received signal. Because error protection coding requires the use of more bits to allow correction of a signal received in error, compared to less bits to merely detect (but not correct) error(s), some wireless technologies only use error detection bits and not error protection bits, particularly for signals which can be repeated if necessary. These wireless systems use portables telephones which typically operate very close to a base station which results in a strong signal which does not experience much distortion.

Control signals such as power control, timing advances, frequency handoff must also be merged into the digital information to be transmitted. The control information may have a more reliable type of error protection and detection process which is different than the speech data. This is because control messages are more important to the operation of the wireless telephone than voice signals. The tradeoff for added error protection and detection bits is the reduced amount of data that is available for voice signals or control messages. The ability to detect and correct errors is a big advantage of digital coding formats over analog formats but it does come at the cost of the additional data required.

In Figure 9.8, the channel coding section adds error protection bits to the speech data signal, mixes the information signal with control signals, and formats the composite digital signal to the necessary burst structure for radio transmission.

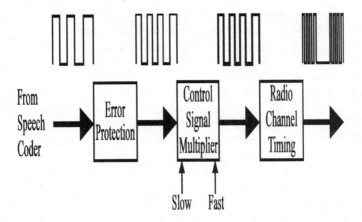

Figure 9.8, Channel Coding

Audio Processing

Analog wireless telephones require audio processing. Audio signals are processed (shaped) to prepare them for radio transmission. The audio processing section must also allow the combining of control signals and the muting (blanking) of audio signals when interference or non-voice signaling occurs.

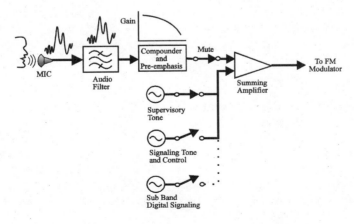

Figure 9.9, Analog Audio Processing

289

Figure 9.9 shows a typical analog audio processing section. The microphone converts acoustic signals (sound) into audio-frequency electrical signals via a microphone. The audio signal is filtered so unwanted frequencies (high frequency parts) are not converted. The audio signal is processed by a compandor circuit and pre-emphasis circuit which reduces the gain for loud acoustic signals and increases the gain for high frequency parts. This processed audio signal is then combined by a summing amplifier with a supervisory control signal (typically 4 kHz to 6 kHz) and a signaling tone (typically 8 kHz or 10 kHz) or control signal when necessary. The audio signal can be disconnected (muted) from the summing amplifier when control messages are being sent or the wireless telephone has detected interference. The combined audio signal is then sent to an FM (or phase) modulator. The modulator converts the audio signal into proportional frequency or phase changes of an RF signal. The Motorola NAMPS analog wireless telephones use digital sub band signaling to send control messages instead of a supervisory control tone or a signaling tone. When using digital sub band signaling, messages can be sent continuously along with the voice signal.

Logic Section

In early applications, the logic section contained discrete logic components (such as AND and OR gates) for each processing section [4]. Today, logic sections usually contain a microprocessor operating from stored program memory. The logic section coordinates the overall operation of the transmitter and receiver section by allowing the insertion and extraction of control messages. Control signals can be analog (such as Signaling Tones) or digital (a FACCH control message). Control messages must be continually inserted and removed from the transmitter and receiver sections. The logic section encodes and decodes control signals and performs the call processing procedures.

The logic control section inserts and extracts special control messages that coordinate the transceiver and control head (keypad and display). The receiver extracts base station commands from the received radio signal and routes them to the logic section. In accord with the base station commands, the logic section then controls the transmitter and updates the control head display information if necessary. When the user initiates commands via the keypad (for example, dials digits), the commands are transferred to the logic section. The logic section then controls the transmitter assembly.

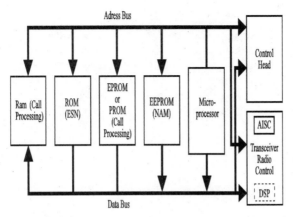

Figure 9.10, Logic Section

There are usually four types of memories used in a transceiver: Electrically Erasable programmable Read Only Memory (E2PROM), [Erasable] Programmable Read Only Memory ([E]PROM), Read Only Memory (ROM), and Random Access Memory (RAM). E2PROMS can have information written to, read from , and can be erased electrically although E2PROMS cannot be written to an unlimited number of times and are slow in the writing cycle. EPROMS can be written to, read from , and erased but must be erased by an ultraviolet light. PROMS can be written to and read from, but only can be programmed (written to) one time. ROM has information stored in physical links stored at the manufacturing facility and can only be read from. RAM can be written to and read from rapidly, but loses information when power is removed.

E2PROMs are used to store data such as the wireless telephones phone number and other seldom changing long term information. EPROMS or PROMS hold the stored programs for call processing. ROMs hold permanent data that cannot be changed such as the Electronic Serial Number (ESN). RAM is storage for data which is temporarily stored such as the radio channel number for which a call is in progress.

Although some wireless telephones have powerful DSPs for speech processing, the basic control of most wireless telephones is performed by a simple 8 bit or 16 bit microprocessor. The microprocessor coordinates the logic section. When the microprocessor controls the radio section, it may perform this through an ASIC or a DSP. Figure 9.10 shows key parts of a wireless telephone logic section.

Subscriber Identity

Each mobile telephone and end customer must have an identification code to allow cellular systems to deliver and process calls along with collecting billing information. Customer information is stored in a mobile telephone or a removable device. For early analog cellular phones, the subscriber identity was stored in a Programmable Read Only Memory (PROM) chip which required a separate chip programmer. The PROM chip would be installed by the programmer and was not convenient to remove.

There are two basic types of identification use by the digital cellular technologies. North American cellular systems use a Number Assignment Module (NAM) and GSM systems use a Subscriber Identity Module (SIM) that can be packaged on a removable card. In addition to the information contained in the NAM and SIM, additional equipment identification codes such as Electronic Serial Numbers (ESNs) are included with the physical equipment.

Power Supply

All wireless telephones require power to operate. Power may be supplied by an internal battery, power line voltage, or other source such as a car battery or cigarette lighter socket. Initially, almost all mobile telephones were wired directly to a car battery. Some car telephones were converted to transportable phones that operated from a 12 volt battery or were plugged into a cigarette lighter. A majority of wireless telephones today are portable telephones which operate from small rechargeable batteries. The first few portable telephones operated from 7.2 volt batteries and the trend is towards lower voltage batteries.

Battery Technology

As the use of portable phones has increased, the importance of battery technology has increased. There are several types of batteries used in wireless telephones; Alkaline, Nickel Cadmium (NiCd), Nickel Metal Hydride (NiMH) and Lithium (Li). A new type of battery technology that has increased energy storage capacity, Zinc Air, is being explored.

With the introduction of portable cellular telephones in the mid 1980s, battery technology became one of the key challenges for cellular phones. In the mid 1990s, over 80% of all cellular phones sold were portable or transportable models rather than fixed installation car phones. Battery technology is a key factor in determining portable phones' size, talk time and standby time.

Batteries are categorized as primary or secondary. Primary batteries must be disposed of once they have been discharged while secondary batteries can be discharged and recharged for several cycles. Primary cells (disposable batteries) which include Carbon, Alkaline, and primary Lithium cells see limited use in wireless telephones. Although they are readily available, disposable battery packs must be replaced after several hours of use and are more expensive than the cost of a rechargeable a battery. Disposable batteries have the advantages of a long shelf life and no need for a charging system.

One of the most common batteries used in portable wireless telephones is the rechargeable NiCd cell which consists of two metal plates made of nickel and cadmium placed in a chemical solution. The package is vented to prevent explosions due to improper charging or discharging. A NiCd cell can typically be cycled (charged and discharged) 500 to 1000 times and is capable of providing high power (current) demands required by the radio transmitter sections of portable wireless telephones. While NiCd cells are available in many standard cell sizes such as AAA and AA, the battery packs used in wireless telephones are typically uniquely designed for particular models of wireless telephones. The internal batteries of the battery pack typically cannot be replaced by the user. Some NiCd batteries can develop a "memory" of their charging and discharging cycles and their useful life can be considerably shortened if they are not correctly discharged. This is known as the "memory effect," where the battery remembers a certain charge level and won't provide more energy even if completely charged. Some NiCd batteries use new designs that reduce the "memory effect."

NiMH batteries use a hydrogen adsorbing metal electrode instead of the cadmium plate. NiMH batteries can provide up to 30% more capacity than a similarly sized NiCd battery. However, for the same energy and weight performance, NiMH batteries cost about twice as much as NiCd batteries [5].

Li-Ion batteries are the newest technology that is being used in portable cellular phones. These provide increased capacity versus weight and size. A typical Li-Ion cell provides 3.6 volts versus 1.2 volts for NiCd and NiMH cells. This means that one third the number of cells are need to provide the same voltage.

Zinc-air batteries have shown they can provide more storage capacity than lithium-ion batteries. They use oxygen from the air to enable reactions that generate electricity. They also have no memory effect and a low self discharge, which allows for a longer shelf life after charging. Zinc-air batteries are being produced for portable computers and will require more testing and study before they are available for wireless telephones. Figure 9.11 shows the relative capacity of different battery types.

The charge and discharge characteristics of batteries are important because portable wireless telephones contain some type of charging system and battery status indicator. Customers typically desire to charge and discharge at any time. The charging system must determine when a battery is fully charged. The charging system stops charging the battery (or may change to a very low trickle charge rate) at a suitable cutoff point. The cutoff point can be determined by temperature, maximum voltage level, a change in voltage (delta voltage), or a combination of these factors. The battery status indicator is typically a form of a voltage meter. Some portable telephones actually indicate the voltage level of the battery but most choose to display a level (similar to a gas gauge). NiCd batteries have the most difficulty to accurately displaying battery levels because of their relatively flat discharge curves, typically displaying little or no change in voltge over the mid range discharge.

Some of the future trends in batteries are new packaging and higher mAh ratings for the same size and weight. Flat cells will allow denser packaging of batteries. Today between four and six round cells are placed into a single battery pack resulting in a lot of wasted "dead space" between the cells. As consumers use their phones more and more they are demanding longer standby and talk times. Because a 0.6 watt portable phone has to transmit at power levels determined by the radio system properties, the battery technology and more efficient power amplifiers are the only ways to increase talk time.

Accessories

Accessories are optional devices that may be connected to wireless telephones to increase their functionality. Accessory devices include

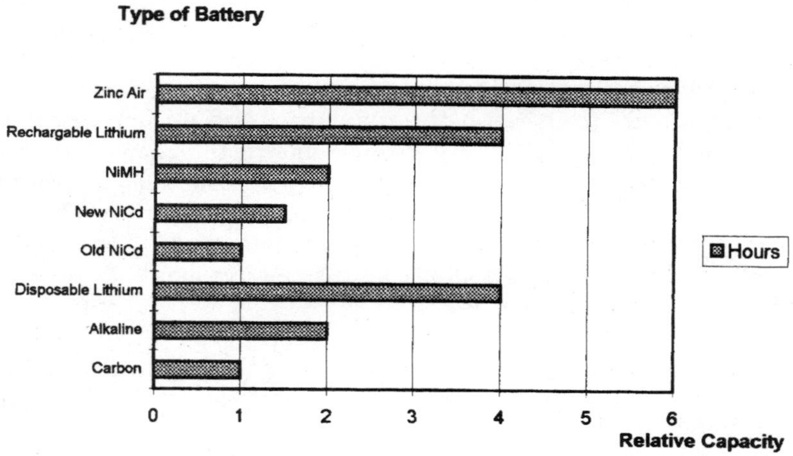

Type of Battery

Figure 9.11, Battery Storage Capacity

hands-free speakerphone, data transfer adapter (modems), voice activation, battery chargers, high gain antennas, and many others.

Hands Free Speakerphone

For safety and convenience reasons, the car telephone may have an option allowing the subscriber to use hands-free operation [6]. A hands-free system consists of: a speaker, usually located in the cradle assembly; a remote microphone, usually located near the visor; and interface circuitry which connects the audio paths and allows for sensing when the user requests hands free mode. Because digital cellular systems take time to process and convert audio signals, multiple delayed echoes caused by hands free operation can be very annoying to the user. The wireless telephone and /or cellular system may contain sophisticated echo cancellors to help reduce the effects of delayed echo.

Data Transfer Adapters

Subscribers sometimes want to send digital information via their wireless telephones. The wireless telephone can offer optional connections for a facsimile, modem, or a standard plain old telephone service (POTS) dialtone to the wireless phone. On the PSTN network, voice or data information can be sent reliably under good to excellent radio channel conditions if it is within the 300-3000 Hz frequency range. While the audio frequency range on the wireless radio channel may be the same as the PSTN, the varying nature of the RF channel is not well suited for efficient standard data transfer. Special error correction modems exist to increase the reliability and efficiency of data transfer via the cellular system [7]. A standard telephone interface may also allow a cellular phone to operate with a touch tone or rotary phone [8]. This interface simulates a dial tone and interprets and translates touch tone signals into a signal compatible with cellular signaling, and a call is initiated without the requirement of pre-origination dialing.

Digital wireless transmission allows new possibilities for data transmission. Instead of converting digital bits (pulses) into audio tones that are sent on the radio channel, the digital bits are sent directly into the phase modulator of the Mobile telephones digital transmitter. This allows much higher data transmission rates. Unfortunately, when the digital bits are received on the other end at the base station, they must be converted to a format that is able to be transmitted through the public switched telephone network (PSTN).

For the analog cellular system (for example, AMPS and TACS), digital information is converted by a modem to audio signals. For the digital cellular system, digital information is only buffered and delayed in time or has some error protection coding added to it for direct transmission. When the modem signal on the analog cellular system is received by the base station, it is ready to be sent to the PSTN. When the digital information is received on the digital cellular system by the Base Station, it must be converted to a signal that can be sent on the PSTN. This is typically performed by a modem and convertor installed at the base location. Modems are not necessary if the PSTN has the capability to directly send digital information such as Integrated Services Digital Network (ISDN).

This data interface plugs into the bottom of a portable cellular phone, uses a cellular modem to convert between audio and digital signals, and connects to a serial port on a computer.

Voice Activation

Another optional feature includes voice activation which allows calls to be dialed and controlled by voice commands. It is recommended that a call should not be dialed by a handset while driving [9], but a call can be initiated via voice activation without significant distraction.

Two types of speech recognition exist—speaker dependent and speaker independent. Speaker dependent recognition requires the user to store his voice command to be associated with a particular command. These recorded commands are used to match words spoken during operation. Speaker independent recognition allows multiple users to control the telephone without the recording of a particular voice. Speaker independent recognition is generally less accurate than speaker dependent recognition, and often may require the user to repeat the command if not recognized the first time. To prevent accidental operation of the cellular telephone by words in normal conversation, key words such as "phone start" are used to indicate a voice command [10].

Battery Chargers

There are two types of battery chargers, trickle and rapid charge. A trickle charger will slowly charge up a battery by only allowing a small amount of current to be sent to the wireless telephone. The battery charger may also be used to keep a charged battery at full capacity if the wireless telephone is regularly connected to an external power source (such as a car's cigarette lighter socket). Rapid chargers allow a large amount of current to be sent to the battery to fully charge it as soon as possible. The limitation on the rate of charging is often the amount of heat generated, the larger the amount of current sent to the battery the larger the amount of heat.

The charging algorithm can be controlled by either the phone itself or by circuits in the charging device. For some types of batteries, rapid charging reduces the total number of of charge and discharge cycles or the useful life of the battery. A charger will charge for a period of time, until a voltage transient occurs (called a knee voltage) and checks for temperature of the battery. The full charge is indicated by a couple of different conditions. Either the temperature of the battery can reach a level where the charging must be turned off, the voltage level will reach its peak value for that battery type, or the voltage level will stop increasing. Most chargers will then enter a trickle charge mode to keep the battery fully charged. Some chargers for NiCd batteries discharge the battery before charging to reduce the memory effect. This is called battery reconditioning.

Software Download Transfer Equipment

Some wireless telephones store their operating software in re-programmable memory (flash memory). Having this type of re-programmable type of memory allows new or upgraded software to be downloaded to the telephone to add feature enhancements or to correct software design errors. The new operating software sometimes can be downloaded with a service accessory that normally contains an adapter box connected to a portable computer. The new software is transferred from the computer through the adapter box to the wireless telephone. Optionally, an adapter box can contain a memory chip with the new software which eliminates the need for the portable computer. These types of devices have been very useful for the newer technologies where the systems and specifications change often. Changes can be made easily in the field without opening up a wireless telephone. Once a system is mature, wireless telephones will probably contain a non-re-programmable type memory device to help reduce costs. Figure 9.12 shows a software transfer system.

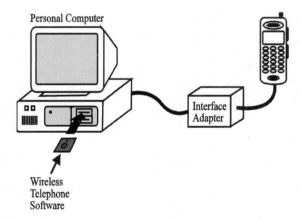

Figure 9.12, Software Transfer Device

Antennas

Antennas convert radio signal energy to and from electromagnetic energy for transmission between the mobile telephone and base Station. While there are several factors that will affect the performance of an antenna, only a fixed amount of energy is available for conversion. Antennas can improve their performance by focusing

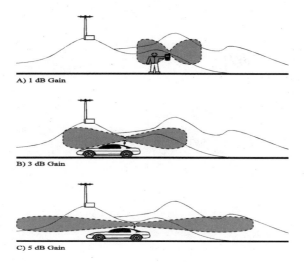

Figure 9.13, Antenna Gain

energy in a particular direction which reduces energy transmitted and received in other directions. The amount of gain is specified relative to a unity (omnidirectional) gain antenna. Car mounted antennas typically use 3 dB or 5 dB gain antennas. Portable antennas commonly use 0 to 1 dB gain antennas because people may turn the phone to many angles or leave the phone laying flat on the table.

Figure 9.13 shows the antenna patterns of antennas with different gains. In Figure 9.13 (a) a portable wireless telephone has 1 dB of signal gain. The radio pattern is almost even all around the telephone in any horizontal direction. Figure 9.13 (b) shows a car with an antenna gain of 3 dB. The gain comes from focusing the radio energy along the roadway and little energy is transmitted above and below the antenna. When using a 5 dB gain antenna, this allows the signal to transmit further along the roadway while transmitting less energy above and below the horizontal direction. As figure (c) shows, having more gain than necessary may cause the transmitted signal to miss the cell site because the radio energy is transmitted in a beam which is narrow and which misses the cell site.

References:

1. Sinnema, William, "Electronic Transmission Technology," Prentice Hall, NJ, 1979, pp.201-244.
2. MRT editorial staff, "Why Cellular Mobiles Use `High-Gain' Antennas," Mobile Radio Technology, Volume 5, Issue 5, May 1987, pp. 44-46.

3. Cellular Telecommunications Industry Association, Narrow AMPS Forum, Chicago, IL, 9 December 1990.

4. The Bell System Technical Journal, Vol58, No1, American Telephone and Telegraph Company, Murray Hill, New Jersey, January, 1979.

5. SuperPhone conference, Institute for International Research, New York NY, January 22-24, 1997.

6. CTIA Winter Exposition, "Safety," San Diego, February 17, 1991.

7. O'Sullivan, Harry M., U.S Patent 4,697,281, Cellular Telephone Data Communication System and Method, 1987.

8. West, William L., U.SPatent 4,658,096, System for Interfacing a Standard Telephone Set with a Radio Transceiver, 1987.

9. CTIA Winter Exposition, "Safety," San Diego, February 17, 1991.

10. Zeinstra, Mark , U.SPatent 4,827,520, Voice Actuated Control System for Use in a Vehicle, 1989.

Chapter 10
Wireless Networks

Wireless Networks

Wireless networks inter-connect base stations to each other and to various other telephone networks. Creating and managing a wireless network involves equipment selection and installation, implementation methods, inter-connection to the public switched telephone network (PTSN), other network interconnections (such as IS-41), and system planning.

These wireless networks consist of wireless base stations, communication links, switching center and network databases and a link to the PSTN. In a typical wireless system, a proprietary interface provides the intra-system connections that link base stations to the Mobile service Switching Center (MSC). The MSC coordinates the overall allocation and routing of calls throughout the cellular system. Inter-system connections can link different wireless network systems to allow wireless telephones to move from system to system. Inter-system connections can be proprietary, or they may either use the standard ITU GSM (European) or the EIA/TIA interim standard IS-41 (North American) interface. The design and layout of a wireless system is a continually changing and complex process. New technologies allow for variations in the type and use of cellular system equipment.

Figure 10.1 illustrates the fundamental interconnections in a cellular system network. Base stations convert radio signals from wireless telephones to a form suitable for transfer to the MSC. The radio signals may be analog (for example, AMPS or TACS) radio technology or digital radio transmissions (for example, TDMA or CDMA). If the cellular system is dual mode (for example, digital and AMPS), the base station radio equipment can process more than one type of radio signal. Calls are transferred between cells within the system (intra-system hand-off) or to cell sites in an adjacent system (inter-system hand-off). Currently, the MSC coordinates these processes, although this may change with the increased use of distributed switching. However, regardless of the type of radio transmission, the MSC routes calls to and from cell sites and the PSTN.

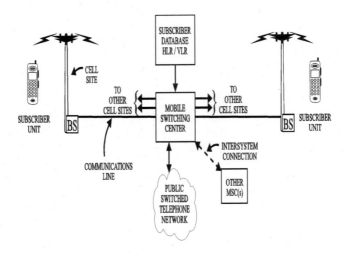

Figure 10.1, Cellular System Network Block Diagram

Base Stations

Base Stations may be stand alone transmission systems which hang on a power line or be part of a cell site which is composed of an antenna system (typically a radio tower), building, and base station radio equipment. Base station radio equipment consists of RF equipment (transceivers and antenna interface equipment), controllers, and power supplies. Base station transceivers have many of the same functional elements as a wireless telephone.

The radio transceiver section is divided into transmitter and receiver assemblies. The transmitter section converts a voice signal to RF for transmission to wireless telephones and the receiver section converts RF from the wireless telephone to voice signals routed to the MSC. The controller section commands insertion and extraction of signaling information.

Unlike the wireless telephone, the transmit, receive, and control sections of a base station are usually grouped into equipment racks. For example, a single equipment rack may contain all of the RF amplifiers or voice channel cards. For some analog or early-version digital cellular systems, one transceiver in each base station is dedicated for a control channel. In most digital cellular systems, control channels and voice channels are mixed on a single radio channel.

Figure 10.2 illustrates the components of a base station. Note that each assembly (equipment rack) contains multiple modules, one for each RF channel. Base station components include the following: voice cards (sometimes called line cards), radio transmitters and receivers, power supplies, and antenna assemblies.

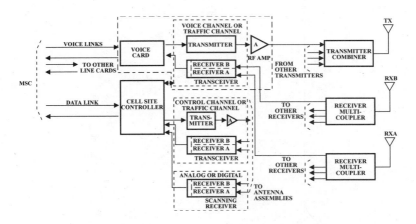

Figure 10.2, Base Station Block Diagram

Analog base stations are equipped with a radio channel scanning receiver (sometimes called a locating receiver) to measure wireless telephones' signal strength and channel quality during hand-off. The digital hand-off process has the advantage of using wireless telephones to provide radio signal strength and channel quality information back to the switching system. This information greatly improves the MSC's hand-off decisions.

Radio Antenna Towers

Wireless base station antenna heights can vary from a few feet to more than three hundred feet. Radio towers raise the height of antennas to provide greater area coverage. There may be several different antenna systems mounted on the same radio tower. These other antennas my be attached to a paging transmitter, a point to point microwave link, or Specialized Mobile Radio (SMR) dispatch radios. Shared use of towers by different radio systems in this way is very common, due to the economies realized by sharing the cost of the tower and shelter. However, great care must be taken in the installation and testing to avoid mutual radio interference between the various systems.

Figure 10.3 Microcell Base Station
Source: Hughes Network Systems
Photography: John Keith Photography

Figure 10.3 shows a miniature (Microcell) base station. Most base stations have multiple antennas. One antenna is used for transmitting and two are used for reception for each radio coverage sector. In some cases, where space or other limitations prevent the use of three separate antennas, two antennas may be used, with one of the two serving as both a transmit and receive antenna, and the other as a receive antenna only. Special radio frequency filters are used with the shared antenna to prevent the strong transmit signal from causing deleterious effects on the receiver.

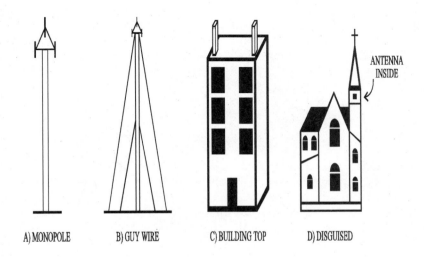

A) MONOPOLE   B) GUY WIRE   C) BUILDING TOP   D) DISGUISED

Figure 10.4 Cell Site Radio Towers

The basic antenna options are monopole mount, guy wire, free standing, or man made structures such as water towers, office buildings, and church steeples. Monopole heights range from 30-70 feet; free standing towers range from 20-100 feet; and guy wire towers can exceed 300 feet. Cell site radio antennas can also be disguised to fit in with the surroundings. Figure 10.4 illustrates several antenna systems. Figure 10.4 (a) is a free standing single pole called a monopole Figure 10.4 (b) is a tower supported by guy wires. Free standing antennas (not shown) are self-supporting structures with three or four legs. Cellular system antennas can also be located on building tops (Figure 10.4 (c)) to focus their radio energy to specific areas, or disguised inside a building as shown in Figure 10.4 (d).

Radio Equipment

A base station transmitter contains audio processing, modulation, and RF power amplifier assemblies. An audio processing section converts audio signals from the communications link to frequency and audio levels optimized for FM (analog) or phase shift modulation (digital). The transmitter audio section also inserts control information such as power control messages to the mobile set. A modulation section converts the audio signals into proportional phase shifts at the carrier frequency. The RF power amplifier boosts the signal to much higher power levels (typically 40 to 100 watts) than the wireless telephone (typically less than 1 Watt). Once initially set, the transmitter power level is normally fixed unless it is changed to vary cell boundaries or slightly increased to compensate for radio signal absorption by foliage in the spring and summer seasons, either manually or automatically (by control section software).

Receiver

The base station receiver sections consist of an RF amplifier, demodulator, and audio processor. The RF amplifier boosts low level signals received from wireless telephones to a level appropriate for input to the demodulator. The demodulator section converts the RF to audio or digital voice signals. Audio processing converts the optimized audio to its original frequency and amplitude levels. Receiver audio processing also extracts control information and converts the output audio level for transmission on the voice channel communication links to the MSC. In addition to all of these functions, most receivers are also able to select or combine the strongest radio signals that are received on the two different receive antennas at the Base Station, a process called diversity reception.

Controller

The controller sections perform control signal routing and message processing. Controllers insert control channel signaling messages, set up voice channels, and operate the radio location/scanning receiver. In addition, controllers monitor equipment status and report operational and failure status to the MSC. Typically, there are three types of controllers: base station controller, base station communications controller, and transceiver communications controller.

The base station controller coordinates the operation of all base station equipment on commands received from the MSC. The base station communications controller buffers and rates-adapt voice and data communications from the MSC. The transceiver communications controller converts digital voice information (PCM voice channels) from the communications line to RF for radio transmission and routes signals to the wireless telephones. The transceiver controller section also commands insertion and extraction of voice information and digital signaling messages to and from the radio channel.

RF Combiner

The RF combiner has certain similarities to the duplexer in a wireless telephone (see chapter 9). Each of a base stations many radio channels are usually served by a dedicated RF amplifier. The RF combiner allows multiple RF amplifiers to share one antenna without their signals interfering with each other. RF combiners are narrow bandpass filters with directional couplers that allow only one specified frequency to pass through. The filtering and directional coupling prohibits signals from one amplifier from leaking into another. A narrowband RF filter is a tuned chamber or cavity which is designed to produce a special internal electromagnetic field pattern called a standing wave, with dimensions that allow only a narrow range of frequencies (and their corresponding wavelengths) to pass. , To change the resonant frequency, the chamber's dimensions can be changed by moving a threaded metal rod inside it either manually by a screw device or automatically by a servo motor (auto-tune). Turning the rod to extend it further into the cavity produces resonance and a peak bandpass frequency which is lower (and corresponds to a longer wavelength), while shortening the length of the rod inside the cavity raises the resonant frequency (corresponding to a shorter wavelength). A later section in this chapter discusses an application of a broadband linear amplifier that may eliminate the need for multiple amplifiers and RF combiners.

Receiver Multi-Coupler

To allow one antenna to serve several receivers, a receiver multi-coupler must be attached to each receiving antenna. Figure 10.5 illustrates a receiver multi-coupler assembly. Because a receiver multi-coupler output is provided for each receiver antenna input, the split-

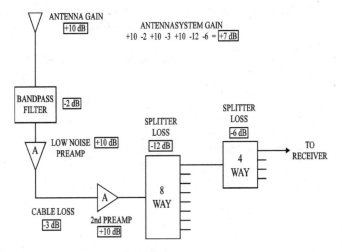

Figure 10.5, Receiver Multi-Coupler

ting of received signal reduces its total available power to each individual receiver. By increasing the number of receivers, the signal to noise ratio to each receiver section is reduced. Low noise RF preamplifiers are included with the multi-coupler to boost the low level received signals prior to the RF multi-coupler splitter.

Communication Links

Communication links carry both data and voice information between the MSC and the base stations. Options for the physical connections include wire, microwave, or fiber optic links. Alternate communication links are sometimes provided to prevent a single communication link failure from disabling communication [1]. Some terrain conditions may prohibit the use of one type of communication link. For example, microwave systems are not usually used in extremely earthquake-prone areas because they require precise line-of-sight connection Small shifts in the earth can mis-align microwave transceivers to break communications.

Regardless of the physical type of communication link, the channel format is usually the same. Communication links are typically digital time-multiplexed to increase the efficiency of the communication line. The standard format for time-multiplexing communication channels between cell sites in North America is the 24 channel T1 line, or multiple T1 channels. The standard format outside of North America is the 32 channel (30 useable channels) E1 line.

Figure 10.6 illustrates T1 (North American) and E1 (European) standard communication links. The T1 communication link is divided into time frames which contain 24 time slots plus a framing bit. To allow for control signaling in standard telephone systems, the last of the 8 bits are "stolen" from various slots in one out of every 6 consecutive frames. This has a negligible effect on the quality of speech transmitted. In cellular base station links, signaling is usually accomplished by setting aside one channel for control signaling instead of voice, so the "stolen" or "robbed" bit method need not be used. An E1 communication link is divided into time frames which contain 32 time slots. Two of the E1 time slots are dedicated as synchronization and signaling control slots, leaving 30 for voice.

Each slot contains eight bits of information. For standard land line voice transmission, each analog voice channel is sampled 8000 times per second and converted to an 8 bit PCM digital word. The 8000 samples x 8 bits per sample results in a data rate of 64 kb/s, and it is called a DS0 or PCM (one channel).

For T1 communication lines, 24 DS0's plus a framing bit are time multiplexed onto the high speed T1 channel frame. Therefore, with a frame length of 193 bits, the data rate is 8000 x 193 = 1.544 Mb/s. For

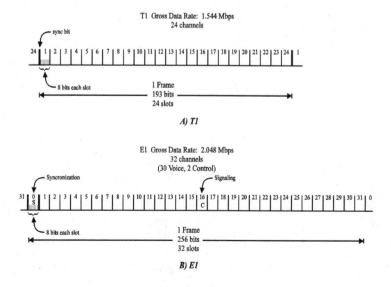

Figure 10.6, T1 and E1 Communication Links

E1 communication lines, 32 PCM channels are time multiplexed, resulting in a gross data rate of 2.048 Mb/s.

Antenna Assembly

When a wide area cellular system is first established, base station antenna assemblies usually employ horizontally omni-directional antennas [2]. As the system matures, directional (sectored) antennas replace the original antennas to reduce interference. An antenna assembly in each sector usually consists of one transmitting antenna and two receiving antennas.

Separate transmit and receive antennas are used to keep excessive amounts of the transmitter's RF energy from being coupled into receive antennas. The few feet of separation between the antennas provides more than 40 dB of isolation. In some installations, where antenna tower platform space is limited, and three antennas cannot be used, one antenna must also be used for transmitting. In this case, a very deep notch isolation filter is used to prevent the transmit signal from leaking into receivers on the shared antenna.

Two receive antennas are used for diversity reception to minimize the effects of Rayleigh signal fading (discussed later). Using two receiving antennas enables a technique that employs two receivers per channel to select the antenna which is receiving the stronger signal or allows the combination of the RF energy of both antennas. The technique improves power reception up to 6 dB or more, improves the signal to noise ratio (S/N) up to 3 dB or more, and reduces the effects of fading signals.

Scanning or Locating Receiver

A scanning or locating receiver measures wireless telephone signal strength for hand-off decisions. It can tune to any channel and measure the received signal strength, from which it determines a wireless telephone's approximate distance from the base station.

When the signal strength falls below a level determined by the serving base station transceiver, the base station signals the MSC that a hand-off will be necessary soon. The MSC then commands one or more adjacent cell sites to tune their scanning receiver(s) to monitor the wireless telephone's radio channel and continually measure and report the signal strength. The MSC (or other base station controller) compares the reported signal strength with other signal levels to decide hand-off.

All of the new cellular technologies provide for hand-off decisions assisted by the wireless telephone. Using signal and interference levels from wireless telephones, the cellular system can better decide when a hand-off may be necessary. The information from wireless telephones also reduces the data communications traffic burden of communications between adjacent base stations while coordinating (passing signal quality information) the hand-off process. The digital wireless telephone's hand-off information can possibly eliminate the need for a base station scanning or locating receiver.

Power Supplies and Backup Energy Sources

Power supplies convert the base station power source to regulated and filtered AC and DC voltage levels required by the base station electronics assemblies. Batteries and generators are used to power a base station when primary power is interrupted. Backup power is also needed for radio equipment and cooling systems. In 1989, a hurricane destroyed almost all land line communications in parts of Puerto Rico. Due to good planning, cellular communications were unaffected, and became the primary communication link [3].

In most climates, the heat generated by the radio equipment in the base station shelter will require installation of air-conditioning equipment to prevent excessively high temperature in the shelter. If the base radio equipment is permitted to operate at higher than rated temperature (for example, due to a failure of the cooling equipment), it may malfunction temporarily (until the shelter temperature is lowered) and the "burn out" lifetime of many of the components, particularly the RF power amplifiers, is significantly shortened.

Maintenance and Diagnostics

Base station radio equipment must be maintained and repaired as equipment assemblies fail, and cell sites are often remote and scattered throughout a cellular system. In response to maintenance needs, advanced maintenance and diagnostic tools have been created.

Base stations require software to operate and can be installed at the factory or loaded by the maintenance technicians after the cell site's radio equipment has been installed. As base station software improves, the updated version of the software must be installed into the base station controllers. Some systems do not require technicians to visit the cell site, but instead download new software via the communication links. During the software download process, one or more communication link voice channels (time slots) are dedicated to transfer software programs instead.of voice

The base station continuously monitors the status of its own operation and performance and sends report messages to the MSC. If spare equipment is available when base station equipment fails, the controller can reconfigure to continue service. Other maintenance tasks include routine testing to detect faults before they affect service. Routine test functions operate in a background mode and are suspended when faults are detected. Diagnostics begin and status reports may be continuously printed to inform system operators that maintenance may be required.

If a system fails, the suspect equipment assemblies must be tested to isolate faults. To verify equipment performance and monitor operational status, test signals are inserted at various points. Loop-back testing inserts test signals on one path (such as the forward direction) of the system and monitors the response of the signals on a return path.

Figure 10.7 illustrates two processes of loop-back test paths that are used to test a base station from the MSC. To test the communication link between the MSC and base station, the MSC sends an audio test signal to the base station on a voice channel (path 1). The line card then returns the test signal to the MSC via another voice path. If the return is unsuccessful, the fault is in one of the two voice channels or the voice channel card interface. To determine if the radio transmitter and receiver are working correctly, a second test path (path 2) routes a test audio signal through transmission and reception equipment. The test samples a portion of the output signal before the 45 MHz frequency shift and transmission via the antenna section, then directs the signal to the receiver section. If the MSC receives the test signal, the RF transmission and reception equipment are operational. Other loop-back paths can isolate other network equipment assemblies.

Many base stations of both the analog and digital type are normally equipped with a test transceiver. This test transceiver can perform the same functions as a mobile set, but it is remotely controlled from a central test control location, usually at the MSC. A voice channel usually used for one of the normal base transceivers is temporarily reassigned to permit a conversation to be routed through the test transceiver. A technician or craftsperson at the test desk can then place or receive a call at the cell site in question without the delay or requirement to first send a person to that site with a mobile set. This will quickly indicate if there is no base transmitter signal, or if a particular base transmit carrier frequency is not functioning, and similar problems can be quickly diagnosed. When the test transceiver is operated remotely (possibly via a standard telephone voice channel), test signals which are transmitted by the Base Station can be sampled and verified (path 3).

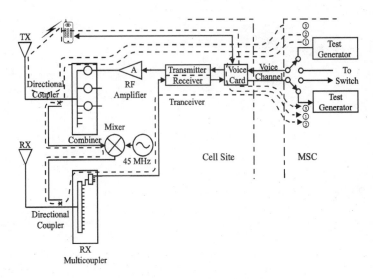

Figure 10.7, Cellular System Loop-Back Testing

Switching Center

A switching center coordinates all communication channels and processes. Formerly called Mobile Telephone Switching Office (MTSO), the Mobile service Switching Center (MSC) processes requests for service from wireless telephones and land line callers, and routes calls between the base stations and the PSTN. The MSC receives the dialed digits, creates and interprets call processing tones, and routes the call paths.

Figure 10.8 illustrates an MSC's basic components: system and communication controllers, switching assembly, operator terminals, primary and backup power supplies, wireless telephone database registers, and, in some cases, an authentication (subscriber validation) center.

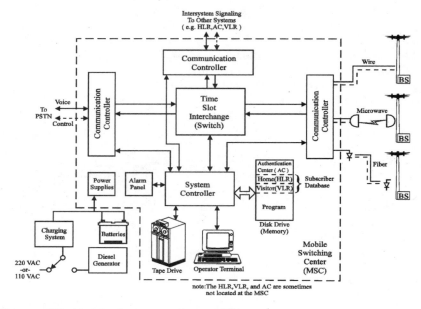

Figure 10.8, Mobile Switching Center Block Diagram

A system controller coordinates the MSC's operations. A communications controller adapts voice signals and controls the communication links. The switching assembly connects the links between the base station and PSTN. Operator terminals are used to enter commands and display system information. Power supplies and backup energy sources power the equipment. Subscriber databases include a home location register (HLR), used to track home wireless telephones, and a visitor location register (VLR) for wireless telephones temporarily visiting or permanently operating in the system. The authentication center (AC) stores and processes secret keys required to authenticate wireless telephones.

Switching Equipment

The switching assembly connects the base stations and the PSTN with either a physical connection (analog) or a logical path (digital).

314

Early analog switches required a physical connection between switch paths. Today's digital cellular switches use digital communication links.

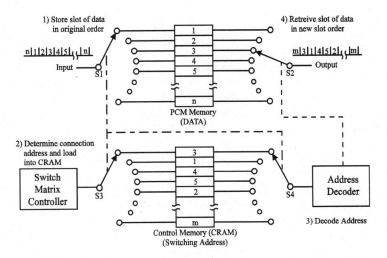

Note: Switches (S1-S4) are electronic and do not use mechanical parts.

Figure 10.9, Time Slot Interchange (TSI) Block Diagram

A switching assembly is a high speed matrix memory storage and retrieval system that provides connections between the base station voice channels and the PSTN voice channels. Figure 10.9 illustrates a simplified switching matrix system. Time slots of voice channel information are input through switch S1 to be sequentially stored in the PCM data memory. Time slots that are stored in the PCM memory are retrieved and output through S2 to the slots which are routed to another network connection (such as a particular PSTN voice channel). Switch 1 (S1) is linked to switch 3 (S3) and switch 4 (S4) so that each moves to predetermined memory locations together (for example, all at position 1). The address in the control memory determines the position of switch 2 (S2). The switch matrix controller determines which addresses to store in control memory slots. This address system matches input and output time slots.

Communication Links

Controllers

Controllers coordinate base stations, MSC switching functions, and PSTN connections. A system controller (or a subsection) creates and interprets commands between the MSC and the base stations, controls the MSC switch, validates customers requesting access, maintains air time and PSTN billing records, and monitors for equipment failures. Communication controllers process and buffer voice and data information between the MSC, base stations, and PSTN [4].

The communication controllers combine the channels from multiple voice channels on T1 or E1 lines in one high speed data channel. They also reverse the process by separating the voice paths from the high speed data channel and routing them to T1 or E1 lines. The high speed data channel contains time slots representing each voice path. The time slot for each voice path is switched in or out of a memory location in the switching matrix. The communication controllers also route call control commands (for example, hang up, dialed digits) to the control assembly.

Communication Links

Communication links are dedicated lines which transfer several channels (usually 64 kb/s voice channels) of information between the base station and switching system. In addition to the primary purpose of carrying voice traffic, one channel is typically reserved for control information (data) which allows the MSC to command the base station. To ensure reliability, alternate communication links can route channels through different network points if one or more communication links fails or is unavailable. Most communication links between cell sites and MSC's use a T1 or E1 Time Division Multiplex (TDM) Pulse Coded Modulation (PCM) digital transmission system.

When many communication channels are required, high speed links can be used. High speed communication links are made up of combined lower speed links. Even though MSCs may connect to a PSTN with hundreds of voice channels, MSC's use several T1 or E1 lines to provide the many voice channels needed. Operator Terminals

Operator terminals control most maintenance and administrative functions. The operator terminal is usually a computer monitor and keyboard dedicated to controlling equipment (for example, turn on a radio channel) and modifying the subscriber database. Operator terminal(s) might or might not be at the MSC, and there may be more than one.

Backup Energy Sources

Backup energy sources are required to operate the cellular network system when the primary power is interrupted. Backup energy to power switching equipment, subscriber databases, and cooling systems is usually a combination of batteries and diesel generators. During normal operations, batteries are charged with a charger using primary power. The batteries are directly connected to the cellular system, and when outside power is interrupted, they immediately and continuously power the system. After a short period of power loss, a diesel generator automatically begins to power the battery charger.

Network Databases

Home Location Register

The home location register (HLR) is a subscriber database containing each customer's Mobile Identification Number (MIN) and Electronic Serial Number (ESN) to uniquely identify each customer. Each customer's user profile includes the selected long distance carrier, calling restrictions, service fee charge rates, and other selected network options. The subscriber can change and store the changes for some feature options in the HLR (such as call forwarding). The MSC system controller uses this information to authorize system access and process individual call billing.

The HLR is a magnetic storage device for a computer (commonly called a hard disk). Subscriber databases are critical, so they are usually regularly backed up, typically on tape, to restore the information if the HLR system fails.

Visitor Location Register

The visitor location register (VLR) contains a subset of a subscriber's HLR information for use while roaming, and also contains the same type of information regarding home subscribers when they are at home and active in the system. The VLR eliminates the need for the visited MSC to continually check with the visitor's HLR each time access is attempted. The visitor's information is temporarily stored in the VLR memory, and then erased either when the wireless telephone registers in another system or after a specified period of inactivity.

Billing Center

A separate database, called the billing center, keeps records on billing. The billing center receives individual call records from the HLR. The billing records are converted into automatic message accounting (AMA) format to collect and process the information. The billing records are then transferred via tape or data link to a separate computer (typically "off-line") to generate bills and maintain a billing history data base.

Authentication Center

The authentication center (AC) stores and processes information required to authenticate a wireless telephone. During authentication, the AC processes information from the wireless telephone and compares it to previously stored information. If the processed information matches, the wireless telephone passes. The details of this process are explained in another section.

Public Switched Telephone Network

The public switched telephone network (PSTN) is the land line telephone system connects a wireless telephone to any telephone connected to the PSTN. Wireless telephones, land line plain old telephone service (POTS) dialtone, and other networks such as private automatic branch exchanges (PABX), all have different capabilities. Unfortunately, some control messages (such as calling line indicator) cannot be sent between the wireless telephone and different telephone networks, prohibiting some advanced features that digital systems could offer.

Figure 10.10 is an overview of the PSTN system. Two types of connections are shown: voice and signaling. End office (EO) and tandem office (TO) switches route voice connections. The end office (EO) switch is nearest to the customer terminal (telephone) equipment. Tandem office (TO) switching systems connect end office (EO) switches when direct connection to an end office is not economically justified. Tandem office switches can be connected to other tandem office switches. Signaling connections are routed through a separate signaling network called Signaling System number 7 (SS7). The SS7 network is composed of signaling transfer points (STPs) and signaling control points (SCPs). An STP is a telephone network switching point that routes control messages to other switching points. SCPs are databases that allow messages to be processed as they pass through the network (such as calling card information). SS7 messages are usually send in a reserved voice grade channel (64 kb/s) between the STP, SCP and other parts of the system. The messages in SS7 have the form of data packets similar to data packets used in other data communications systems such as X.25. Part of the similarity is the use of a special "flag" bit pattern (01111110) which is sent (at least once) at the end of each packet as a separator to indicate where one packet ends and another begins. A special reversible process called bit stuffing is used to modify the contents of a packet before transmission so it never contains the flag bit pattern. This bit stuffing process is undone at the receiving end of the data communication link. Each packet contains extra bits after the actual message which are used to provide an error detection code. Correctly received packets are acknowledged by a transmission of a special message in the opposite directions, from the receiving to the transmitting end of the data communications link. If the receiving end does not acknowledge within a pre-arranged time, the packet transmission is repeated. SS7 messages permit control of a wide variety of network functions, and are normally very reliable in their operation.

Two types of land line networks are shown: a local exchange carrier (LEC) and inter-exchange carrier (IXC) network. LEC providers furnish local telephone service to end users. An Inter-Exchange Carrier (IXC) is the long distance service provider. In some countries, local and long distance providers are operated by the same company or government agency. In the US, the LEC and IEX businesses were legally separated due to an anti-trust settlement which took effect in 1984. Then the Telecommunications Act of 1996 has again permitted the same company to operate both IXC and LEC services, with a more

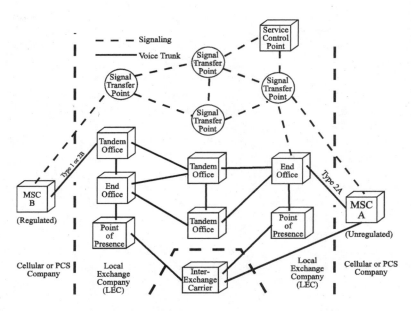

Figure 10.10, Public Switched Telephone Network (PSTN)

open competition in both types of markets. Before that new law becomes effective in the US, government regulations prohibited directly connecting an IXC to some end-user equipment, requiring a point of presence (POP) connection. The POP connection is a location within a local access and transport area (LATA) designated to connect a local exchange carrier (LEC) and an inter-exchange carrier (IXC). A LATA is typically a geographic service area boundary for the local exchange carriers (LECs).

The MSC is the gateway from the cellular system to the public switched telephone network (PSTN). Various types of connections can connect the MSC to the PSTN: Type 1, Type 2A or Type 2B. Type 1 connects the MSC with an end office (EO). A Type 2A ties the MSC into a tandem office. Type 2B occurs in conjunction with a Type 2A, and connects to an end office to allow alternate routing for high usage.

Wireless Network System Interconnection

Subscribers can only visit different cellular systems (ROAM) if the systems communicate with each other to hand-off between systems, verify Roamers, automatically deliver calls, and operate features uniformly. Fortunately, cellular systems can use standard protocols to directly communicate with each other. These inter-system communications use brief packets of data sent via the X.25 packet data network (PDN) or the SS7 PSTN signaling network. SS7 and X.25 are essentially private data communication networks. SS7, which is used by the telephone companies, is available only to telephone companies for direct routing using telephone numbers. The X.25 network does not route directly using telephone numbers. Some MSCs also use other proprietary data connections. No voice information is sent on the SS7 or X.25 networks. Only inter-system signaling (such as IS-41) is sent via these networks to establish communication paths. Voice communications are routed through the PSTN.

Ideally, inter-system signaling is independent of cellular network radio technology, but this can be difficult between systems where radio technologies differ. Consider inter-system hand-off between a CDMA-capable and a TDMA-capable cell site (assuming the wireless telephone were capable of both). The CDMA system uses soft hand-off while TDMA does not. As new features in cellular systems change, inter-system signaling messages must change to support them.

Communication between MSCs is performed either by a proprietary or standard protocol. Standard protocols such as IS-41 allow MSCs of different makes to communicate with few or no changes to the MSC. Regardless of whether a standard (for example, IS-41) protocol or a manufacturers private (proprietary) protocol is used, the underlying data transferred via inter-system signaling is the same. If changes are required to communicate with a different protocol, an interface (protocol converter) changes the proprietary protocol to standard protocol. The interface has a buffer which temporarily stores data elements being sent by the MSC and reformats it to the IS-41 protocol. Another buffer stores data until it can be sent via the control signaling network.

Inter-System Hand-off

Inter-system hand-off links the MSCs of two adjacent cellular systems during the hand-off process. During inter-system hand-off, the MSCs involved continuously communicate their radio channel parameters. Figure 10.11 illustrates inter-system hand-off between two different manufacturers' MSCs. The process begins when the serving base station (#1) informs the MSC (system A) that a hand-off is required. The MSC determines that a base station in an adjacent system is a potential candidate for hand-off. The MSC requests the adjacent MSC (system B) to measure the wireless telephone's signal quality. Both base stations (#1 and #2) measure the wireless telephone's signal quality until hand-off. In many cases, hand-off may be immediate. The serving MSC (system A) compares its measured signal strength with the signal strength that the MSC in system B measures. When the system B MSC measures a sufficient signal, the system A MSC requests the

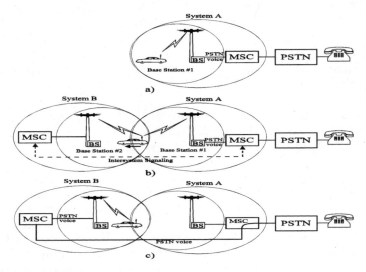

Figure 10.11, IS-41 Inter-System Hand-off

hand-off. Base station #1 issues the hand-off command, informing the wireless telephone to tune to another frequency, and base station #2 begins communicating to the wireless telephone on the new frequency. The voice path is then connected from the anchor (original) MSC to the system B MSC, and the call continues. After the anchor MSC receives a confirmation message that the wireless telephone is successfully operating in system B, the radio resources in the original cell site become available.

During inter-system hand-off, adjacent MSCs are typically connected by a T1 or E1 link, providing both inter-system messaging and voice communications. An external land line connection between the MSCs is technically possible, but the setup time between them is slower.

Roamer Validation

Roamer validation is the verification of a wireless telephone's identity using registered subscriber information. Validation is necessary to

322

limit fraudulent use of cellular service. The two types of roamer validation are "post-call" and "pre-call." Post-call validation occurs after a call is complete, and pre-call validation occurs before granting access to the system.

During the early deployment of cellular systems, the limited connections between systems resulted in delays of minutes or even hours before roaming cellular subscribers could be validated. To allow customers to use the phone immediately, early systems used post-call validation. Pre-call validation became possible when improved inter-system interconnection greatly reduced validation time.

Figure 10.12 illustrates roamer validation. When a wireless telephone initiates a call in a visited system (step 1), the cellular system attempts to find the wireless telephone's ID in its visitor location register (VLR). Let us consider the case in which the visited system determines that the wireless telephone is not registered in its system (step 2). Using the wireless telephone's ID (phone number), the visited cellular system sends a message to the wireless telephone's home system requesting validation (step 3). The HLR compares the ESN and MIN to determine if it is valid (step 4). If the subscriber proves valid, the HLR responds to MSC-A to indicate that validation was successful (step 5). After MSC-A receives confirmation that the visiting

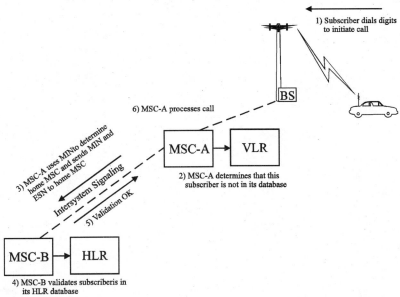

Figure 10.12, Roamer Validation

wireless telephone is valid, the call is processed (step 6). MSC-A's VLR may then temporarily store the wireless telephone's registration information to validate the subscriber's identity rather than requesting validation from the home system again for the next call. After a predetermined period of wireless telephone inactivity, the information stored in the VLR will be erased. If the wireless telephone was recently operating in another cellular system, the home system informs the old visited system that the wireless telephone has left. This allows the old visited system to erase the wireless telephone's identification information.

Authentication

Authentication is the exchange and processing of stored information to confirm a wireless telephone's identity. Authentication is significant because roamer validation cannot detect illegally cloned (telephones not owned by the authentic customer but containing duplicated identification) wireless telephones.

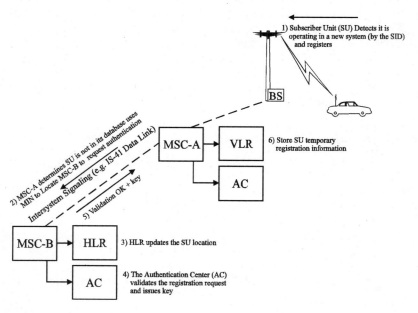

Figure 10.13, Authentication

New technologies offer a new authentication process to verify a subscriber's identity. The new process transfers stored information between the wireless telephone and an authentication center (AC).

The two primary options for inter-system authentication are: 1) the visited MSC can use a temporary key, or 2) the MSC can request the authentication center (AC) to validate wireless telephones each time. If the AC validates the SU[SU never defined!!!] each time, the visited MSC must send all the authentication parameters. If the AC provides a temporary key, the visited MSC can use the key while the subscriber is in the system without validation from the home system for each call.

Figure 10.13 illustrates a wireless telephone's authentication process. When a wireless telephone detects a new cellular system (new system identifier), it attempts to register with the system (step 1). The visited system searches for the wireless telephone's ID in it's visitor location register (VLR) and determines that the wireless telephone is not yet registered. The visited cellular system uses the wireless telephone's ID (phone number) to request authentication of identity (step 2) from the subscriber's home cellular system. If the home system information processes correctly, the authentication center (AC) validates the registration request (step 4). The AC either confirms validation or creates a key for future authentication using information received from the home system (step 5). MSC-A's VLR then temporarily stores the subscriber's registration information (step 6) for future authentication without contacting the subscriber's home HLR for the next call. After a predetermined period of inactivity, the temporary authentication information stored in the VLR will be erased.

Automatic Call Delivery

Ideally, call delivery is completely automatic whether the wireless telephone is in its home system or visiting another system. Such automatic delivery requires the home system to continuously track the wireless telephone's location. Roamer validation is the means for providing this information back to the home system.

To enable voice connection between the home system and visited system, a temporary location directory number (TLDN) is assigned for each automatic call delivery request.

Figure 10.14 illustrates basic inter-system call delivery. When a home system (MSC-A) receives a call for its subscriber, the MSC checks its home location register (HLR) to determine if the wireless telephone is operating in another cellular system (step 1). The home MSC then

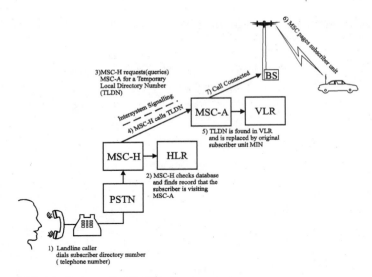

Figure 10.14, Automatic Call Delivery

sends a request to the visited MSC for a TLDN (step 2). The TLDN is cross-referenced with the MIN in the VLR (step 3). The home system (MSC-H) then initiates a call to the TLDN (step 4). The visited system (MSC-A) receives the call and finds the TLDN number is listed in it's VLR (step 5). It pages the wireless telephone using the MIN which was previously stored in the VLR (step 6). When the subscriber answers, the call is connected (step 7). After a predefined interval of inactivity or notification that the roaming subscriber has entered another system, the TLDN is dis-associated from that particular MIN. The TLDN can be placed back into a pool of TLDNs to be used for other calls.

Subscriber Profile

Features such as call forwarding, three-way calling, and call waiting activation options may operate differently in different cellular systems. Although EIA/TIA IS-53 and other standards have standard system feature specifications to resolve many of these issues, it may be undesirable to change established feature controls (such as dialing *69). Also, some features are linked to special services that go beyond functional operation (such as speed dialing). For subscribers to experience the same feature operation while roaming, system feature operations can be transferred to the visited system using inter-system messaging.

Figure 10.15 illustrates how subscriber profile features operate in a visited system. During initial registration in the visited system, the wireless telephone's feature profile can be transferred to the visited system's VLR. When the roaming subscriber activates a feature by

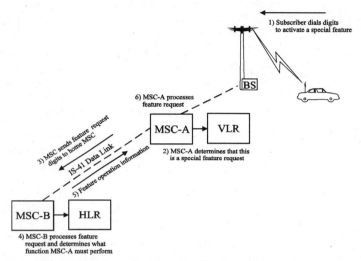

Figure 10.15, Subscriber Feature Profile Transfer

dialing a digit pattern (step 1), the VLR determines that the subscriber has selected a feature that its home system must process (step 2). MSC-A sends the dialed digit feature request to the subscribers home system (HLR) for processing (step 3). The HLR reviews the subscriber's feature profile, processes the request (for example, call forwarding), and determines what tasks the visited MSC must perform (step 4). MSC-B then sends any additional operation instructions to MSC-A (step 5). Finally, MSC-A completes processing the feature request (step 6).

Inter-system specifications continue to evolve as new features are added. If the inter-system signaling protocol does not include the profile feature request (such as IS-41 revisions 0 through B), a failure tone is given and the advanced service are not provided.

Wireless System Planning

Cellular system design is tedious, ongoing, and unique for each cellular system. A cellular system is built upon radio coverage areas created by RF transmission from interconnected base stations. Each separate radio coverage area must be planned to work together for optimization of the whole system. The selection of frequencies and power levels at each base station determines the cell site and system serving capacity limits. Transmission power levels are influenced by antenna directivity, gain, and type, and by surrounding terrain.

Radio propagation factors, strategic planning (for example, key heavy traffic locations), system testing and validation, and system expansion (capacity) requirements all affect cellular system design. Radio propagation characteristics, which change with the seasons, affect radio signal quality. Strategic planning requirements include data acquisition, MSC and cell site location selection, equipment procurement, and validation. System capacity expansion involves optimizing the existing system and adding/dividing radio service coverage areas.

Radio Propagation

Cellular system radio propagation has unique attenuation, signal fading characteristics, and signal quality requirements. Attenuation varies as a function of distance and terrain. Fading characteristics result in variations in received signal levels over short distances. Signal quality is limited by both received level and interfering signal strength.

Path Loss

For RF energy radiating into free space, power density decreases at more than 20 dB/decade of distance. At 10 times the distance from the transmitter (for example, 10 meters to 100 meters), received power decreases by at least a factor of 100. RF energy its partially transmitted through obstructions such as buildings, but average attenuation through such obstructions is about 40 dB per decade [5], and attenuation in heavy foliage can be more than 60 dB per decade [6]. Therefore, as seasons change and leaves fall, decreased attenuation can increase cell site RF coverage areas.

Fading Characteristics

Radio signal fading can cause weak or dead spots in radio coverage areas so that subscribers hear a noticeable drop in audio quality and volume. In a deep signal fade, transmission can become distorted or interrupted, and subscribers must move to an area with better signal strength. At UHF radio frequencies, multipath fading causes weak signal strength spots every few inches. The theoretical mathematical description of this phenomenon is called Rayleigh fading.

Multipath fading results from multipath propagation, the reception of rays of radio power which arrive at the receiving antenna both by direct line-of-sight paths from the transmitter, and also by zig-zag paths involving reflection from surfaces of buildings, cliffs, the ground, sides of trucks and other vehicles, and so forth. The longer diagonal or zig-zag paths have greater time delay. When several replicas of the same signal are received at slightly different times, some have a different out of phase relationship to each other. The oscillating carrier frequency electric field from one ray may be repeatedly pushing the electrons in the antenna *up* while another ray with different delay and phase is pushing the same electrons *down*. When two radio signals are added in phase, their combined maximum power is higher than either individual signal. Adding two equally strong signals that are out of phase by 180 degrees can nearly eliminate the radio signal. The theoretically ideal statistical variation of such fluctuations due to many different rays having different individual signal strength is called Rayleigh fading. Rayleigh fading variations occur approximately every 1/2 wavelength. At 840 MHz (US cellular frequency band), deep fades (weak spots) are about 7 inches (18 cm) apart. At 1900 MHz (PCS/PCN frequency band), deep fades are about 3 inches (8 cm) apart.

Radio signal fades are unpredicable and are frequency and time dependent. It is unlikely that a signal fade on one frequency will be occurring on another frequency at the same time. It is unlikely that a signal fade *at one location* on one frequency will be occurring *at another location* on the same frequency at the same time This latter situation is the theoretical basis for using two separated receive antennas at the base station to improve the composite signal via receive diversity. The new digital technologies overcome some of the effects of Rayleigh signal fading either by using a very wide radio channel, by slow frequency hopping, or by handing the call off to another radio channel that is not experiencing the fading characteristics.

Signal Quality

Signal quality varies throughout a cellular system coverage area, and may be degraded by a low carrier-to-noise (C/N) ratio or a low carrier-to-interference (C/I) ratio. In a C/N limited environment, communications quality decreases as the wireless telephone signal strength declines and approaches the thermal noise (natural background noise) inherent in the radio channel. A C/N limited (noise limited) environment would only occur near the outer edges of the outermost cell of a cellular system, where the interference from other cells in the system is very small. In a C/I limited environment, communications quality decreases as the wireless telephone moves nearer to a strong interfering signal. A C/I limited (interference limited) environment occurs in all of the inner cells of a cellular system. In a single cell system where the wireless telephone can wander out of the service coverage area completely, signal strength available to overcome the background noise (C/N ratio) is the limiting factor. In mature cellular systems, as most in the US are rapidly becoming, the limiting factor is signal strength available to overcome interference from other cell sites (C/I ratio).

To reuse frequencies, cellular systems rely on path loss and adequate spacing between base stations and cells which use the same frequency to prevent interference. The minimum allowable distance between cell sites is determined by the minimum permitted carrier to interference ratio (C/I). The permitted C/I ratio (more accurately, C/(I+n)) is different for different radio technologies (analog FM, various types of digital modulation) and also depends upon the ratio of the radio bandwidth to the audio bandwidth (for FM) or the digital bit rate. For example, for standard North American 30 kHz bandwidth analog FM cellular, the permitted C/I ratio is 18 dB. For NAMPS, with a 10 kHz bandwidth, the permitted C/I ratio is 23 dB. From knowledge of the permitted C/I ratio and the path loss, it is possible to calculate the nearest distance to other cells having the same carrier frequency. This is expressed as the D/R (distance to cell radius) ratio. Once the /R ratio is calculated, frequencies are selected to avoid interference among cell sites. For AMPS cellular systems, with 18 dB C/I and 40 dB/decade path loss, the reuse distance D/R ratio is typically 4.6. This means that if a cell site with a one-mile radius is assigned channel 424, then channel 424 could not be reused at any cell site nearer than 4.6 miles.

Strategic Planning

Strategic planning for a cellular service provider involves setting company goals, such as subscriber growth, quality of service, and cost objectives. It also involves making plans for obtaining those goals. Building and expanding a cellular system requires collecting demographic information, targeting key high traffic locations, selecting potential cell and MSC sites, conforming to government regulations, purchasing equipment, construction, and testing validation.

Most providers gather physical and demographic information first. For example, transportation thoroughfares, industrial parks, convention centers, railway centers, and airports may be identified as possible high usage areas. Estimates of traffic patterns help to target coverage areas for major roadway corridors. Terrain maps, marketing data, and demographic data are all used to divide the cellular system into RF coverage areas. The object is to target gross areas where cell site towers may be located. The raw data needed might include system specifications, road maps, population density distribution maps, significant urban center locations, marketing demographic data, elevation data, and PSTN and switch center locations.

For the US market, government regulations include quality of service (typical limiting the blocked call attempt ratio to 2%, or P02 grade of service) and time intervals for service offerings [7]. While business considerations may indicate that radio coverage is not necessary (for example, an unpopulated rural area), government regulations may require that area to be covered within a specified period.

After systems are planned, equipment manufacturers and their systems are reviewed and purchase contracts are signed. During various stages of equipment installation, validation testing is performed to ensure that all of the planning goals are being realized.

After the system is planned and cell site locations are selected, RF simulation begins. Calculations based on antenna elevation and terrain data are used to estimate expected signal strengths and quality levels.

The results of such calculations are rendered graphically onto transparent overlays which can be placed over standard topographical maps (published by the US Coast and Geodetic Survey in the US, and by similar government agencies in other countries). Typically, different colors used on these overlays indicate different values of signal strength levels. System simulations may predict estimated signal coverage and performance levels, but to be certain, temporary cell sites are often tested using a crane to lift a temporary antenna to the planned tower height. Theoretical calculations are often imprecise everywhere by a relatively uniform dB error, which can be determined only by comparing theory and experimental measurement. Once the proper dB correction factor is known from this comparison, the theoretical calculation can be used for evaluation of other base antenna locations in the cell with considerably improved precision.

Frequency Planning

When deploying narrowband radio channels, nearby cell sites cannot use the same frequency. Therefore, the frequencies to be used in each cell site must be planned to account for differing cell site boundaries and terrain conditions that enhance or reduce interference from nearby cell sites.

Radio channels in the same cell site can be separated by as little as 90 kHz for AMPS and30 kHz for North American TDMA, only 10 kHz for NAMPS channels, and 200 kHz for GSM. However, it is usually preferred to *not* use adjacent carrier frequencies (30 kHz separation for North American TDMA or 200 kHz for GSM) in the same cell when that can be avoided by means of a different frequency plan. Use of adjacent frequencies in adjacent cells is not supportable in AMPS due to excessive adjacent carrier frequency interference under most conditions. Use of adjacent carrier frequencies in North American TDMA or GSM is usually permissible except when one cell is significantly larger than its neighbor in radius. Proponents of CDMA state that no carrier frequency separation is required for IS-95 CDMA in adjacent cells, and they plan to use the same carrier frequency in all cells in the system. Opponents of CDMA claim that this can be done only by severely reducing the number of simultaneous conversations in each cell, and the discrepancy of these two claims will be better understood after more field experience has been accumulated with working CDMA systems. However, frequency planning requirements typically specify that narrowband radio channels attached to the same antenna be separated by several radio channel frequency carriers. For analog (AMPS, TACS, and NAMPS) and US TDMA (IS-54 and IS-136), the separation is typically 7 to 21 carrier channels. For GSM, separation is typically 4 carrier channels. For IS-95 CDMA, the same radio channel can be used in each cell according to the plan of CDMA proponents.

Sectorization

Sectorization adds radio channel capacity by dividing a cell site's radio coverage area into sectors. The number of radio channels used in the cell site is multiplied by the number of sectors. For example, when three 120-degree sectors are used, antennas focused in one direction interfere less with antennas in the opposite direction. Therefore, frequencies in other cell sites can be reused in more closely placed cells than with omnidirectional antennas.

System Testing and Verification

System testing and verification determine if RF coverage quality is optimal good and if the system is operating correctly. Signal quality is tested for individual cells , then adjacent cells are measured to determine system performance. System operation is verified by measuring the signal quality level at hand-off, blockage performance, and the number of dropped calls.

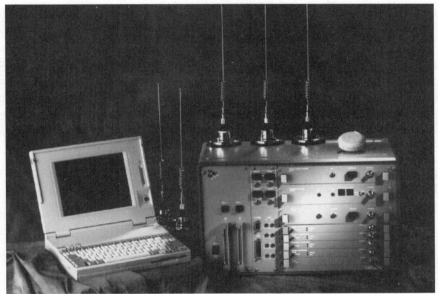

Figure 10.16, Cellular System Monitoring Equipment
Source: LCC

RF coverage area verification confirms that a minimum percentage (usually 90% of the cell area is the goal) of coverage area is being served. Received signal strength and interference levels are measured to locate holes where terrain and obstructions cause the signal levels to fall below an acceptable level. Such coverage holes may require another cell site or repeaters to amplify the existing signal and refocus the energy into the dead spot, or in or behind an obstacle (for example, a parking garage).

System operation can be verified by recording the signal levels received by a test wireless telephone during hand-off and access. Blocking probability can be estimated from the number of access attempts rejected by the system. The test equipment shown in Figure 10.16 monitors and records the signal quality levels as the test wireless telephone moves throughout the system area. The resultant data can then be plotted on a map overlay to determine where signal levels and hand-off thresholds need to be increased and decreased.

System Expansion

Cellular systems are expanded to allow more customers to obtain service in a given area. This can be performed either by the addition of cell sites (adding radio coverage areas) or by increasing radio spectrum efficiency. All new technologies increase radio spectrum efficiency by allowing multiple users to share a single radio channel or by allowing more radio channels into the same frequency spectrum.

Cellular systems can expand by adding cell sites (called cell division), but at a high cost. For example, replacing a 15-mile radius cell site with 1/2-mile radius cell sites would require more than 700 new cell sites. All of the new cell sites would need to be interconnected, and hand-off switching would increase dramatically. Figure 10.17 illustrates how cellular systems can expand through dividing the coverage area. This is economically feasible when the revenue producing traffic immediately increases in proportion to the added capital and operating cost due to installing new smaller cells, but traffic tends to grow more slowly than is desired by the system operator.

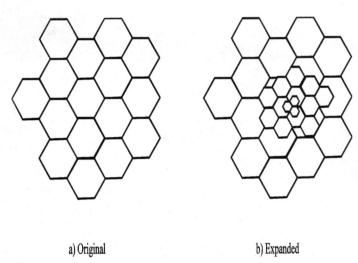

a) Original   b) Expanded

Figure 10.17, Cellular System Expansion Via Cell Division

Cellular system capacity can be expanded with new radio technologies. By adding or replacing low capacity radio channels (such as AMPS) with higher capacity radio technologies (such as TDMA or CDMA), each cell site can serve more customers. Unfortunately, such expansion requires new hardware and network software for new technologies.

Network Options

While industry specifications may define how radio channels and messages are coordinated, manufacturers and cellular system operators have many options. New technologies can be integrated or set up as separate overlay systems. Use of various types of technological upgrades in different parts of the system have various different effects on the overall system technological and economic performance, some more significant than others. Integrated systems use a single switching system to control different types of radio channels while overlay systems use two or more switches to control the different types of radio channels. Communication channels between the MSC and individual cells can be sub rate multiplexed to decrease communication link costs. For example, the normal 64 kb/s digital speech coding used on a T1 communication link channel can be replaced with 2 channels using 32 kb/s ADPCM speech coding, which supports 48 rather than 24 channels on the existing T1 link, thus reducing the operating cost per voice channel. Broadband linear amplifiers allow the use of different types of radio channel cards without changing RF amplifier equipment, and can also reduce wasted power at the base station. Distributed switching can reduce the burden on an MSC and provide for advanced features. Repeaters can extend the range of any type of cellular radio system.

Integrated and Overlay Systems

Dual-mode digital cellular systems support two implementation systems: integrated and overlay. The integrated approach makes one system responsible for assigning both analog and digital radio channels. The overlay approach uses two separate systems to assign radio channels. For example, in an overlay approach, the analog system may coordinate AMPS radio channels and the overlay system could coordinate digital radio channels. Overlay systems permit continued use of older cellular equipment that may not be replaced or upgraded to digital service.

In addition to the use of two completely separate cellular systems with separate MSCs and base stations, a "halfway" form of overlay is also used. In this method, two different types of base stations are used, but both types are controlled by and connected to the same MSC. This has the advantage of simpler and smoother inter-mode handoff, for example between TDMA and analog on a dual mode mobile set.

Both types of overlay systems give the operator the economic advantage of installing a limited number of base radios of the new type (usually digital) in a pre-existing system of the old type (usually analog). The new digital cells have higher altitude antennas and more base transmitter power, so they cover the area of a large number of smaller analog cells. These larger cells are sometimes called "umbrella" cells because they cover a large area. Because the amount of traffic from the new digital mobile sets is smaller in the beginning of their use, the investment in new base stations can be limited to an amount which is not disproportionate to the initial revenue from these new digital mobile sets. Later, as the number of digital mobile sets increase, the digital base equipment can be installed in ordinary size cells as well, always keeping the capital cost and the added revenue in proportion.

Sub Rate Multiplexing

Whatever transmission medium is used between cell sites and the MSC, maintaining the network's voice and data communication links is costly. Various methods, such as sub rate multiplexing, can improve the efficiency of these links. Sub rate multiplexing combines multiple compressed voice channels on a single communication channel (such as a 64 kb/s DS0).

Several digital cellular equipment configuration options can optimize sub rate multiplexing. One option is where to locate the speech coder and channel coder: in the cell site or MSC. Significant economies can be achieved by placing the low bit-rate digital speech coder required for the radio link (for example, VSELP for TDMA, or QCELP for CDMA) at the MSC rather than the base station. Data transfer on a radio channel includes error protection bits along with the compressed voice data bits. The error protection bits protect against the radio signal distortions common in cellular radio transmissions. T1 or E1 communication links between the base station and MSC are either wire, fiber, or fixed microwave, so distortion is low and few error protection bits are needed. The low requirement for error protection allows for the removal of cellular radio channel coding bits (error protection bits) on the link between the base station and the MSC. The resulting low

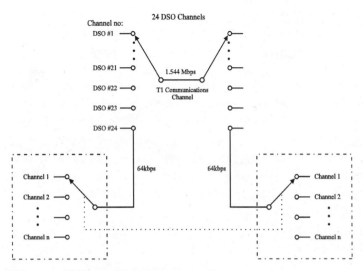

Figure 10.18, Sub Rate Multiplexing

bit rates (1.2 - 9.6 kb/s) allow several compressed digitally coded voice channels to be shared on a single DS0 64 kb/s communications channel. Figure 10.18 illustrates how several sub rate multiplexed channels can share a single DS0 channel.

Since the cost of the communication links between the MSC and cell sites can be a significant part of the system cost, using each DS0/PCM channel in a T1 or E1 communication link for multiple conversations is an important potential system cost reduction. For analog cellular systems, the using 32 kb/s ADPCM digital voice coding on each voice communications channel combines 2 conversations on each DS0/PCM channel. This increases the number of voice channels per communication link to 48 for T1 links and 60 for E1 links. Other systems use the VSELP coder at the MSC end to encode each voice channel at 13 kb/s, and thus three voice channels can be combined in one DS0 channel over the T1 link (with some unused bits left over in each DS0 channel). It is not advisable to use ADPCM on the MSC-BS link and then convert it into another form of low bit rate coding such as VSELP, since the quality of the speech is degraded due to the double digital coding conversion. This loss of quality is called trans-coding loss. However, 32 kb/s ADPCM on the MSC-BS link is quite good for use with analog cellular.

Broadband Linear Amplified Base Stations

With a single linear RF amplifier, it is technically possible to mix any combination of cellular radio signal types (AMPS, NAMPS, TDMA, CDMA). This linear amplifier allows any of the cellular technologies to by installed by adding transceiver cards into the equipment frames. This mix of radio channels can then be controlled by the cellular sys-

Figure 10.19, City RFx in-building distributed antenna system
Source: ADC

tem. In a sectored cell, it is also possible to switch the RF signal from the input of the linear RF amplifier serving one sector to the input of another linear RF amplifier serving another sector when additional traffic capacity is needed in the second sector. This technique of flexibly allotting traffic capacity to different sectors is sometimes called "face switching." It is a very beneficial capability which overcomes the problem of fluctuating traffic demand in different sectors of a sectored cell, which cannot be handled using the technique of connecting a fixed number of high power RF outputs to an antenna via a combiner.

Suitably designed wideband linear RF power amplifiers can also be more efficient overall than using a two stage combiner. A two stage combiner is only 25% efficient. It converts 75% of the transmitter RF power into heat in the directional coupler parts of the combiners. The only major drawback of the use of a wideband linear RF power amplifier relates to reliability. For a higher level of reliability, a wideband linear RF amplifier needs to be installed with a back-up amplifier which can be switched into its place, because the failure of a single amplifier (without backup) will kill *all* carrier frequency transmissions in that cell or sector. This problem is called a "single point of failure" and is the main drawback to use of wideband linear RF power amplifiers.

Distributed Switching

Distributed switching is the transfer of some coordination of network switching to locations distributed throughout a cellular system network. Distributed switching may also allow advanced features such as 4 digit dialing.

Small, easily installed cell sites called microcellular systems have increased the use of distributed switching. Microcells can be installed in high usage areas such as airports or office parks to expand cover-

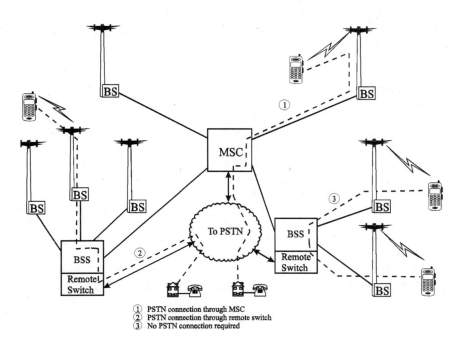

① PSTN connection through MSC
② PSTN connection through remote switch
③ No PSTN connection required

Figure 10.20, Distributed Switching

age. Many of these systems relieve the burden on the MSC through distributed switching. Since many calls on a private system are within the system and the MSC does not need to be involved, private cellular systems often distribute switching to reduce the work of the MSC.

Figure 10.20 illustrates how distributed switching allows calls to be placed between wireless telephones without the central MSC. In communication path 1, the wireless telephone communicates to a land line customer through the MSC. Using communications path 2, the wireless telephone communicates to a land line telephone customer through a distributed switch. The distributed switch is connected directly to the PSTN. Using communication path 3, two wireless telephones are connected with each other via a distributed switch. No connection to the PSTN is required.

Cell Site Repeaters

Repeaters are vital in providing cost-effective cellular service in rural areas by making it possible to extend the range of analog and digital channels. Repeaters can also be used to fill in RF coverage behind obstacles or inside buildings, tunnels, or parking garages within the normal geographic boundaries of a cell. Repeaters receive a radio signal from a nearby cell site, amplify that signal, and re-transmit it in a new direction at the same radio frequency. With this same-frequency repeater, it is important to install the two antennas so that the path loss between them is greater in magnitude than the gain of the repeater amplifier, to prevent self oscillation. Also, for border areas which receive radio signals both from the repeater output antenna and directly from the base transmitter, there is an artificially enhanced multipath fading and signal dispersion problem which causes problems to some digital cellular systems. Some other repeaters receive, decode, and re-transmit the radio waves on a new frequency. These are called "frequency shifting" repeaters. They have two advantages: 1) they cannot cause self oscillation regardless of where or how the input and output antennas are installed, and 2) they cannot cause enhanced multipath fading. However, their proper use requires careful modification of the call setup and handoff commands used in that area, since the carrier frequency transmitted by the base station is not the actual physical carrier frequency seen by the mobile set in the repeater radio areas, and the commands which direct the mobile station to change to a designated carrier frequency must be properly composed by the call processing software to take this into account.

Figure 10.21, Cellular System Repeater
Source: Antenna Specialists

It is possible to extend the range over a hundred miles using repeaters, but the increased range may introduce propagation delays and phase distortion. The propagation delays from repeaters may exceed the maximum delay offset that digital wireless telephones can accommodate. In addition, repeaters that use class C amplifiers introduce phase distortion and high or bit error rates (BERs) for TDMA and CDMA digital radio channels. Figure 10.21 illustrates a cell site repeater that can repeat analog and digital radio channels.

Diverse Routing of Network Communication Links

Cellular systems are interconnected via communication links. Figure 10.22 illustrates a sample cellular system communication link structure. The links between cell sites typically have alternate routing (for example, via another cell site) for redundancy. If one link is damaged or otherwise inoperative, most or all of the traffic it formerly carried can be routed via the second diverse link. The purpose of diverse geographical routing is to avoid local disasters, such as a contractor with a backhoe cutting a buried cable, or a truck which hits a telephone pole and breaks the telecommunication wires and cables supported on that pole.

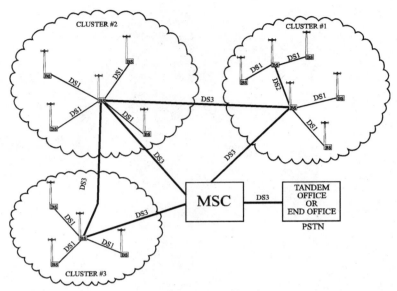

Figure 10.22, Sample Cellular System Communication Link Structure

References:

1. CTIA Winter Exposition, "Disaster Experiences," Reno Nevada, February 6, 1990.
2. Lee, William, "Mobile Cellular Telecommunications Systems," McGraw Hill, NY, 1989.
3. CTIA Winter Exposition, "Disaster Experiences," Reno Nevada, February 6, 1990.
4. Interview, Richard Levine, Beta Laboratories, 9 January 1996.
5. Felix, Kenneth, U.S. Patent 4,887,265, Packet-Switched Cellular Telephone System, December, 1989.
6. Lee, William, "Mobile Cellular Telecommunications Systems," McGraw Hill, NY, 1989, p. 116.
7. Ibid.
8. FCC Regulations, Part 22, Subpart K, "Domestic Public Cellular Radio Telecommunications Service," 22.903, June 1981.

Chapter 11

Wireless Economics

Wireless Economics

Wireless telephone wholesale costs have dropped by approximately 20% per year over the past 7 to 10 years [1]. While the technology and mass production cost reductions for wireless telephones and systems are mature, new digital wireless telephones are more complex and typically do not have the large sales volume that promotes cost savings through mass production. System equipment costs for digital cellular equipment must also compete against a mature, competitive analog cellular (for example, AMPS, TACS, and NMT) equipment market, which already has the advantage of cost reductions due to large production runs.

The economic goal of a wireless network system is to effectively serve many customers at the lowest possible cost. The ability to serve customers is determined by the capacity of the wireless system. Two key factors are used in determining the capacity of the system: the size of the cell sites and the spectral efficiency of the radio channels. When using any radio access technology, system capacity is increased by the addition of smaller cell site coverage areas, which allows more radio channels to be reused in a geographic area. If the number of cell sites remains constant, the efficiency of the radio access technology (for example, the number of users that can share a single radio channel) determines the system capacity.

Wireless service providers usually strive to balance the system capacity with the needs of the customers. Running systems over their maximum capacity results in blocked calls to the customer, while running systems that have excess capacity results in the purchase of system equipment that is not required, which increases cost. Any wireless system (including AMPS) can be designed for very high capacity use through very small cell site coverage areas. The objective of the new digital wireless technologies is to achieve cost-effective service capacity, using techniques such as narrow radio channels, digital voice compression, or voice activity.

Purchasing and maintaining wireless system equipment is only a small portion of the cost of a wireless system. Administration, leased facilities, and tariffs may play significant roles in the success of cellular systems.

The wireless marketplace is undergoing a change. New service providers, such as Specialized Mobile Radio (SMR) and Personal Communications Service (PCS) providers, are entering the market. This is likely to increase wireless services competition. Sales and distribution channels may become clogged with a variety of wireless product offerings. Advanced wireless digital technologies offer a variety of new features that may increase the total potential market and help service providers to compete. These new features may offer added revenue and provide a way to convert customers to more efficient digital service. The same digital radio channels that provide voice services may offer advanced messaging and telemetry applications.

Wireless Telephone Costs

The initial digital cellular mobile telephones introduced in 1992 were approximately two times the cost of their analog equivalents, while Narrowband AMPS (NAMPS) mobile telephones were introduced at approximately the same cost as their analog equivalents. Time Division Multiple Access (TDMA) and Code Division Multiple Access (CDMA) digital mobile telephones reaching the market in 1996 were approximately two to three times the cost of comparable analog units. However, the average wholesale cost of analog mobile telephones (AMPS) has dropped from $307 in 1992 to approximately $104 in 1996 [2]. The high cost of digital mobile telephones is due to the following primary factors: development cost, production cost, patent royalty cost, marketing, post-sales support, and manufacturer profit.

Development Costs

Development costs are non-recurring costs that are required to research, design, test, and produce a new product. Unlike well-established FM technology, non-recurring engineering (NRE) development costs for digital wireless telephones can be high, due to the added complexity of digital design. Several companies have spent millions of dollars developing digital wireless products. Figure 11.1 shows how the non-recurring development cost per unit varies as the quantity of production varies from 20,000 to 100,000 units. Even small development costs become a significant challenge if the volume of production of the digital wireless telephones is low (below 20,000 units). At this small production volume, NRE costs will be a high percentage of the wholesale price.

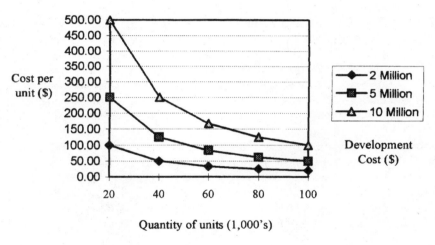

Figure 11.1, Mobile Telephone Development Cost

The introduction of a new technology presents many risks in terms of development costs. Some development costs that need to be considered include: market research; technical trials and evaluations; industrial, electrical, and software design; prototyping; product and FCC testing; creation of packaging, brochures, user and service manuals; marketing promotion; sales and customer service training; industry standards participation; unique test equipment development; plastics tooling; special production equipment fabrication; and overall project coordination.

When a new product is created which is not a revision of an existing product, a cost-effective design is sometimes compromised by using readily available components. Cost effective design is achieved by integrating multiple assemblies into a custom chip or hybrid assembly. Custom integrated circuit chip development is used to integrate many components into one low-cost part. Excluding the technology development effort, custom application specific integrated circuit (ASIC) development typically requires a development setup cost that ranges from $250,000 to $500,000. There may be more than one ASIC used in a digital wireless telephone.

Cost of Production

The cost to manufacture a mobile telephone includes the component parts (bill of materials), automated factory assembly equipment, and human labor. Digital wireless telephones are more complex than the older analog units. A digital mobile telephone is composed of a radio transceiver and a digital signal processing section. The primary hardware assemblies that affect the component cost for digital mobile telephones are digital signal processors (DSPs) and radio frequency assemblies. A single DSP, and several may be used, may cost between $15 and $28 [3]. The radio frequency (RF) assemblies used in digital wireless telephones often include more precise linear amplifiers and fast switching frequency synthesizers. These RF components cost approximately $15-30. Other components that are included in the production of a mobile telephone include printed circuit boards, integrated circuits and electronic components, radio frequency filters, connectors, plastic case, a display assembly, a keypad, a speaker and microphone, and an antenna assembly. In 1995, the bill of materials (parts) for a digital mobile telephone was approximately $100.

The assembly of wireless telephones requires a factory with automated assembly equipment. Each production line can cost between two and five million dollars. Typically, one production line can produce a maximum of 500-2,000 units per day (150,000-600,000 units per year). The number of units that can be produced per day depend on the speed of the automated component insertion machines and the number of components to be inserted. Typically, production lines are often shut down one day per week for routine maintenance and two weeks per year for major maintenance overhauls which leaves about 300 days per year for the manufacturing line to produce products. Between interest cost (10-15% per year) and depreciation (10-15% per year), the cost to own such equipment is approximately 25% per year. This results in a production facility overhead of $500 thousand to

$1.25 million per year for each production line. Figure 11.2 shows how the cost per unit drops dramatically from approximately $10-25 per unit to $1-3 per unit as volume increases from 50,000 units per year to 400,000 units per year.

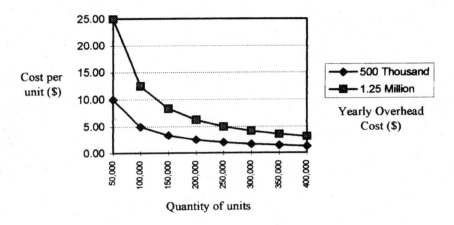

Figure 11.2, Factory Assembly Equipment Cost

While automated assembly is used in factories for the production of mobile telephones, there are some processes that require human assembly. Efficient assembly of a mobile telephone in a modern factory requires 1/2 to 1 hour of human labor. The amount of human labor is a combination of all workers involved with the plant, including administrative workers and plant managers. The average loaded cost of labor (wages, vacation, insurance) varies from approximately $20 to $40 per hour, which is based on the location of the factory and the average skill set of human labor. This results in a labor cost per unit that varies from $10-40. Because digital wireless telephones may have more parts to assemble due to the added complexity, the labor cost may increase.

Patent Royalty Cost

Another significant cost factor to be considered are patent royalties. Cellular technology was originally developed and patented by AT&T [4]. To the author's knowledge, AT&T has never requested a single royalty payment for this fundamental technology, which is not the case for the new digital standards. Several companies have disclosed that they believe they have some proprietary technology that is required to implement the IS-54 and IS-95 standards [5]. Prior to the creation of the GSM standard, companies agreed to freely license the GSM technology to other companies participating in the standards development process.

Some large manufacturing companies exchange the right to use their patented technology with other companies that have patented technology they want to use. For manufacturers who do not exchange patent rights, in 1995, the combined royalties for IS-54 digital cellular phones was up to $70 per unit, and the combined royalties for IS-95 digital cellular phones was up to $100 per unit. Patents from other companies that may be desirable or essential to implement the standard specifications may not have been discovered or disclosed.

Marketing Cost

The marketing cost, which is included in the wholesale cost of the wireless telephone, includes a direct sales staff, manufacturer's representatives, advertising, trade shows, and industry seminars.

Wireless telephone manufacturers typically dedicate a highly paid representative or agent for key customers. Much like the sales of other consumer electronics product, manufacturers employ several technical sales people to answer a variety of technical questions prior to the sale.

There are wireless telephone manufacturers that use independent distributors to sell their products. This practice is more prevalent for smaller, lesser known manufacturers who cannot afford to maintain a dedicated direct sales staff. These representatives typically receive up to four percent of the sales volume for their services.

Advertising programs used by the wireless telephone manufacturers involve broad promotion for brand-recognition and advertisements targeted for specific products. The budget for brand recognition advertising typically ranges from less then one percent to over four percent. Product-specific advertising is often performed through co-operative advertising. Co-operative advertising involves dedicating a percentage of the sales invoice (typically 2-4%) as an advertising allowance. When the advertising allowance meets the manufacturers requirements, the allowance is paid back to the customer. This approach allows customers to determine the best type of advertising for their specific markets. The typical advertising budget for mobile telephone manufacturers varies from approximately three to six percent.

Cellular system manufacturers exhibit at trade shows, typically 3 to 4 times per year. Trade show costs are high. Cellular mobile telephone manufacturers exhibiting at trade shows typically have large trade show booths, gifts, and theme entertainment. Medium to large hospitality parties at the trade shows are also common. Wireless telephone manufacturers often bring 15 to 40 sales and engineering experts to the trade shows to answer customer questions.

To help promote the industry and gain publicity, wireless telephone manufacturers participate in a variety of industry seminars and associations. The manufacturers typically have a few select employees who write for magazines and speak at industry seminars.

All of these costs and others result in an estimated marketing cost for mobile telephone manufacturers of 10 to 15 percent of the wholesale selling price.

Post Sales Support

The sale of cellular mobile telephones involves a variety of costs and services after the sale of the product, including warranty servicing, customer service, and training.

A customer service department is required for handling distributor and customer questions. Because the average customer for a wireless telephone is not technically trained in radio technology, the amount of non-technical questions can be significant. Fortunately, customer questions can typically be answered during normal business hours.

Distributors and retailers require training for product feature operation and servicing. The post sales support cost for wireless telephones is typically between 4 to 6 percent.

Manufacturer's Profit

Manufacturers must make a profit as an incentive for producing products. The amount of profits a manufacturer can make typically depends on the risk involved with the manufacturing of products. As a general rule, the higher the risk, the higher the profit margin.

The wireless telephone market in the early 1990's became very competitive due to the manufacturers' ability to reduce cost through mass production. To effectively compete, manufacturers had to invest in factories and technology, which increased the risk and the required profit margin. In 1995, the estimated gross profit in the wireless telephone manufacturing industry was 15 to 35 percent [6].

System Equipment Costs

The cost for wireless system equipment is due to the following primary factors: development cost, production cost, patent royalty cost, marketing, post sales support, and manufacturer profit.

Development Costs

These wireless network system equipment development costs are much higher than wireless telephone development costs. When a completely new technology is introduced, wireless network system development costs can exceed $500 million because the complexity of an entire wireless system is significantly greater than a mobile telephone and more testing and validation is required.

While the base station radios perform similar to a wireless telephone, the coordination of all the wireless telephones involves many additional electronic subsystems. Additional assemblies include communication controllers in the base station and switching center, scanning locating receivers, communication adapters, switching assemblies, and large databases to hold subscriber features and billing information. All of these assemblies require hardware and very complex software.

Unlike mobile telephones, when a cellular system develops a problem, the entire system can be affected. New hardware and features require extensive testing. Testing cellular systems can require thousands of hours of labor by highly skilled professionals. Introducing a new technology is much more complex than adding a single new feature.

Cost of Production

The physical hardware cost for wireless digital network system equipment should be more expensive than analog network system equipment due to the added technological complexity. However, the physical hardware cost for digital network system equipment may actually be less than older analog equipment. This is a result of a more competitive market and economies of scale.

The costs to manufacture a wireless network system include the component parts, automated factory equipment, and human labor. The number of wireless network system assemblies produced is much smaller than the number of wireless telephones. Setting up automated factory equipment is time consuming. For small production runs, much more human labor is used in the production of assemblies because setting up the automated assembly is not practical. The production of system equipment does involve a factory with automated assembly equipment for specific assemblies. However, because the number of units produced for system equipment is typically much smaller than wireless telephones, production lines used for cellular system equipment are often shared for the production of different assemblies, or remains idle for periods of time.

With over 150 countries developing and expanding their wireless systems, the demand for wireless system equipment is increasing exponentially. This increased demand allows for larger production runs, which reduce the average cost per unit. Large production runs also permit investment in cost-effective designs, such as using ASICs to replace several individual components.

The telecommunications industry has begun to standardize interfaces between system equipment. Different manufacturers can develop and sell parts of the wireless system without the cost burden of investing in the development of all the wireless network system components. It is possible in the Global System for Mobile Communications (GSM) system for one manufacturer to supply the cellular system base station control equipment and another manufacturer to supply the base station radios. The ability for a manufacturer to focus on the high demand components without a large development investment increases competition and reduces prices to the wireless system customer.

The maturity of digital technology is promoting cost reductions through the use of cost-effective equipment design and low-cost commercially available electronic components. In the early 1990s, many technical system equipment changes were required due to changes in radio specifications. Manufacturers had to modify their equipment based on field test results. For example, complex echo cancellors were required due to the long delay time associated with digital speech compression. Manufacturers typically did not invest in cost-effective custom designs because of the rapid changes. As the technology has matured, the investment in custom designs is possible with less risk. In the early 1990s, it was also unclear which digital technologies would become commercially viable, which limited the availability of standard components. Today, the success of digital systems has created a market of low-cost digital signal processors and RF components for digital cellular systems.

Like the assembly of wireless telephones, the assembly of system radio and switching equipment involves a factory with automated assembly equipment. The primary difference is the smaller production runs, multiple assemblies, and more complex assembly.

The number of equipment units that are produced is much smaller than the number of mobile telephones produced because each radio channel produced can serve 20-32 subscribers. The result is much smaller production runs for wireless network system equipment. While a single production line can produce a maximum of 500-2,000 assemblies per day [7], several different assemblies for radio base stations are required. A change in the production line from one assembly process to another can take several hours or several days. Wireless system radio equipment requires a variety of different connectors, bulky RF radio parts, and large equipment case assemblies. Due to the low-production volumes and many unique parts, it is not usually cost effective to use automatic assembly equipment. For unique parts, there are no standard automatic assembly units available. Because of this more complex assembly and the inability to automate many assembly steps, the amount of human labor is much higher than for wireless telephones.

Each automated production line can cost two to five million dollars. The number of units that can be produced per day varies depending on the speed of the automated component insertion machines, the number of components to be inserted, the number of different electronic assemblies per equipment, and the amount of time it takes to change/setup the production line for different assemblies. If we assume there are four electronic assemblies per base station radio equipment (for example, controller, RF section, baseband/diagnostic processing section, and power supply), the automated production cost for base station equipment should be over four times that of mobile telephones.

Figure 11.3 shows how the production cost per unit drops dramatically from approximately $400-1,000 per unit to $50-125 per unit as the volume of production increases from 5,000 units per year to 40,000 units per year. This chart assumes production cost is four times that of wireless telephones due to the added complexity and the use of multiple assemblies.

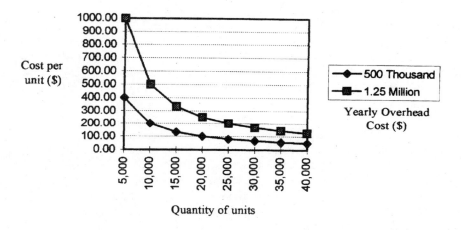

Figure 11.3, Factory Assembly Equipment Cost

While automated assembly is used in factories for the production of wireless telephones, there are some processes that require human assembly. Efficient assembly of base station units in a modern factory requires between 5 and 10 hours of human labor. The amount of human labor includes all types of workers from administrative workers to plant managers. The average loaded cost of labor (wages, vacation, insurance) varies from approximately $20-40 per hour, which is based on the location of the factory and average workers skill set. The resultant labor cost per unit varies from $100-400.

Patent Royalty Cost

There are only a few large manufacturers that produce wireless network system equipment due to the fact that the use of many different technologies is involved. Large manufacturers have a portfolio of patents that are commonly traded. Cross-licensing is common and tends to reduce the cost of patent rights. When patent licensing is required, the patent costs are sometimes based on the wholesale price of the assemblies in which the licensed technology is used.

Marketing Cost

The marketing costs that are included in the wholesale cost of wireless system equipment includes a direct sales staff, sales engineers, advertising, trade shows, and industry seminars. Wireless system manufacturers often dedicate several highly paid representatives for key customers. Wireless system sales are much more technical than the sale of wireless telephones. Manufacturers employ several people to answer a variety of technical questions prior to the sale.

Advertising used by the cellular system equipment manufacturers involves broad promotion for brand recognition and advertisements targeted for specific products. The budget for brand recognition advertising is typically small and is targeted to specific communication channels because the sale of wireless system equipment involves only a small group of people who typically work for a wireless service provider. Product-specific advertising is also limited to industry specific trade journals. Much of the advertising promotion of wireless system equipment occurs at trade shows, industry associations, and direct client presentations. The advertising budget for wireless system equipment manufacturers is typically less than two percent.

Wireless system manufacturers exhibit at trade shows typically 3 to 4 times per year. The trade show costs are sometimes much higher than the trade show costs for wireless telephone manufacturers. Wireless system equipment manufacturers exhibiting at trade shows often have large hospitality parties that sometimes entertain thousands of people. Wireless system manufacturers often bring 60-100 sales and engineering experts to the trade shows to answer customer questions.

To help promote the industry and gain publicity, wireless system manufacturers participate in many industry seminars and associations. These manufacturers use trained experts to present at industry seminars.

All of these costs and others result in an estimated marketing cost for system equipment manufacturers of approximately 8-10% of the wholesale selling price.

Post-Sales Support

The sale of wireless systems involves a variety of costs and services after the sale of the product. This includes warranty servicing, customer service, and training. A 24-hour customer service department is required for handling customer questions. Customers require a significant amount of training for product operation and maintenance after a system is sold and installed. The post-sales support costs for wireless system equipment is typically 3-5%.

Manufacturer Profit

Standardization of systems and components, particularly GSM, has led to a rapid drop in the wholesale price of system equipment. While the increased product volume of wireless system equipment has resulted in decreased manufacturing costs, the gross profit margin for wireless system equipment has decreased. The estimated gross profit in the wireless telephone manufacturing industry is 10-15% [8].

Network Capital Costs

The wireless service provider's investment in network equipment includes cell sites, base station radio equipment, switching centers, and network databases. One of the primary objectives of the new technologies was to decrease the network cost per customer which was made possible because the new technologies can serve more customers with less physical equipment.

In theory, existing analog cellular technology can serve almost an unlimited number of subscribers in a designated area by replacing large cell site areas with many Microcells (small cell coverage areas). However, expanding the current analog systems in this way increases the average capital cost per subscriber due to the added cost of increasing the number of small cells and interconnection lines to replace a single large cell. For example, when a cell site with a 15 km radius is replaced by cell sites with a 1/2 km radius, it will take over 700 small cells to cover the same area.

One of the reasons that digital cellular technologies were developed was to allow for cost-effective capacity expansion. Cost-effective capacity expansion results when existing cell sites can offer more communication channels, which allows more customers to be served by the same cell site. As systems based on such new technologies expand, the average cost per subscriber decreases.

Cell Site

The cell site is composed of a radio tower, antennas, a building, radio channels, system controllers, and a backup power supply. The cell site radio tower is typically 100-300 feet tall. The cost ranges between 30 thousand and 300 thousand dollars. While some of the largest towers can cost $300 thousand, an average cost of $70 thousand is typical because, as systems expand, smaller towers can be used.

Many cell sites can be located on a very small area of land. Land is either purchased or leased. In some cases, existing tower space can be leased for $500-$1,000 per month. If the land is purchased, the esti-mated cost of the land is approximately $100 thousand.

A building on the cell site property is required to store the cell site radio equipment. This building must be bullet proof, have climate con-trol, and various other non-standard options. The estimated building cost is $40 thousand.

Cell sites are not usually located where high-speed telephone commu-nication lines are available. Typically, it is necessary to install a T1 or E1 communications line to the cell-site which is leased from the local phone company. If a microwave link is used in place of a leased com-munication line, the communications line installation cost will be applied to the installation of the microwave antenna. The estimated cost of installing a T1 or E1 communications line is approximately five thousand dollars.

The land where the cell site is to be located must be cleared, founda-tions poured, fencing installed, building and tower installed. A con-struction cost of $50 thousand is estimated. Table 11.1 shows the esti-mated cost for a typical cell site without the radio equipment.

In addition to the tower and building cost, radio equipment must be purchased. The cost of the radio equipment usually varies based on technology and the number of customers that use the system.

After the total investment of each cell site is determined, the cell site capital cost per customer can be determined by dividing the total cell site cost by the number of subscribers that will share the resource (cell

Item	Cost x $1,000's
Radio Tower	$70
Building	$40
Land	$100
Install Communication Line	$5
Construction	$50
Antennas	$10
Backup Power Supply	$10
Total	$285

Table 11.1, Estimated Cell Site Capital Cost Without Radio Equipment

site). Because not everyone uses every radio channel at the same time, cellular systems typically add 20-32 subscribers per voice channel. For the different cellular technologies, each RF channel can supply one or more voice paths. The number of voice paths per radio channel is multiplied by the number of radio channels per cell site. If 20 subscribers are added to the system for each voice path, the average number of subscribers per cell site varies from 1,000 to 19,200.

Since a single RF channel can serve many users, either the number of base station RF channel equipment assemblies is decreased, or the system capacity is increased. For example, a cell site that has 24 analog RF channels can convert 12 channels to full-rate TDMA (three users to one 30 kHz RF channel) and support 48 voice channels. With the introduction of the half rate TDMA channels (six users to one 30 kHz RF channel), the same cell site with 24 analog RF channels could convert 12 analog RF channels to half rate TDMA and support 84 voice channels. A future consideration is the acceptance of digital speech interpolation (DSI) to allow the transmission of speech data only when a person is actively speaking. DSI doubles the capacity (assuming 50% voice activity) by allowing the same 24-channel cell site to convert 12 channels to digital and have up to 156 voice channels.

Table 11.2 shows a sample of system equipment costs as digital technology evolves. In column 1, we see analog FM technology that supports one voice channel per carrier. Because each subscriber will only access the cellular system for a few minutes each day, approximately 20 subscribers (customers) can share the service of a single radio channel. NAMPS radio channels also provide one voice channel per carrier. The advantage of NAMPS is that because the radio channels are one-third the width of AMPS channels, more channels can be placed in each cell site. This increases the total number of subscribers per cell site to approximately 3,060. For the IS-54/IS-136 system, each RF channel supports up to three users (full rate), so each RF channel cost is shared by 60 subscribers. The evolution of IS-136 will include a half rate system where six users can share each radio channel. The ETDMA system (discussed in chapter 11), uses DSI to allow up to 10 users to share a single IS-136 radio channel. For IS-95 CDMA where each RF channel supports approximately 20 users, each RF channel cost is shared by up to 400 subscribers. For GSM, each RF channel can provide service to 8 simultaneous users which allows up to 160 subscribers to share the radio channel cost. The GSM phase 2 specification (discussed in chapter 7) allows up to 16 users to share each radio channel (half rate). While it is not suggested that all of the available RF channels can be converted to digital in the near future, the following table shows target costs that project the reasonable costs of a cellular system equipment in the future.

	AMPS/ ETACS	NAMPS	IS-54/IS-136 TDMA	IS-95 CDMA	GSM
Cost per RF Radio Channel	10,000	10,000	15,000	45,000	15,000
Number of Radio Channels per Cell Site (3 sector)	51	153	51	24	30
Total Radio Channel Cost	510,000	1,530,000	765,000	1,080,000	450,000
Tower and Building Cost	285,000	285,000	285,000	285,000	285,000
Total Cell Site Cost	795,000	1,815,000	1,050,000	1,365,000	735,000
Number of Voice Paths per Radio Channel	1	1	3	20(est)	8
Number of Voice Paths per Cell Site	51	153	153	480	240
Number of Subscribers per Voice Channel	20	20	20	20	20
Number of Subscribers per Cell Site	1020	3060	3060	9600	4800
Cell Site Capital Cost per Subscriber	$779	$593	$343	$142	$153

Table 11.2, Cell Site Capital Cost per Subscriber

The multiplexing of several radio channels through one RF equipment reduces the number of required RF equipment assemblies, power consumption, and system cooling requirements. Multiplexing in this way typically reduces cell site size and backup power supply (generator and battery) requirements, and ultimately, cost.

Mobile Switching Center

Cell sites must be connected to an intelligent switching system (called the "switch"). An estimate of $25 per subscriber is used for the cellular switch equipment and its accessories,based on one Mobile Switching Center (MSC) costing $2.5 million that can serve up to 100,000 customers.

The switching center must be located in a long-term location (10-20 years) near a local exchange carrier (LEC) public switched telephone network (PSTN) central office connection. The building contains the switching and communication equipment. Commonly, a customer database called the home location register (HLR) is located in the switching center. The switching center software and associated cellular system equipment typically contain basic software that allows normal mobile telephone operation (place and receive calls). Special software upgrades that allow advanced services are available at additional cost.

Operational Costs

The costs of operating a cellular system includes leasing and maintaining communication lines, local and long distance tariffs, billing, administration (staffing), maintenance, and cellular fraud. The operational cost benefits of installing digital equipment includes a reduction in the total number of leased communication lines, a reduction in the number of cell sites, a reduction in maintenance costs, and a reduction of fraud due to advanced authentication procedures.

Leasing and Maintaining Communications Lines

Cell sites must be connected by leased communication lines between radio towers, or by installing and maintaining microwave links between them. The typical cost for leasing a 24-channel line between cell sites in the US in 1992 was $750/month [9]. Microwave radio equipment can cost from $20 thousand to $100 thousand.

The number of subscribers that can share the cost of a communication line (loading of the line) varies with the type of service. For cellular-like subscribers who typically use the phone for two minutes per day, approximately 480 customers can share a T1 (20 subscriber per voice path x 24 voice paths per communication line) or 600 customers per

E1 (20 subscribers per voice path x 30 voice paths per communication line). For residential-type service, where customers use the phone for approximately 30 minutes per day, approximately 120 customers can be loaded onto a T1 or 150 per E1. For office customers who use the phone for approximately 60 minutes per day, approximately 60 customers can be loaded onto a T1 or 75 for E1.

The monthly cost per subscriber is determined by dividing the monthly cost by the total number of subscribers. Table 11.3 shows the estimated monthly cost for interconnection charges. The estimated monthly cost is based on 100% use of the communication lines. If the communication lines are not used fully (it is rare that communication lines are used at full capacity), the average cost per line increases.

Digital signal processing for all the proposed technologies allow for a reduction in the number of required communications links through the use of sub-rate multiplexing. Sub-rate multiplexing allows several users to share each 64 thousand bit per second (kb/s) communications (DS0/PCM) channel. This is possible because digital cellular voice information is compressed into a form much smaller than existing communication channels. If 8 kb/s speech information is sub-rate multiplexed, up to 8 voice channels can be shared on a single 64 kb/s channel, which can reduce the cost of leased lines significantly.

Service	Line Cost per Month	No Chan	Load	Total Cost per Month
Cellular	750	24	20	1.56
LEC (residential)	750	24	5	6.25
Office	750	24	2.5	12.5

Table 11.3, Monthly Communications Line Cost

Local and Long Distance Tariffs

Telephone calls in cellular systems are often connected to other local and long distance telephone networks. When cellular systems are routed to existing landline telephone customers, they are typically connected through the wired telephone network (usually via the LEC). The local telephone company typically charges a small monthly fee and several cents per minute (approximately 3 cents per minute) for each line connected to the cellular carrier. Because each cellular subscriber uses their mobile telephone for only a few minutes per day, the cellular service provider can use a single connection (telephone line) to the PSTN to service hundreds of subscribers.

In the United States and other countries that have separate long-distance service providers, when long distance service is provided through a local telephone company (LEC), a tariff is paid from its cellular service provider to the LEC. These tariffs can be up to 45% of the per minute charges. In 1993, MCI paid $5.3 billion in the United States for local exchange tariffs out of the $11 billion in revenues [10]. Due to government regulations limiting the bundling of local and long-distance service, it is necessary for some cellular service providers to separate their local and long-distance service. In the future, regulations may permit cellular carriers to bypass the LEC and save these tariffs.

Billing Services

Cellular systems exist to provide services and collect revenue for those services. Billing involves gathering and distributing billing information, organizing the information, and invoicing the customer.

As customers initiate calls or use services, records are created. These records may be provided in the customer's home system or a visited system. Each billing record contains details of each billable call, including who initiated the call, where the call was initiated, the time and length of the call, and how the call was terminated. Each call record contains approximately 100-200 bytes of information [11]. If the calls and services are provided in the home system, the billing records can be stored in the company's own database. If they are provided in a visited system, the billing information must be transferred back to the home system. In the mid 1980s, cellular systems were not interconnected, which required the use of a clearing house. The clearing house was used to accumulate and balance charges between different cellular service providers. Billing records from cellular systems were typically transferred via tape in a standard automatic message accounting (AMA) format.

In the 1990s, standard intersystem connection allowed billing records to be transferred automatically, which made advanced billing services such as advice of charging or debit account billing possible. Advice of charging provides the customer with an indication of the billing costs. Debit account billing allows a cellular service provider to accept a prepaid amount from a customer (perhaps a customer that has a poor credit rating) and decrease his or her account balance as calls are processed.

With the introduction of advanced services, billing issues continue to become more complicated. The service cost may vary between different systems. To overcome this difficulty, some service providers have agreed to bill customers at the billing rate established in their home system. In the United States, Cellular Digital Packet Data (CDPD) services are billed at the home subscribers rate [12].

Each month, billing records must be totaled and printed for customer invoicing, invoices mailed, and checks received and posted. The estimated cost for billing services is $1 per month. This billing cost includes routing and summarizing billing information, printing the bill, and the cost of mailing. To help offset the cost of billing, some wireless service providers have started to bundle advertising literature from other companies along with the invoice. To expedite the collection, some wireless service providers offer direct billing to bank accounts or charge cards.

Operations, Administration, and Maintenance

Running a wireless service company requires people with many different skill sets. Staffing requirements include executives, managers, engineers, sales, customer service, technicians, marketing, legal, finance, administrative, and other personnel to support vital business functions. The present staffing for local telephone companies is approximately 35 employees for each 10,000 customers. Wireless telephone companies have approximately 20-25 employees per 10,000 customers, and paging companies employ approximately 10 employees per 10,000 customers [13]. If we assume a loaded cost (salary, expenses, benefits, and facility costs) of $40,000 per employee, this results in a cost of $3.33 to $11.66 per month per customer ($40,000 x (10-35 employees) /10,000 customers/12 months).

Maintaining a wireless system requires calibration, repair, and testing. System growth involves frequency planning, testing, and repair. When a frequency plan is changed, manual tuning of 20-30 radio channels per cell site is required. Large urban systems may have over 400 cell sites. Because it is desirable to perform frequency re-tuning in the same evening, some wireless service providers borrow technicians from neighboring systems when their small staff cannot perform this task. All of the technologies offer potential automatic RF frequency planning (see MRI, MAHO, or CDMA frequency planning in chapters 4-6).

During the year, the geographic characteristics of a wireless system change (for example, leaves fall off trees). This changes the radio coverage areas. Typically, a wireless carrier tests the radio signal strength in its entire system at least four times per year. Testing involves having a team of technicians drive throughout the system and record the signal strength.

Maintenance and repair of wireless systems is critical to the revenue of a wireless system. In large systems, a staff of qualified technicians are hired to perform routine testing. Smaller wireless systems often have an agreement with another wireless service provider or a system manufacturer to provide these technicians when needed. Wireless systems have automatic diagnostic capabilities to detect when a piece of equipment fails. Most wireless systems have an automatic backup system, which can provide service until the defective assembly is replaced.

Land and Site Leasing

In rural areas, exact locations for cell site towers are not required. The result is that land-leasing is not a significant problem. In urban areas, and as systems mature, more exact locations for cell sites are required. This results in increased land-leasing costs. By using a more efficient RF technology, one cell site can be used to serve more channels, which limits the total number of required cell sites.

Land leasing is typically a long-term lease for a very small portion of land (40-200 square meters) for approximately 20 years or longer. The cost of leasing land is dependent on location. Premium site locations such as sites on key buildings or in tunnels can exceed the gross revenue potential of the cell site.

Another leasing option involves leasing space on an existing radio tower. Site leasing on an existing tower is approximately $500/month. Site leasing eliminates the requirement of a building and maintaining a radio tower.

Cellular Fraud

It is estimated that cellular fraud in the United States during 1994 was in excess of $460 million [14]. This was approximately three percent of the $14 billion yearly gross revenue received [15]. Each of the new cellular standards has an advanced authentication capability, which limits the ability to gain fraudulent access to the cellular network.

The type of cellular fraud seen has changed over the years. Initially, cellular fraud was subscription fraud. With advances in technology available to distributors of modified equipment, cellular fraud has changed to various types of access fraud. Access fraud is the unauthorized use of cellular service by changing or manipulating the electronic identification information stored inside of a mobile telephone.

Subscription fraud occurs when a cellular phone is registered by a person using false identification. After the required documentation is provided, unlimited service is often provided by the cellular service provider. When the bill is unpaid, the fraudulent activity is determined and service is disconnected. Some cellular service providers now require valid identification and credit checks prior to service activation, which reduces subscription fraud.

In the mid 1980s, roamer fraud was possible. Roamer fraud occurs when a mobile telephone is programmed with an unauthorized telephone number and home system identifier so that it looks like a visiting customer. Because some of the cellular systems in the mid 1980s were not directly connected to each other, these systems could not immediately validate the visiting customer. In the 1990s, intersystem connection provides validation of the phone number, which limits (or eliminates) roamer fraud.

To allow fraudulent access to valid cellular customer accounts, criminals began to modify the Electronic Serial Number (ESN) of mobile telephones to match a valid subscriber's ESN. This duplication of subscriber information is called "cloning." To enable the cloning process, mobile telephones have to be modified to accept a new ESN and a valid ESN must be acquired. ESNs are typically stored in a hard-to-get (secure) memory area of a mobile telephone. It typically takes a very technical person to be able to override the security system in the mobile telephone to modify the ESN. Obtaining valid ESNs is possible by reading the ESN on the label or by using a commercially available test set that commands the mobile telephone to send its ESN. Cellular carriers are able to detect changing patterns of use when a cloned phone has been created. If the subscriber's billing account jumps dramatically, the cellular customer can be contacted to have their phone number changed. The original ESN is then marked invalid.

To overcome the barriers of ESNs becoming invalid, criminals designed their systems to change the ESN during each call. This changing ESN process is called "tumbling." Tumbling uses valid ESNs that are pre-stored in the mobile telephone or captured from the radio channels during a valid subscriber's regular access. These valid ESNs are used by the modified mobile telephones either during each call or randomly each time they send or receive calls.

Most of these methods can be detected and blocked by the use of Mobile telephone authentication information. Authentication is a process of using previously stored information to process keys that are transferred via the radio channel. Because the secret information is processed to create a key, the security information is not transferred on the radio channel. The secret information stored in the wireless telephone can be changed at random either by manual entry, by the customer, or by a command received from the wireless system. Authentication is supported in all of the digital technology specifications.

Marketing Considerations

The cellular service providers commissioned the development of digital cellular technology to allow cost-effective system capacity expansion and to provide more revenue from advanced services. To obtain this cost savings and new feature service revenue, it is necessary to have a percentage of subscribers who have digital-capable equipment that can access the advanced technology. Digital cellular marketing programs have focused on converting existing subscribers to digital service, enticing new customers to purchase digital over analog, and targeting new customers for advanced services. The key marketing factors that may determine the success of digital cellular includes the type of new services, system cost savings, pricing of voice and data service, mobile telephone cost, consumer confidence, new features, retrofitting existing customer equipment, availability of equipment, and distribution channels.

Service Revenue Potential

At the end of 1995, there were over 74 million cellular telephone customers in the world. In the United States, there were over 24 million customers with an average monthly service bill of $58.65 [16]. While the average cellular telephone bill is not much higher than the average wired residential telephone bill, the amount of usage for a cellular telephone is approximately one tenth of residential usage.

The average cellular telephone bill in the US has declined eight to nine percent each year over the last five years [17]. The average charge per minute has not decreased much; however, the amount of usage has because new customers entering into the market are consumers that do not use their wireless telephone very much.

The number of subscribers on the US cellular systems have been increasing by over 45% per year over the past five years [18]. Some of this growth is due to new system service areas and the decreased price of wireless telephones. It is not reasonable to assume a continued 45% yearly growth period as the total number of wireless telephones would exceed the total world population in only 12 years. Figure 11.4 shows the combined wireless subscriber growth in Europe, Asia, and the Americas over the past five years.

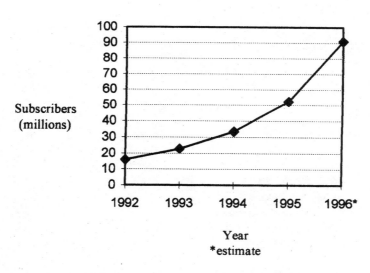

Figure 11.4, World Cellular Subscriber Growth
Source: US Dept of Commerce

The main revenue for wireless service providers is derived from providing telecommunications service. In 1995, a majority of the service revenue came from voice services. Digital wireless systems provide for increased service revenue which comes from a variety of sources such as advanced services and system cost reduction.

System Cost to the Service Provider

One of the advantages of digital service is to allow more customers to share the same system equipment. Cellular system equipment costs account for approximately 10-15 % of the service provider's revenue. Digital wireless systems can offer a reduction of approximately 60 % of system equipment cost per customer. Some of the advanced features of digital wireless (such as authentication to reduce fraud) also provide for reductions in operations, administration, and maintenance (OA&M) costs. These system cost reductions offered by digital wireless technology may be necessary to allow existing cellular service providers to effectively compete against other wireless service providers. As more companies (such as PCS companies) begin to offer cellular-like services, the potential for a surplus of voice channel time exists.

Voice Service Cost to the Consumer

Over the past few years, the average cost of airtime usage to a cellular subscriber has not changed very much. To help attract subscribers to digital service, some cellular carriers have offered discounted airtime plans to high usage customers. This discount provides a significant incentive to the high usage customers. By shifting a small portion of these customers to digital service, the loading on the older analog radio channels is reduced.

Data Service Cost to the Consumer

There are two types of data services that are available to customers: continuous (called "circuit switched data") or brief packets (called "packet switched data"). Typically, continuous data transmission is charged at the same rate as voice transmission. Packet data transmission is often charged by the packet or by the total amount of data that has been transferred.

Numeric paging systems can serve 40,000-100,000 paging customers per radio channel [19], while cellular can serve only 20-32 customers per radio channel. In the United States during 1994, the average revenue for a paging customer was approximately $10 per month and the average revenue for a cellular customer was $56.21 per month [20]. If paging is viewed as packet data service, the total revenue potential per radio channel is significantly higher than voice. CDPD is a packet data service that is offered on cellular. Using the pricing of 7 cents to 23 cents per kilobyte of data [21], the average cost to the customer is $3 to $10 per minute.

Wireless Telephone (Mobile Phone) Cost to the Consumer

In 1984-85, cellular mobile telephone prices varied from $2000-$2500 [22]. By 1991, you could get a free cellular phone with the purchase of a hamburger at selected Big Boy Restaurants in the United States [23]. One of the primary reasons for the continued penetration of the cellular market is the declining terminal equipment costs and stable airtime charges [24]. In 1995, the wholesale price of digital wireless telephones was higher than analog wireless telephones. The wireless service providers often subsidize the sale of end-user telephones, which reduces revenue. Wireless service providers do not usually anticipate revenues from the sale of telecommunications equipment.

Many of the service providers are not concerned with the profit on wireless telephone equipment because their goal is to gain monthly service revenue. In the Americas, it is common for the cellular service provider to subsidize the sale of cellular subscriber equipment. It is not as common for such subsidies in Europe. To help introduce digital wireless telephones into the marketplace, subsidies may be higher for digital wireless telephone equipment than analog wireless telephones. Some wireless telephone manufacturers use groups of serial numbers to identify mobile telephones that have digital capability.

The wireless telephone service activation subsidy and the type of distribution channels used usually affects the retail price paid by the consumer. Figure 11.5 shows the wholesale mobile telephone cost in the United States over the past seven years [25].

Consumer Confidence

To effectively deploy a new technology, the consumer must have confidence that the technology will endure. The next generation of wire-

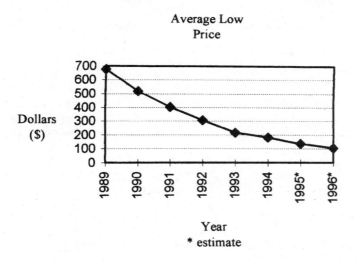

Figure 11.5, Wholesale Mobile Telephone Cost in the United States
Source: Herschel Shosteck Associates, Wheaton, Maryland, USA

less technology was required to be compatible the old and new systems to maintain this consumer confidence [26]. In some parts of the world, there are multiple digital technologies. These technologies offer different features, services, and radio coverage areas. Consumers must be willing to choose a technology that may not exist in future years.

New Features

Customers purchase mobile telephones and wireless service based on their own value system, which estimates the benefits they will receive. New features provide for new benefits to the consumer. These features can be used for product differentiation and to increase service revenue. New features available for digital cellular include longer battery life, caller ID, and message services. These new features are sometimes used to persuade customers to convert to digital or to pay extra for these new advanced services.

While the first FM cellular telephone weighed over 80 pounds and required almost all available trunk space, the first digital wireless telephones were only slightly larger than their analog predecessors. The size of digital wireless telephones continues to be reduced as production volumes allow for custom ASIC development, which integrates the analog and digital processing sections. There is no reason that the digital wireless telephones cannot approach the same size as analog cellular telephones [27].

New features allow for different types of customers. With advanced data capabilities, cellular service may focus its products, services, and applications on non-human applications (many non-human applications were discussed in Chapter 1). Without a change in focus, wireless service providers could become limited to the voice services market which may eventually reach saturation.

Retrofitting

Retrofitting involves mobile telephone hardware conversion and equipment exchange programs. Due to the limited capability of analog mobile telephones, they cannot be easily converted to digital capability. While it may be possible to replace the analog mobile telephone (mobile car phone) transceiver with a digital transceiver, the retrofit market is considered small.

One of the easiest ways to reduce system blockage (system busy signals) and quickly increase system efficiency is by retrofitting high-usage subscribers [28]. Retrofitting high-usage subscribers can increase the total demand for digital wireless telephones over several years. There have been some retrofit incentive programs created that allow a customer to exchange their analog mobile telephones for a digital mobile telephone. These programs have required the customer to exchange or designate another customer to receive the old analog unit prior to receiving a digital unit. In return for this exchange, the digital mobile telephone would be provided to the customer at reduced cost or no cost.

Higher commissions have been paid for digital mobile telephones to entice new customers to select digital mobile telephones over their analog equivalents. Some manufacturers group ESN numbers to allow the cellular service provider to know if the mobile telephone has digital capability that allows the carrier to differentiate activation commissions between analog and digital mobile telephones.

Churn

Churn is the percentage of customers that discontinue cellular service. Churn is usually expressed as a percentage of the existing customers that disconnect over a one-month period. Churn is often the result of natural migration (customers relocating) and switching to other service providers. Because some wireless service providers contribute an activation commission incentive to help reduce the sale price of the phone, which can be a significant cost if the churn rate is high. The percentage of churn in North America over the last five years has remained relatively constant at approximately 2.8% per month [29].

Wireless carriers and their agents have gone to various lengths to reduce churn. This includes programming in lockout-codes to lengthy service agreements. Programming lockout-codes (typically 4-8 digits) can be entered by the programmer into some mobile telephones to keep the mobile telephone from being re-programmed by another wireless service provider. Wireless service providers sometimes require the customer to sign a service agreement, which typically requires them to maintain service for a minimum of one year. These service agreements have a penalty fee in the event the customer disconnects service before the end of the one year period.

Availability of Equipment

The design and production of digital wireless equipment requires significant investment by a manufacturer. Digital telephones are more complex, and portable digital wireless telephones are typically larger and more expensive. There are fewer manufacturers that offer digital wireless equipment which has reduced competition and maintained higher prices than their analog equivalents.

The first cellular portable, introduced in 1984 by Motorola, weighed 30 oz and had approximately 30 minutes of talk time. In early August 1991, Motorola released the Micro-Tac Light which weighed 7.7 oz [30] and had 45 minutes of talk time. The additional DSP circuitry that is required for a TDMA dual mode portable weighs less than 3oz using commercially available DSPs [31], which allowed the introduction of a 10 oz IS-54 TDMA dual mode portable phone in 1993. In 1995, a 6.9 oz digital cellular phone was introduced for GSM and IS-54 TDMA. In 1996, the minimum size and weight of digital cellular handportable mobile telephones were close to analog cellular handportables.

Distribution and Retail Channels

Products produced by manufacturers are distributed to consumers via several distribution and retail channels. The types of distribution channels include: wholesalers, specialty stores, retail stores, power retailers, discount stores, and direct sales.

Wholesalers purchase large shipments from manufacturers and typically ship small quantities to retailers. Wholesalers will usually specialize in a particular product groups, such as pagers and cellular phones.

Specialty retailers are stores that focus on a particular product category such as a cellular phone outlet. Specialty retailers know their products well and are able to educate the consumer on services and benefits. These retailers usually get an added premium via a higher sales price for this service.

Retail stores provide a convenient place for the consumers to view products and make purchases. Retailers often sell a wide variety of products, but a salesperson may not have an expert's understanding of or be willing to dedicate the time to explain the features and cellular service options. In the early 1990s, mass retailers began selling cellular phones. Mass retailers sell a very wide variety of products at a low profit margin. The ability to sell at low cost is made possible by limiting the amount of sales time spent providing customer education on new features.

Power retailers specialize in a particular product group such as consumer electronics. Power retailers look for particular product features that match their target market. Power retailers carry only a select group of products. Because there are only a few products for the consumer to select from, the demand for a single product is higher than if several different models were on display. This tends to increase sales for a particular product, which leads to larger quantity purchases and discounts for the power retailer.

Discount stores sell products at a lower cost than their competitors. They achieve this by providing a lower level of customer service. Because there is limited customer service, many of the wireless telephones sold in discount stores are pre-programmed or are debit (prepaid) units.

Some wireless service providers employ a direct sales staff to service large customers. These direct sales experts can offer specialty service pricing programs. The sales staff may be well-trained and typically sell at the customer's location.

Distribution channels are commonly involved in the activation process. The application for cellular service can take a few minutes to several hours, and programming of the mobile telephone for the consumer must be performed. With several new cellular service providers such as Personal Communications Service (PCS) and Personal Communications Network (PCN), the complexity of sale and activation can increase. To simplify this process, some wireless service providers have streamlined the application process and pre-programmed the mobile telephones to make the process more like a typical retail sale.

Because there are several new technologies and different models of phones, access to particular distribution channels is limited. In 1995, each retailer carried (stocked) approximately 3-4 different manufacturer's brands. Retailers can only dedicate a limited amount of shelf space for each product or service, which may limit the introduction of new digital products and services into the marketplace.

References:

1. Herschel Shosteck, "The Retail Market of Cellular Telephones," Herschel Shosteck Associates, Wheaton, Maryland, USA, 1st Quarter, 1996.
2. Ibid.
3. Cellular Integration Magazine, "Tech-niques," Argus Business , January 1996.
4. United States Patent 3,663,762, "Mobile Communication System," May 1992.
5. Letter to the TIA voting members from Eric Schimmel, 21 November, 1990.
6. Schlesinger, Jeffrey, personal interview, UBS Securities, New York, February 12, 1996.
7. Glen, Bob, personal interview, Sparton Electronics, Raleigh NC, January 1996.
8. Schlesinger, Jeffrey, personal interview, UBS Securities, New York, February 12, 1996.
9. EMCI, "Digital Cellular, Economics and Comparative Analysis," Washington DC, 1993.
10. Paine Webber Conference, New York City, 1993.
11. D.M. Balston, D.M., "Cellular Radio Systems," Artech House, MA, 1993, p. 223.
12. Wireless Internet conference, Council for Entrepreneurial Development, Raleigh, NC USA, September 1995.
13. Hamilton, Elliott, personal Interview, EMCI Consulting, Washington DC, 25 February 1996.
14. North Carolina Electronics Information Technologies Association, "Cellular Fraud," Raleigh, NC, November, 1995.
15. Cellular Telecommunications Industry Association, "Wireless Factbook," Washington DC, Spring 1995.
16. Cellular Telecommunications Industry Association, Mid-Year Results, Washington

DC, 1995.

17. Cellular Telecommunications Industry Association, "Wireless Factbook," Washington DC, Spring 1995.

18. Ibid.

19. Horsman, Lee, personal interview, Allen Telecom, February, 1996.

20. Cellular Telecommunications Industry Association, "Wireless Factbook," Washington DC, Spring 1995.

21. Sprint, Wireless Data Symposium, Raleigh, NC, December, 1995.

22. Calhoun, George, "Digital Cellular Radio," Artech House, MA, 1988, p. 69.

23. Crump, Stuart F., Cellular Sales and Marketing, Creative Communications Inc., Vol. 5, No. 8, Washington DC, August 1991, p. 2.

24. Chan, Hilbert, "The Transition to Digital Cellular," IEEE 1990 Vehicular Technology Conference, 1990, p. 191.

25. Shosteck, Herschel, "The Retail Market of Cellular Telephones," Herschel Shosteck Associates, Wheaton, Maryland, USA, 1st Quarter, 1996.

26. Stupka, John, CTIA Winter Exposition, "Technology Update," Reno Nevada, 1990.

27. Telecommunications Industries Association, Transition to Digital Symposium, "New Services and Capabilities," TIA, Orlando, FL, Sep 1991.

28. Ibid.

29. Hamilton, Elliott, personal interview, EMCI, Washington DC, 6 March 1996.

30. Crump, Stuart F., Cellular Sales and Marketing, Creative Communications Inc., Vol5, No8, Washington DC, August 1991, p. 1.

31. Telecommunications Industries Association Transition to Digital Symposium, "New Services and Capabilities," TIA, Orlando, FL, Sep 1991.

Chapter 12
Future Wireless Technologies

Future Wireless Technologies

As the wireless technology continues to evolve, new systems are being developed, tested, and marketed each year. There have been new wireless technologies identified that will better serve customer needs. These technologies may efficiently replace wired telephone systems or offer cost efficient data message transmission. The systems to make this possible include Integrated Dispatch Enhanced Network (iDEN), Cellular Digital Packet Data (CDPD), Extended Time Division Multiple Access (E-TDMA), two-way voice paging, Spatial Division Multiple Access (SDMA), and satellite cellular systems.

iDEN^{TM}

Integrated Dispatch Enhanced Network (iDEN^{TM}) is a digital only TDMA radio system, formerly called Motorola Integrated Radio System (MIRS). iDEN^{TM} is primarily being used for 800 MHz Specialized Mobile Radio (SMR) channels which range from 806 MHz to 821 MHz and 851 MHz to 866 MHz. Originally, MIRS technology was developed by Motorola for a company called Fleet Call, now known as Nextel Since it began purchasing or partnering with other SMR companies in 1988, Nextel and its partners had radio coverage in almost all of North America in 1995.

The iDEN^{TM} technology is a TDMA system which divides the frequency band into 25 kHz radio channels, and then the radio channel is divided into frames which contain 6 slots. When used for cellular like service, three users can share each radio channel. Should a lower voice quality be acceptable, then a maximum of 6 users could share each radio channel. The iDEN radio channel uses 16 level quadrature amplitude modulation (QAM) to allow up to 64 kb/s of data on a single 25 kHz radio channel, and also has the capability to combine voice and data transmissions. The iDEN^{TM} technology may also be referred to as Digital Integrated Mobile Radio System (DIMRS).

In addition to the ability of the iDEN^{TM} system to operate similar to a cellular system, the iDEN^{TM} system also allows most of the original dispatch type services. This includes dispatch and messaging services along with direct two-way wireless communications between each handset.

CDPD

Cellular Digital Packet Data (CDPD) was created to address the need of subscribers to be able to rapidly transmit a small amount of data and not tie up a cellular radio channel for a long period of time. Cellular Digital Packet Data is a wireless digital packet transfer system that overlays on top of an existing cellular system. Because the CDPD system transfers packets of data only when they are needed, the CDPD system does not maintain a constant connection between the two users. This type of system is referred to as "connectionless" because there are no pre-determined time periods or dedicated resources for packet transmission.

Digital data can be sent on an existing cellular radio voice/traffic channel. Unfortunately, cellular radio channels were not designed for rapid initialization to send small amounts of data. CDPD has changed the way a Mobile telephone accesses the network to help minimize the setup and transfer time for each data transfer period.

The process of CDPD divides a subscriber's data information into small packets which contain their destination address, and the packets are then sent to their destination by the best path possible at the time of transfer. The travel time for each packet between the origination and destination may be different. This is because the packets of information are often sent on different routes due to the availability

Figure 12.1, CDPD Radio Transceiver
Source: Cincinnati Microwave

of communications paths. As the packets are received, they are reassembled in the proper order by using a packet index number decoded at the receiving end of the transmission. Figure 12.1 shows a CDPD radio transceiver.

The CDPD network and radio specifications were announced in 1992 by a consortium of cellular carriers. In 1995, there were over 12,000 cell sites in the United States which had CDPD capability [1]. The technology is based upon industry standards that are maintained by the CDPD forum, which is an organization of cellular carriers and others who are interested in promoting CDPD [2].

System Elements

While the names of the equipment assemblies are different in a CDPD system, the CDPD equipment assemblies perform similar functions as those found in a cellular system. A CDPD Mobile telephone is called a mobile end station (MES)and the base station is termed a mobile data base station (MDBS). The switch is now called a mobile data intermediate system (MD-IS). Also, the public switched telephone network (PSTN) is replaced with various packet and other data connection networks.

377

System Operation

The management of channels to coexist with the AMPS systems is one of the most important aspects of a CDPD system. The radio resource management (RRM) entity exists in CDPD systems to handle this activity. There are three general categories of channel management used by CDPD systems. The most desired method of channel management is a cooperative exchange of radio channel activity information between the AMPS system and the CDPD system. In this method the AMPS system notifies the CDPD system of channels that are available or likely to be available in the near future. Independent management of the channels by the CDPD system is the second method of channel management. This method requires the use of an RF sniffer that can detect activity on the AMPS channels. The sniffer is used to examine both used and unused channels to provide a list of available channels of the CDPD system. The difficulty of this method is finding a suitable algorithm that can quickly detect used channels as well as predict which channels are likely to be available for use in the future. The third method of channel management is the use of dedicated channels. This entirely prevents interference between CDPD and AMPS systems but has the drawback of reducing the number of channels available for AMPS on systems that have assigned most of their available radio channels. Figure 12.2 shows an overview of shared and dedicated CDPD radio channel management.

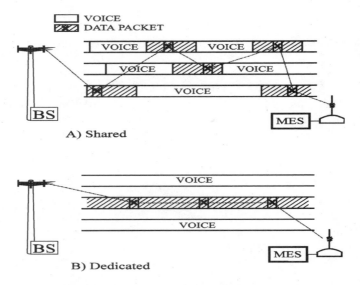

Figure 12.2, CDPD Radio Channel Management

Due to the fact that there are no dedicated control channels for CDPD, Mobile telephones can make access attempts on the data channel each time a data transfer is required. This minimizes the amount of overhead on the cellular system and allows CDPD data transmission to be more cost effective for small data transfers than the traditional circuit switched data.

Figure 12.3 shows the basic operation of the CDPD system network. As a first step, the MES scans cellular radio channels for the CDPD pilot signal. After it has locked onto a CDPD channel, it will begin to transmit data (step 2). If the radio channel that the MES is currently operating on becomes active with a cellular voice channel, the MES will move to the next CDPD radio channel and continue to send data. The data control is coordinated and gathered by the MDBS which converts the information and routes it to the MD-IS. The MD-IS is often located in the cellular switching center so the information may be sent on the existing communication channels linking the Base Station and the MSC (step 4). The MD-IS adapts and routes the data to the designated external network (for example, PPDN or Internet) and the external network routes the data to the address given by the MES.

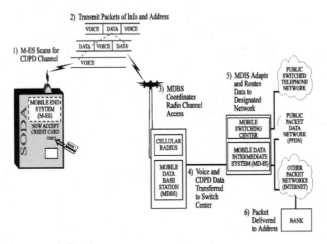

Figure 12.3, CDPD System

The CDPD frequency band is identical to the AMPS cellular system with the AMPS control channels being removed from the CDPD band. RF power levels are also the same as AMPS cellular. Output power in CDPD is controlled by the MES which calculates its output power level by measuring the received signal strength. A lower received signal results in a higher transmit power level. GMSK modulation is used for CDPD to transfer data at a 19.2 kb/s. Because some of the data bits are used for packet addressing and system control, the average data transfer rate is approximately 7 to 8 kb/s.

Similar to a cellular Mobile telephone's unique ESN identifier, each MES has its own unique serial number. referred to as the Network Entity Identifier (NEI). While the NEI is used to validate subscriber identity, one or more Temporary Equipment Identifiers (TEI) are used to address the MES when it is operating in a particular service area. Multiple TEI's may be used to provide different services. For example, one TEI may be used for point to point (single recipient) and another may be used for broadcast services (multiple receivers to a single message).

Every MES in a CDPD system belongs to a home system location. Part of the home system's responsibility is to maintain a data base containing the location of all of its MES. Every message sent to a MES first goes to its home system location where it is then forwarded to the current location. The home system keeps track of the MES through a registration process that tracks each cell that a MES enters.

After the MES determines that it is not desirable to maintain connection (probably due to poor signal strength), it will terminate transmission and seek a new CDPD channel. Once it has established a link on a new channel (typically a new cell site), transmission can continue. The handoff process is completely controlled by the MES in CDPD systems.

CDPD has authentication and data link confidentiality (transmission privacy) capability similar to digital cellular systems. Each MES can be validated by the system and the MES can also validate system authentication inquiries. Once the validation occurs, the key is used to encrypt the data transferred by the radio packets.

E-TDMA

ETDMA (Extended TDMA) was proposed by Hughes Network Systems in 1990 as an extension to the existing IS-54 TDMA standard. ETDMA uses the existing TDMA radio channel bandwidth and channel structure and are tri-mode as they can operate in AMPS, TDMA, or ETDMA modes. While a TDMA system assigns a Mobile telephone fixed time slot numbers for each call, ETDMA dynamically assigned time slots on an as needed basis. The ETDMA system contains a half-rate speech coder (4 kb/s) that reduces the number of information bits that must be transmitted and received each second. This makes use of voice silence periods to inhibit slot transmission so other users may share the transmit slot. The overall benefit is that more users can share the same radio channel equipment and improved radio communications performance. The combination of a low bit rate speech coder, voice activity detection, and interference averaging increases the radio channel efficiency to beyond 10 times the existing AMPS capacity.

ETDMA Operation

ETDMA radio channels are structured into the same frames and slots structures as the standard IS-54 radio channels. Some or all of the time slots on all of the radio channels are shared for ETDMA communication, which is similar to IS-54 and IS-136 radio channels, or slots can be shared on different frequencies. When a Mobile telephone is operating in extended mode, the ETDMA system must continually coordinate time slot and frequency channel assignments. The ETDMA system performs this by using a time slot control system. On an ETDMA capable radio channel some of the time slots are dedicated as control slots on an as needed basis. ETDMA systems can assign either an AMPS channel, a TDMA full-rate or half-rate channel, or an ETDMA channel. The existing 30 kHz AMPS control channels are used to assign analog voice and digital traffic channels

Figure 12.4 shows an ETDMA radio system. Some of the radio channels include a control slot which coordinates time slot allocation. This usually accounts for an estimated 15% of available time slots in a system. The control time slots assign an ETDMA subscriber to voice time slots on multiple radio channels. You will notice that the voice path in Figure 12.4 moves between voice time slots (shaded) on multiple radio channels.

ETDMA uses the following process to allocate time slots from moment to moment as needed. The cellular radio maintains constant communications with the Base Station through the control time slot (marked "C" in Figure 12.4). When a conversation begins, the cellular radio uses the control slot to request a voice time slot from the Base Station. Through the control slot, the Base Station assigns a voice time slot and sets the cellular radio to transmit in that assigned voice time slot. During each momentary lull in phone conversation, the transmitting cellular radio gives up its voice time slot, which is then placed back into the Base Station's pool of available time slots.

When a cellular radio is ready to receive a voice conversation, the Base Station uses the control slot to tell it which voice time slot has the conversation being sent. The cellular radio receiver then tunes to the appropriate slot. Through the control slot, the Base Station constantly monitors the cellular radio to determine whether it has given up a slot or needs a slot. In turn, the cellular radio constantly monitors the control slot to learn which time slot contains voice conversation being sent to it.

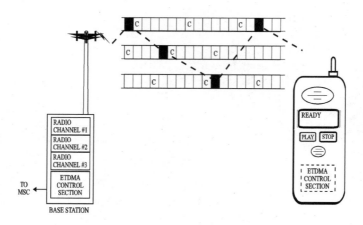

Figure 12.4, ETDMA System

Two-Way Voice Paging

A new development for cellular systems is the dedicated two-way voice pager. The difference between a standard wireless telephone and a two-way voice pager is the voice pager stores voice messages in its internal memory. The operation of a voice paging system allows for unique non-real time message delivery. Figure 12.5 shows a cellular voice pager.

Figure 12.6 shows a voice messaging system which is connected to a cellular system. In step 1, the caller dials a voice pager telephone number. After a recorded message, the caller leaves a message which is stored in a voice mailbox (step 2). The voice mail system then dials the Cellular Voice Pager's (CVP) phone number and attempts to deliver the message (step 3). After the CVP has answered the call, the voice messaging system checks to see if enough memory is available in the CVP for the new message(s). If there is enough memory, the voice

382

Figure 12.5, Cellular Voice Pager
Source: ReadyCom

mailbox will transfer the message (step 4) and it will be stored in the CVP voice message memory (step 5). After the message is completely delivered, the CVP will acknowledge the successful receipt of the message (step 6) and it can be deleted from the voice mail system's memory. The subscriber will be alerted by visual and/or audio that a message is waiting. The message can then be played at any time from Cellular Voice Pager's memory (step 7).

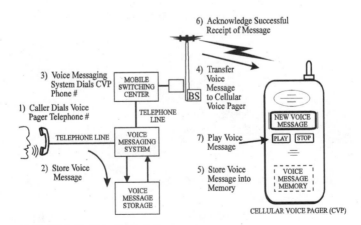

Figure 11.b Cellular Voice Paging System

Figure 12.6, Voice Paging System

The acknowledgment feature is an important part of voice paging. Message acknowledgment allows a new service called confirmation paging to be provided. After the message is successfully delivered, the voice message system can deliver confirmation to the caller via a pre-recorded confirmation message call back, fax message, or email.

Another advantage of the voice message system is the ability to compress voice messages for more efficient delivery. Figure 12.7 shows a sample of some of the compression steps. The first step removes the pauses between words. Next, the stored voice message is played back 2-4 times faster than normal voice. Both of these types of voice compression are not possible for normal (real time) cellular conversations.

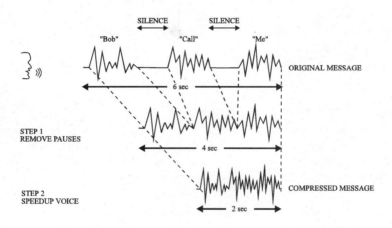

Figure12.7, Voice Paging Compressed Voice

By providing non-real time messaging it allows for the use of extended sleep mode. Similar to numeric paging systems, non-real time messages can be delayed for several minutes without a significant impact. This is not possible with standard cellular Mobile telephones where the sleep period cannot exceed a few seconds. This is because a caller who dials a Mobile telephone is probably only willing to only wait a few rings before hanging up the phone. The battery life could be extended to weeks by allowing the CVP to wake up for very short periods (tenths of seconds) and sleep for several minutes,

SDMA

Spatial Division Multiple Access (SDMA) is a technology which increases the quality and capacity of wireless communications systems. Using advanced algorithms and adaptive digital signal processing, Base Stations equipped with multiple antennas can more actively reject interference and use spectral resources more efficiently. This would allow for larger cells with less radiated energy, greater sensitivity for portable cellular phones, and greater network capacity.

Figure 12.8 shows a typical 120 degree sectored antenna (an antenna which covers 1/3 of the cell site area). Within the coverage area denoted, the antenna can communicate with only one subscriber per traffic channel in the radio coverage area at any given time. The performance of the system is constrained by the levels of interference present.

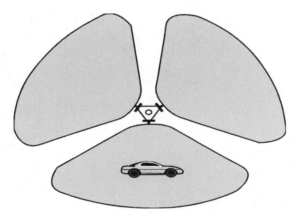

Figure 12.8, Conventional Sectored Cell Site Radio Coverage

One company, ArrayComm, of San Jose California US, has implemented SDMA into its IntelliCell™ Base Stations. Because of the IntelliCell Base Station's ability to reject interference on the uplink and control transmission patterns on the downlink, Spatial Channels™ can be created. Figure 12.9. ArrayComm has simultaneously operated 3 Mobile units on the same traffic channel in the same local area. Using an 8 element antenna array to track multiple users, the IntelliCell system exhibited carrier to interference (C/I) ratios which consistently exceeded 30 dB.

An IntelliCell Base Station which implements Spatial Channels will track the movement of each user in the system. If one user moves too close to another and the users are on the same conventional traffic channel, the system will automatically hand off one of the users to another traffic channel frequency.

The SDMA technology can be applied to any of the wireless systems in use today (AMPS, TDMA, CDMA, etc.). Its benefits include larger cells, less interference to other cells, and lower mobile/portable transmit power.

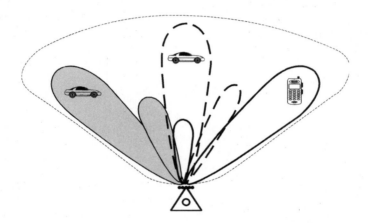

Figure 12.9, Spatial Division Multiple Access

Satellite Cellular Systems

There are several new wireless and network technologies that will impact future cellular telephone systems; therefore, new frequency allocations are being assigned to provide for these advanced services.

In 1995 the United States began licensing bands of frequencies for advanced wireless services. It is perceived that these technologies will combine the services of paging, cellular, data, and cordless telephones into one portable device. Personal Communications Services (PCS) will occur on these bands of frequencies that have been recently auctioned by the federal government. While the technologies may offer some unique services, PCS companies will provide similar services as cellular carriers.

Integrated Dispatch Enhanced Network (iDEN) is a wireless technology which is being deployed in 800 MHz Specialized Mobile Radio (SMR) channels. Originally called Motorola's Integrated Radio System (MIRS), iDEN is a new digital TDMA technology that provides both voice and dispatch services. Nextel (originally called Fleet Call) has been acquiring and partnering with SMR operators to begin to offer nationwide voice and dispatch services using iDEN technology.

Extended Time Division Multiple Access (ETDMA) is a technology that was created to extend the capability of the IS-54 and IS-136 TDMA systems. ETDMA provides for an increase in the number of subscribers that can be served by an IS-54 TDMA system.

Voice paging is a new service that is being deployed in the cellular system to transfer voice messages to a cellular voice pager. Cellular system voice paging uses store and forward messaging to allow the queuing of messages and increased information compression.

The packet data system that overlays on a cellular system to allow for efficient delivery of short digital messages if referred to as Cellular Digital Packet Data (CDPD).

A new technology that can be applied to any of the wireless technologies and frequencies to improve the performance and ability to serve more subscribers is Spatial Division Multiple Access (SDMA). This technology allows an increase in the number of subscribers that can be served by any cellular system by using smart antennas to focus the radio energy into narrow radio coverage beam widths.

Satellites can provide voice and data services to a wide regional geographic area which terrestrial (land based) services do not reach. These satellite voice systems have been available for many years, however; these systems required large antennas and expensive subscriber communications equipment. There are new satellite cellular systems planned for deployment in the late 1990's which will allow low cost handheld Mobile telephones.

Satellite communication systems can provide service to any unobstructed location on the Earth. In the late 1990s, several mobile-satellite systems will begin commercial operation [3], [4]. These new satellite systems will supplement low cost terrestrial (land based) wireless communications systems [5].

The satellite systems are much more expensive than cell sites and have a limited lifespan. To help with the distribution of high costs over many customers, some satellite systems offer frequency reuse similar to cellular technology to serve thousands of subscribers on a limited band of frequencies. These systems use sophisticated antenna systems to focus their energy into small radio coverage areas.

Satellite systems are often characterized by their height above the Earth and type of orbit. Geosynchronous Earth Orbit (GEO) satellites hover at approximately 36,000 km, Medium Earth Orbit (MEO) satellites are positioned at approximately 10,000 km, and Low Earth Orbit Satellites are located approximately 1 km above the Earth.

The higher the satellite is located above the Earth, the wider the radio coverage area and this results in a reduction of the number of satellites required to provide service to a geographic area. GEO satellites rotate at the same speed as the Earth allowing the satellite to appear to be stationary over the same location. This allows fixed position antennas (satellite dishes) to be used. Because the GEO satellites hover so far above the Earth, this results in a time delay of approximately 400 msec [6] and increases the amount of power that is required to communicate which limits their viability of handportable Mobile telephones. GEO systems would include: AMSC, AGRANI, ACeS, APMT [7].

The closer distance of the MEO satellite reduces the time delay to approximately 110 msec and allows a lower Mobile telephone transmit power. As a result of the lower position of MEO satellites in the Earth's orbit, they do not travel at the same speed relative to the Earth which requires the need for several MEO satellites to orbit the Earth to provide continuous coverage. MEO systems include Intermediate Circular Orbit (ICO) Project 21, and TRW's "Odyssey."

Low Earth Orbit (LEO) satellites are located approximately 1 km from the surface of the Earth and this allows the use of low power handheld Mobile telephones. LEO satellites must move quickly to avoid falling into the Earth, therefore; LEO satellites circle the Earth in approximately 100 minutes at 24,000 km per hour. This requires many satellites (for example, 66 satellites are used for the Iridium system) to provide continuous coverage. The LEO systems include Iridium Inc.'s "Iridium", Constellation Communications Inc.'s "Aries", Loral-Qualcomm's "Globalstar," and Ellipsat's "Ellipso" Figure 12.10 shows the three basic types of satellite systems.

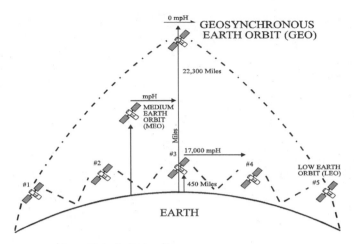

Figure 12.10, Satellite Cellular System

By allowing telephone calls to be routed directly between satellites, some satellite systems communicate directly with each other through microwave links. This eliminates the high cost of routing long distance global calls through other telephone companies. Figure 12.11 shows a picture of a LEO satellite Mobile telephone.

Figure 12.11, GlobalStar Mobile telephone
source: Qualcomm

References:

1. Cincinnati Microwave, "Cellular Digital Packet Data - A Primer On How CDPD Can Change The Way Your Organization Does Business," Cincinnati, OH, July 1995, p1.

2. "Cellular Digital Packet Data System Specification," Volume 1, Release 1.0, Costa Mesa,, CA, CDPD Industry Input Coordinator, July 19, 1993.

3. R. Rusch, "The Market and Proposed Systems for Satellite Communications," Applied Microwave and Wireless, Fall 1995, pp. 10-34.

4. Comparatto, G. M. "Global Mobile Satellite Communications: A Review of Three Contenders," American Institute of Aeronautics and Astronautics, 1994, pp. 1-11.

5. Rydbeck, Nils, "Mobile-Satellite Systems: A Perspective on Technology and Trends," Proceedings of IEEE Vehicular Technology Conference, Atlanta, Georgia, US, April-May 1996,.

6. Ibid.

7. Ibid

APPENDIX I - DEFINITIONS

Abbreviated Alert - An alert order that informs the user that a previous function selected is still active (North American 800 MHz cellular systems).

Access - A process where the mobile telephone competes to gain the attention of the system to obtain service.

Access Channel - Redefined to be Analog Access Channel. Similar to digital Random Access Channel in digital systems.

Access Overload Class (ACCOLC) - A class assigned to a mobile to limit its access attempts in an overloaded system. The overhead message contains an overload field which corresponds to the access overload class assigned to the mobile. When the system becomes overloaded, the overhead message will begin to remove groups of mobiles accessing the system (North American 800 MHz cellular systems).

Activation Commission - An incentive that is paid to a retailer of communications products from a service provider as a reward for establishing a new customer.

Adaptive equalizer - Device to compensate or correct for channel imperfections such as multipath propagation or non-uniform signal delay at different frequencies by using a known portion of the information (such as a training or synchronizing bit pattern) to automatically make adaptive adjustments to produce an internal compensating delay for signals at different frequencies, or by subtracting away signals which arrive too early or too late, thus making the performance equal for all components of the transmitted signal.

Adjacent Channel (or Adjacent Carrier Frequency) - A carrier frequency ± 30 kHz from the current carrier frequency (North American 800 MHz cellular systems), ± 25 kHz from the current carrier (ETACS and other European analog cellular systems), ± 200 kHz from the current carrier for GSM and related systems, or ± 1.23 MHz from the carrier for CDMA IS-95.

Adjacent Channel (or Adjacent Carrier) Interference - This occurs when the spectral power distribution from an adjacent channel creates noise or errors in the channel being used by the mobile or base station.

Advanced Mobile Phone Service (AMPS) - The North American 800 MHz band analog cellular service that is available in the United States and over 33 other countries today. AMPS was originally an AT&T (now Lucent Technologies) trade name.

Advanced Radio Technology Subcommittee (ARTS) - Sponsored by the CTIA to study the industry needs, technology available, and manufacturers support to develop new cellular technology.

Alert - An order sent to the mobile that informs the user that a call is to be received.

Alert with info - An alert message sent on the digital channel; may also contain calling line ID or other information (North American 800 MHz cellular systems).

Alert with info Ack - A message sent back on the digital traffic channel that acknowledges the Alert with info order (North American 800 MHz cellular systems).

Alternate Channel (or Alternate Carrier Frequency) - A cellular carrier frequency ± 60 kHz from the current channel (North American 800 MHz cellular systems), .or (400 kHz for GSM-related systems, or in general 2 carrier frequencies away from the current carrier frequency (see the related term: Adjacent Channel, which is one carrier frequency away).

Analog Access Channel - A cellular system channel that uses frequency shift keying (FSK) to pass data control signals between the mobile telephone and base station. This control channel is used by the mobile to inform the system that it wants to obtain service.

Analog Cellular - An industry term given to today's existing cellular system. It involves transmission of voice information in a method similar to that of a FM radio.

Analog Color Code - A digital code which designates the frequency of one of three audio tones mixed in with the voice channel to distinguish the channel from interfering co-channels.

Analog Control Channel - A channel designated by the cell site to control operations of a analog and dual mode cellular phones (North American 800 MHz cellular and other analog cellular systems). A control channel can be divided into paging and access functions.

Analog Mobile - A mobile unit which is only capable of Analog Cellular service.

Analog Voice Channel - A cellular system channel that operates by using frequency modulation (FSK) to pass voice and data control signals between the Mobile and Base.

Audit - An order generated by the system which determines if the mobile is active within the system.

AUTH1 - An algorithm used to authenticate the mobile set.

Authentication - A process during which information is exchanged between a mobile telephone and a base station to allow a cellular service provider to confirm the true identity of the mobile set, thus inhibiting fraudulent use of the radio system.

Automatic Message Accounting (AMA) - A standard recording system for collection and processing of call billing information.

Backhauling - The completion of a voice connection between a cell site and MTSO often requires the use of the landline network system. A fee often must be paid for the use of these lines. Backhauling is the carrying of cellular calls between the cell sites and the MSC/MTSO via a landline network prior to reaching the PSTN.

Base Station - A controlling transmitting/receiving station that provides service to cellular mobile units. It is sometimes called a Land Station or Cell Site.

Baseband - Audio frequency or digital information to be transmitted via radio, which is distinguished from the modulated radio signal which carries the baseband information.

Basic Trading Area (BTA) - A geographic region where area residents do most of their shopping. The United States has been divided into 493 BTAs. The 1.9 GHz band PCS licenses are granted based on BTA. Derived from Rand McNally commercial maps of trading zones in the United States.

Bearer Services - Telecommunications services that provide facilities for the transport of user information without functional processing of the user data.

Bit Error Rate (BER) - The ratio of bits received in error to the total number of bits transmitted.

Blocking Probability - The percentage of calls that cannot be completed within a busy hour period due to capacity limitations. For example, if within the busiest hour of the day, 100 users attempt accessing the system, and two attempts fail, the blocking probability is 2 percent (often expressed as P02).

Broadband - 1) A signal which has a wider bandwidth than another system with which it is compared; 2) In particular, a signal which has a wider bandwidth than another technology because of a particular process which causes greater bandwidth (for example, the process of combining a low bit-rate data stream with a high bit-rate pseudo-random bit stream in CDMA). Note, wider bandwidth implies that the digital bit rate is greater, when all other parameters are unchanged.

Burst Collisions - The problem of transmit bursts originating from mobiles overlapping in time when received at the base station. This may be due to the propagation time delay difference between a nearby and distant mobile set. Dynamic time alignment was created to solve this challenge.

Busy Idle Status Field - A field sent in the overhead message train which informs the mobile of which access method to use when obtaining service from the system. If the bit is set high (binary 1), the mobile must monitor the Busy-Idle bits to determine if the system is busy. When the bit is set low, the mobile attempts access without monitoring the Busy-Idle bits (North American 800 MHz cellular systems).

Busy-Idle Bits - Bits that are time multiplexed with the forward control channels to indicate if the access channel of the system is busy (North American 800 MHz cellular systems).

Capture Effect - A phenomenon which occurs with FM and PM radio, in which the stronger signal "captures" the attention of the receiver, producing a flawless audio or digital output signal despite the presence of a weaker interfering radio signal. Cellular frequency re-use only can be done because of this.

Capture Ratio - The minimum ratio of desired carrier signal strength to the undesired interference which achieves receiver capture by the stronger signal. For North American analog cellular radio, this ratio (including the effects of fading) is approximately 63/1 or 18 dB.

Carrier - 1) A service provider of cellular service; also called a cellular operator. 2) RF energy emitted from a transmitter assembly, and in some cases, the unmodulated RF sine wave signal.

Carrier-to-Interference Ratio (C/I) - The ratio of the carrier to the combined radio signal power due to adjacent and co-channel interference.

Carrier-to-Noise Ratio - The ration of the carrier to the thermal noise.

CAVE - Cellular Authentication Verification and Encryption algorithm, an algorithm used to produce the authentication and encryption data values in North American 800 MHz digital cellular systems.

Cell Site - A radio antenna assembly, tower or support, and associated equipment that converts phone lines to radio signals for transmission to mobile telephones.

Cell Splitting - A method to increase system capacity by the division of assigned radio areas. Each cell site (base station) can provide a limited number of channels. To increase the total number of channels available in a given area, it is possible to divide an area into a greater number of smaller cells by adding additional radio base stations.

Cellular Geographic Service Area (CGSA) - The area licensed by the FCC to a service provider. It is based on Metropolitan Statistical Areas or New England Metropolitan Areas.

Cellular Mobile Carriers (CMC) - The service providers that are licensed by the FCC to provide service in the CGSA. There are two CMCs authorized for each CGSA.

Cellular Subscriber Station (CSS) - Another name for mobile telephone. This is the preferred term to network providers since it comprises both mobile and fixed stations.

Channel - May have two meanings; logical and physical. In some cases this word is a synonym for a pair of carrier frequencies. The physical channel is composed of 2 carrier frequencies or frequency bands separated by 45 MHz and each with a bandwidth of 30 kHz

(North American 800 MHz cellular systems).For other systems the separation of the two carrier frequencies is 80 MHz (North American 1.9 GHz PCS systems), and/or the channel bandwidth is 25 kHz (European analog cellular systems), 200 kHz (GSM-related systems), or 1.23 MHz (CDMA). The logical channel is a throughput of information on that physical channel. A physical channel can be divided into many logical channels via TDMA or CDMA techniques.

Channel Quality Message (CQM(1,2)) - Messages sent on the digital channel to provide the base station with channel quality information. This information can contain the received signal strength indication (RSSI) or bit error rate (BER).

CHIP (or chip) - 1) The sequence of pseudo-random high bit-rate binary coding bits which are combined with the digital data in CDMA coding; 2) one of these bits. See also PN-PRBS.

Churn - The amount or percentage of customers that disconnect from service. Churn is usually expressed as a percentage of the existing customers that disconnect over a one month period.

Closed User Group (CUG) - A group of users who have access to a defined set of features or channels.

Co-channel - A channel being re-used at another cell site. This is allowed due to the attenuation of the signal because of the distance.

Co-channel Interference - A channel that is used by another cell site on the same frequency which interferes with the current channel. This is caused by insufficient attenuation of distance.

Coder/Decoder (Codec or CODEC) - A device for digitally coding and decoding information.

Coherence Bandwidth - A bandwidth within which either the phases or amplitudes, or the fading characteristics of two received signals have a high degree of similarity.

Color Code: - See Digital Color Code.

Combined Paging and Access - This allows the combining of paging and access functions. This allows one channel to broadcast both paging messages while the corresponding reverse channel allows system access within the same cell site. This leaves the paging channel free to only send system overhead and paging messages. This is the normal mode of operation (North American 800 MHz cellular systems). The antonym (opposite) is separated paging and access channels, which is seldom if ever used.

Combined Paging and Access (CPA) [Field] - This is a bit field in the forward overhead message which informs the mobile if the system has combined paging and access channels. If this bit is set to 1, the mobile must obtain access on the same (corresponding) control channel on which it receives the paging message. If this bit is set to 0, the mobile must tune to a different access channel than the paging channel to obtain service (North American 800 MHz cellular systems).

Combiner - Used in base stations to couple different transmitters to one antenna. The corresponding device for base receivers is called a multi-coupler or an RF pre-amplifier and distributor.

Continuous Transmission - A mode of operation where the mobile does not cycle its power level down when the modulating speech signal amplitude is low.

Control Channel - A cellular system channel dedicated to sending and/or receiving controlling messages between the Base and mobile telephones. (see Sections on Call Processing and RF Channel).

Control Mobile Attenuation Code (CMAC) [Field] - Used in a message sent to the mobile from the base which assigns the mobile to an absolute (specified) power level. This is important in small diameter cells where the mobile must access the system at a low power level to prevent co-channel interfering with other control channels.

Convolutional Coding - A forward error correction method used to provide for correction of data that has been corrupted in transmission. This is accomplished by sending redundant information which allows for reconstruction of the original signal. Convolution coding is mathematically similar to multiplying the binary number value of a message by a pre-defined constant before transmitting, and then transmitting the product value. The received product value is processed in a manner similar to long division using the same pre-defined constant. If the remainder is zero, the implication is that the received data is error free. The precise value of a non-zero remainder can be used to determine if there are errors, and in some cases of limited errors, to correct them.

Coupler - See combiner.

Cross Talk - A problem where the audio from one communications channel is imposed on another channel.

Cyclic Redundancy Check (CRC) Generator - A function used to create a CRC data word. A CRC parity check value is formed by a process similar to computing the remainder derived from dividing the binary number value of the message by a pre-determined constant. This CRC parity check value (remainder) is appended to the message data and sent with it. At the receiving end, the message data part is again processed similarly to produce another CRC parity check value. If the value computed at the receiver agrees with the value received, the received data is presumed error free. If not, the type and location of the bits which differ when comparing the received and re-computed CRC parity check values can be used to determine if there are errors, and in some cases of limited errors, to correct them.

Dead Spot - An area within a service area where the radio signal strength is significantly reduced. This is primarily due to terrain and obstructions. Dead spots are generally eliminated by the use of repeaters or relocation of the cell site.

Dedicated Control Channels - A nationwide allocated set of cellular channels that must be used only for control messages between the Base and mobile telephone. For the wireline or B operator these channels are 333 to 354. For the non-wireline or A operator, the control channels are 332 to 311 (North American 800 MHz cellular systems).

Delay Spread - A result of multipath propagation where symbols become will partially overlap in time due to copies of the same signal being received at a different time. It becomes a significant problem where a high bit rate requires a short bit time interval, and in mountainous areas where signals are reflected at great distances. Corrected by means of an adaptive equalizer.

Digital Cellular - An industry term given to the new cellular technology that transmits voice information in digital form. This differs from Analog cellular in the method for the transmission of voice/data information by digital signals.

Digital Color Code (DCC) [Field] - This is a bit field whose value corresponds to one of the three SAT codes, namely the one assigned to the local cell site. The mobile matches this code to the received SAT frequency to ensure it has locked on to the correct channel (analog cellular only).

Digital Enhanced (or European) Cordless Telephone (DECT) - The DECT system is a dedicated wireless office system which is capable of serving wireless telephones and high bit-rate data devices (32 kb/s or more). The DECT system is a digital system which uses a relatively wide radio bandwidth on each base station, which allows up to 12 simultaneous wireless telephones to share each channel. The DECT system has been adapted to several different frequencies and data rates and is also being used for the US PCS system. DECT and its North American version PWT(E) constitute a third generation standard air interface TDD/TDMA system which has certain technical similarities to the CT-2 and Ericsson CT-3 systems.

Digital Only Unit - Amobile telephone that will only be capable of using the new digital signals. This will not be available for several years.

Digital Signal Processor (or Processing) - see DSP.

Digital Speech Interpolation (DSI) - A voice detecting system which allows transmission only when speech or data information is to be sent. More elaborate DSI systems (used today for satellite and undersea cable but not yet for cellular systems) re-assigns an idle channel to another user for greater capacity utilization of each channel. The simpler DSI used today on cellular systems is simply DTX, which does reduce battery drain in the handset, but does not increase capacity. Extended TDMA (see below) has been demonstrated experimentally in 1989 to use DSI with TDMA cellular.

Digital Voice Color Code (DVCC) [Field] - Provides a unique code from the base station which helps to distinguish its channel from co-channels.

Digital Voice Color Code Status - This is a logical variable contained within the mobile. It can be enabled or disabled dependent on if a match occurs with incoming DVCC. Every burst contains a DVCC field. The mobile must transmit the DVCCs (stored value) even if it does not agree with incoming DVCC (used in North American 800 MHz digital cellular systems IS-54 and IS-136).

Discontinuous Transmission (DTX) - A method where the mobile changes its power level as a result of the input level of its modulating signal. This allows conservation of power when the modulating level is low or no data is to be transmitted.

Discontinuous Transmission (DTX) [Field] - A field included in the overhead message and reverse control channel message which indicates the base and the mobile is capable of discontinuous transmission respectively.

Diversity Reception - The process of combining or selecting one of two stronger of the signals received by two base antennas. These two base antennas are usually separated by a number of wavelengths, typically 5 or more.

Doppler - A frequency offset or change that is a result of a moving antenna relative to a transmitted signal.

Dotting - Used to bit-synchronize the mobile. This also acts like a: "wake up, a message is coming!" warning. .

DSP - Digital signal processing, a method for performing signal processing such as filtering, equalization, or speech coding, in digital form by using a fast microprocessor to perform the appropriate mathematical operations using numerical samples of the input voltage waveform rapidly enough to produce the results in real time. The digital output is converted back into an analog waveform in many applications.

Dual Mode Cellular - The combination of an Analog cellular unit and Digital cellular section into one cellular 'phone. It allows operation of the phone in the existing system as well as the new system when it becomes available.

Electromagnetic Interference (EMI) - electro-magnetic fields or radio signals that interfere with the operation of the mobile telephones.

Electronic Serial Number (ESN) - see Serial Number.

End Office Switching System (EO) - A switch which provides communication paths between originating customer terminal (telephone) equipment and terminating (destination) customer terminal equipment.

Enhanced Specialized Mobile Radio (ESMR) - A name given for the new digital technologies being implemented by the Specialized Mobile Radio operators which provide dispatch, voice, messaging, paging, and wireless data services.

Equalization - A process which modifies the receiver waveform to compensate for changing radio frequency conditions. Primarily used to compensate for multipath propagation (see section on RF channel).

Equalizer - See adaptive equalizer.

Erlang - Amount of voice connection time with reference to one hour. For example, a 6 minute call is 0.1 Erlang (named after Anger K. Erlang, the Danish engineer and mathematician who first analyzed the probability of channel blocking in busy telephone systems).

Extended (or Enhanced) Time Division Multiple Access (E-TDMA) - A time division cellular system which utilizes digital speech interpolation (DSI) which allows transmission only when speech or data information is to be sent. When transmission is inhibited, the same channel and time slot can be shared by or temporarily re-assigned to other users.

Extended Protocol - Optional extended capability of the signaling messages which provide for the addition of new system features (North American 800 MHz cellular systems).

Fast Associated Control Channel (FACCH) - A logical signaling channel which is created by replacing speech data with signaling data for short intervals of time (GSM and IS-54/IS-136).

Field - A dedicated number of bits within a message or data stream which is dedicated to specific functions.

Flash Request - A request to invoke a special processing function by a signal analogous to momentarily depressing the cradle switch on a landline telephone set. Analog cellular only allowed flash requests from the mobile to the base. Dual Mode cellular allows flash requests in both directions.

Flash With Info - A flash message sent over the digital traffic channel. This function allows an indication that the originator needs special processing.

Forward Analog Control Channel (FOCC) - The Analog control channel which is from the base station to the mobile telephone.

Forward Analog Voice Channel (FVC) - The Analog voice/traffic channel which is from the base station to the mobile telephone.

Frame - In is-54 /IS-136, six slots are linked together compose one frame. In GSM related systems, 8 slots are linked to form a frame. Any one or any combination of frames can be assigned to a single user.

Frequency Planning - The selection and assignment of carrier frequencies to cell sites to minimize the interference levels they create with adjacent and alternate cell sites.

Frequency Reuse - The ability to reuse channels on the same frequency. This is possible due to the attenuation of the signals by distance and the "capture effect" of FM or PM when the desired signal is sufficiently stronger than the interfering signal.

Full Duplex - Transferring of voice/data in both directions at the same time. This becomes confusing in a TDMA system because information is reconstructed from short burst of high bit rate into continuous bit streams at a lower bit rate, to allow transfer of voice information in both directions at the same time although actual radio link transmission does not occur simultaneously.

Full Rate - The process where 2 slots per 6 slot frame for IS-54/IS-136 (or 1 slot per 8-slot frame for GSM related systems) is used to convey all speech or data.

Group Identification - A subset of the system identification (SID) which identifies a group of cellular systems.

Guard Time - A time allocated at the end of each TDMA mobile transmit slot so transit time of the signal does not cause collisions between mobiles transmitting on the same frequency.

Half Duplex - The transferring of voice/data in both directions, but not at the same time.

Half Rate - The process where only one slot per six slot frame in IS-54/IS-136 (or one slot per 16-slot double frame in GSM related systems) is used to convey all speech or data.

Handoff (also called Handover in European documents) - A process where a mobile operating on a particular channel will be reassigned to a new channel during a conversation. This may occur for two reasons. First:where the mobile moves out of range of one cell site and is within range of another cell site. Second: where the mobile has requested a cellular channel with different capabilities. For example, this would mean assignment from a digital channel to an analog channel or assignment from an analog channel to a digital channel.

High Tier Systems - Wireless systems that provide communications services through a wide coverage area such as a cellular system.

Home Mobile Station - A mobile that is operating in a cellular system where it has subscribed for service.

Hot Spots - Regions in a cellular service area which have much higher traffic than average.

Hyperframe - The sequence of 2,715,648 frames (2048 repetitions of the superframe cycle of 1321 frames) used for synchronizing encryption calculations and frequency hopping cycles in GSM and related systems.

Improved Mobile Telephone Service (IMTS) - Available in 1964 as the MJ 150 MHz system, it was the first system to offer automatic dialing. IMTS was a technological predecessor of cellular technology.

In Band Signaling - 1) Historically, any signaling that occurs within the voice bandwidth of an analog system. 2) Also used for any system which temporarily or continually uses all or part of the capacity of an analog or digital channel that is normally used for voice, but uses that part for data or information related to controlling the call.

Intercept - A message sent to the mobile to inform the user of an error or that no service could be established when placing a call.

Inter-Exchange Carrier - A long distance service provider (i.e. MCI) that provides inter-exchange service between Local Exchange Carriers (LEC).

Interleaving - 1. A process in which digital data is permuted or rearranged in time order and not sent in direct time sequence to minimize the effect of burst errors. 2. A process of offsetting carrier frequencies in different cells, intended to increase the reuse capacity of a system.

International Mobile Subscriber Identity (IMSI) - This is the GSM subscriber unit's identifying number. It is distinct from the telephone number.

Intersymbol Interference (ISI) - One result of multipath propagation, which results in a time overlap of various copies of the received signal. It can be corrected by an adaptive equalizer.

Landline - Telephone equipment or service of the conventional wired type, as opposed to radio-telephone. .

Local Access and Transport Areas (LATAs) - A legally defined geographic region that is typically the service area for Local Exchange Carriers (LECs) in the United States.

Local Control - A function of the mobile which has been designated to provide special features in addition to those specified by the cellular standard.

Local Exchange Carrier (LEC) - A telephone service provider which furnishes local telephone service to end users.

Low Tier System - A wireless system that provides communications service through a limited coverage area such as a cordless telephone system.

Major Trading Area (MTA) - A geographic region where most of the area's distribution, banking, wholesaling is conducted. The United States has been divided into 51 MTAs. The A and B PCS licensess are granted based on MTA (see also **BTA**).

Malfunction Timer - A timer that is used to turn-off a mobile telephone's transmitter in the case of a malfunction. The malfunction timer runs separate from all other functions. It repetitively counts down and needs to be reset to its high starting value before its count value reaches zero, or it will inhibit the transmitter. If the mobile telephone is operating correctly (without failure) this timer will be reset continually by the software and will not expire.

Metropolitan Statistical Area (MSA) - An area designated by the FCC for service to be provided for by a cellular carriers. There are two service providers for each of the 306 MSA's in the United States.

Mobile Assisted Handoff (MAHO) - A process in which the base requests and the mobile telephone scans and reports signal quality information, which can be used to determine which cell site channel is best suited for handoff. The mobile can provide RSSI information on up to 12 forward channels and BER on its current operating channel (see section on RF channel).

Mobile Identification Number (MIN(1,2)) - This is the mobiles telephone number. It is divided into MIN1 and MIN2. MIN1 is the 7 digit portion of the number. MIN2 is the 3 digit area code portion of the number (North American 800 MHz cellular systems).

Mobile Station (MS) - A receiver transmitter (transceiver) operating in a cellular system. This includes hand held units along with transceivers units installed in vehicles.

Mobile Station Class (MSC) - This is a classification of the power level capability of the mobile. A mobile may be a high, mid-range, or low power station. The mobile identifies to the base its power level capability by using the Station Class Mark field.

Mobile Station Control Message - A message sent from the base station to the mobile over the analog voice channel which allows assignment to a digital traffic channel. This message was modified in IS-54 to contain two words. Two words are necessary to include all the parameters necessary for assignment to the digital channel.

Mobile Telephone Switching Office (MTSO) - Includes switching equipment needed to interconnect mobile equipment with the land telephone network and associated data support equipment.

Mobile-service Switching Center (MSC) - Similar to the Mobile Telephone Switching Office. MTSO is viewed as a trade name by some vendors, while MSC is a legally "neutral" name, not owned by any one vendor.

Multiframe - The sequence of 26 TDMA frames (on a traffic channel) or 51 frames (on a control channel) used in GSM and related systems for scheduling the appearance of different types of information transmission.

Multipath Propagation - Occurs when the same signal transmitted reaches a point via different paths. This is due to signal reflection and refraction (see RF channel).

Multiplexing - The process of combining several resources over a shared medium. This may be in the form of time sharing (time division multiplexing) where one radio channel is divided into time periods and one resource uses the channel for only the dedicated time allowed.

Narrowband Advanced Mobile Phone Service - A cellular system which uses a 10 kHz FM channel bandwidth instead of 30 It utilizes many AMPS operations with relatively minor signaling changes.

Non-Wireline Carriers - Cellular service providers that are not engaged in the business of providing landline telephone service. Also called A operators (North American 800 MHz cellular systems).

Number Assignment Module (NAM) - A 32 byte memory storage area or device which contains user profile data such as Mobile Identification Number (MIN), Electronic Serial Number (ESN), and the registered system identification (North American 800 MHz cellular systems).

Operations, Administration and Maintenance (OA&M) - Supervision of the cellular system and its component parts.

Orders - Messages sent between the mobile and base station.

Out of Band Signaling - 1) Historically, any analog signaling which occurs outside of the 300-3000 Hz audio voice channel bandwidth. 2) Any signaling method which never uses digital bits which are assigned to transmit digitally coded voice, but which instead uses distinct and separate bits.

Overhead Messages - System messages that are sent from the base station to the mobile giving the mobile the necessary parameters to operate in that system.

Overload Class (OLC) [Field] - A field within the global system messages sent to the mobile which indicates if the mobile is authorized to attempt access (North American 800 MHz cellular systems).

Overload Control - A process used by the system to control the access attempts initiated by mobiles. Overload Class (OLC) bits sent in the overhead message inhibit operation of groups of mobiles.

Paging - The process in which the base station sends a message over the control channel informing the mobile that a mobile destination call is incoming.

Paging Channel - A control channel which addresses mobiles directly to alert them of an incoming signal.

Personal Access Communications System (PACS) - A small-cell wireless communication system that evolved from development at Bellcore, coupled with technology from the Japanese Personal Handyphone System (PHS).

Personal Communications Network (PCN) -.1) synonymous with Personal Communications System. 2) For some, the infrastructure or land network part of a PCS system, as distinguished from other parts of the system.

Personal Communications System (or Service) (PCS) -.1) Any type of wireless technology system. 2) A wireless system operating on the North American 1.9 GHz band, in distinction with a system operating on the 800 MHz "cellular" band. 3) PCS-1900 technology, in distinction to other access technologies.

Personal Handyphone System (PHS) -.A short-range TDMA-TDD radio-telephone system developed in Japan. It uses 8 TDMA time slots per carrier.

Phase Locked Loop - A circuit which synchronizes an adjustable oscillator with another oscillator by the comparison of phase between the two signals[1.]

Photogrammetry - A process through which terrain elevation data can be derived from using a stereo image photograph pairs.

PN-PRBS - Pseudo-noise, pseudo-random bit string (or stream), describes the apparently random but actually deterministic pattern of the chip bit stream used in CDMA to encode data before transmission.

Point of Presence (POP) - A location within a Local Access and Transport Area (LATA) that has been designated for connection between a Local Exchange Carrier (LEC) and Interexchange Carrier (IXC) by means of a switch, multiplexer, or other point of demarcation ("demark").

Pop - The population of a cellular service area for a service provider The price paid per pop (for a license or for purchase of a system) is a size-independent measure of the value or worth of a system.

Power Level (PL) - The name for a digital code specifying the mobile telephone's commanded operating power level. .

Pre-origination Dialing - A process in which the dialing sequence takes place prior to the mobile's first communication with the cellular system.

Private Branch Exchange (PBX) - A private switching facility that is used to permit internal call routing and connection to the PSTN.

Protocol Capability Indicator (PCI) [Field] - A bit in the base system overhead information which indicates the cell site has digital channel capability (North American 800 MHz cellular systems).

Protocol Discriminator [Field] - A 2 bit field used on the digital traffic channel. It is reserved for future use for indication of up to 3 different new protocols, and the bits must currently be set to '00'.

RAKE Receiver - An adaptive equalizer for broadband signals such as radar or CDMA. Name derived from appearance of multiple delayed pulses on an oscilloscope screen display like the teeth of a rake or comb.

Rayleigh Fading - A mathematical description of the fading in the the received signal strength, which will vary due to multipath propagation. A fade usually occurs at places separated by approximately a half wavelength (approximately 16 cm on the 800 MHz band and approximately 8 cm on the 1.9 GHz band) (Named for W.J. Strutt, the physicist who analyzed the fading due to multiple reflections of sound waves in the 19th century. He was a hereditary baron in England, and therefore known as Lord Rayleigh).

Read Control Filler (RCF) [Field] - A field in the system overhead information data stream that indicates if the mobile must read the control filler messages prior to attempting access to the system (North American 800 MHz cellular systems).

Refraction - The apparent bending of a radio or light ray as it passes from one material to another, in a case where the two materials have different wave speed for the electromagnetic ray. For example, refraction occurs when a radio ray travels from the higher altitude atmosphere, which has lower density and thus a higher wave speed, into the lower altitude atmosphere, which has higher density and thus a lower wave speed. Similar changes in wave speed can occur due to changes in atmospheric humidity (water vapor content) or other properties. . At lower frequencies than the cellular bands, where radio rays are bent by the ionosphere (a layer in the atmosphere with a high concentration of electrically charged ions and electrons caused by the intense sunlight at high altitude, rays traveling up into the atmosphere bend completely back toward the Earth, resulting in signal "skip." This causes re-appearance of so-called "short wave" radio signals at locations far from the transmitter, with no reception in between the transmitter and the distant ground location. Note that radio skip does not occur on the UHF band used for cellular and PCS systems because of the difference in the effect of the ionosphere (almost no refraction) on UHF radio waves.

Registration - The process where by a mobile telephone identifies itself by sending a message that it is operating in the service area.(analog cellular systems). The processes of attachment and location update in digital systems are similar in purpose.

Registration Identification - A process in which the mobile accesses the cellular network to inform the system it is in its operating area.

Release Request - An order by which the base station or the mobile requests a termination or disconnect of conversation mode.

Re-use Factor - The number of cells in a basic group or cluster which contain all the available carrier frequencies in a system. The most widely used re-use factor is 7. A smaller re-use factor implies that there are more carrier frequencies in each cell, and thus more system capacity. .

Reverse Analog Control Channel (RECC) - The FSK modulated channel that transmits control channel signaling from the mobile telephone to the base station.

Reverse Analog Voice Channel (RVC) - The FM channel that transmits the voice/traffic channel from the mobile telephone to the base station.

Roamer - A mobile telephone that is operation in a cellular system other than its subscribed (home) system.

Roaming - The process by which a mobile operates in a system other than its subscribed (home) system.

Rural Statistical Area (RSA) - A geographic area designated by the FCC for service to be provided for by cellular carriers that falls outside the MSA regions. There are 428 RSAs in the United States.

Scan of Channels - A process where the mobile unit tunes to a defined set of frequencies and locks on to the strongest signal.

Scan Primary set of Dedicated Control Channels {Task} - This is the first task the mobile uses to find which signal is the strongest of the dedicated control channels.

Seamless - 1) Handoff with no loss of information, particularly for TDMA; 2) Inter-system handoff (whether seamless in the sense of meaning 1 or not) so that a user moving from one city to an adjacent city has the perception of geographically seamless service coverage.

Secondary Dedicated Control Channels - A second set of nationwide allocated control channels that may be used (optionally) by a digital cell site switch (North American 800 MHz IS-54 and IS-136 systems).

Sectoring or Sectorizing - A process of dividing cell sites into angular sectors by using directional antennas.

Seizure Precursor - A defined bit stream transmitted by the mobile telephone on the reverse analog control channel used to synchronize the base station receiver (analog cellular systems).

Serial Number - A 32 bit electronic serial number permanently installed in the mobile. For dual mode mobiles, this number is used with the algorithm 'AUTH1' for authentication (North American 800 MHz cellular systems).

Serial Number Response Message - This is a new message that was added, in addition to the requirements of EIA-553, which allows the base station to query the mobile telephone for its un-encrypted serial number (North American 800 MHz cellular systems).

Service Control Point - A signaling processing point in a telephone network that modifies and directs signaling messages to other switching points (Common Channel No. 7 signaling).

Service Provider - An organization that provides cellular service.

Shortened Burst - A shortened transmit burst used by the digital mobile when initial transmit occurs in a cell whose size and thus whose burst timing information has not been established. This is required to overcome propagation delays which may cause burst colli-sions (overlapping received bursts) (IS-54/IS-136 and GSM related systems).

Signaling System #7 (SS7) - A international standard network sig-naling protocol which allows common channel signaling between tele-phone network elements. Also called Common Channel No. 7 Signaling System. Many other abbreviations are in use in addition to SS7.

Signaling Tone (ST) - A 10 kHz tone mixed in with the analog voice signal which is used as a status change signaling device between the mobile telephone and the base station (North American 800 MHz ana-log cellular systems).

Signaling Transfer Point (STP) - A switching point in a telephone network that routes messages to other switching points (Common Channel No. 7 signaling system).

Sleep - The operation of a mobile set with power applied only to an electronic timer, and the transmitter and receiver both powered down. This is part of a scheduled sleep-wake cycle used to conserve battery power in the mobile set when even receiver operation is not required.

Slot - See Time Slot.

Slow Associated Control Channel (SACCH) - Out of Band signal-ing that occurs on the digital traffic channel where messages are transferred by means of 1) a dedicated number of bits (12) assigned within each time slot for North American IS-54 and IS-136 digital cel-lular systems, or 2) by a scheduled transmission of one SACCH burst in 26 time frames, in place of the digitally coded speech burst trans-missions which are used in the other time frames, in GSM-related sys-tems.

Soft Capacity Limit - A system subscriber serving capacity limit which allows more users to receive service at a less than optimal qual-ity (CDMA IS-95 systems).

Special Mobile Group (GSM) - English language translation of the name of the study group created by CEPT to establish a digital cellu-lar standard to be used in Europe. This system while theoretically similar to IS-54 is not compatible with the U.S. 800 MHz band digital cellular system.

Specialized Mobile Radio (SMR) - Private wireless networks which provide dispatch type services. Most SMR operators are licensed to provide service in the portions of the 800MHz frequency band which are not used by 800 MHz cellular service.

Speech Frame - For a digital speech coder used on a full rate traffic channel, this will be a 20 msec period that the speech is sampled. When half rate traffic channel is supported, the sample period will be increased to 40 msec (digital cellular and PCS systems).

Standard Offset Reference (SOR) - This is a time period allocated between the transmit and receive burst that allows the base station to measure radio propagation delay from the mobile to the base It is an interval of 44 symbols (approximately 1.8 msec) for IS-54 and IS-136 systems.

Superframe - The sequence of 1321 TDMA frames in GSM used for scheduling repetitions of the two multiframe patterns.

Supervisory Audio Tone (SAT) - One of several continuous tones that are mixed in with the modulating audio signal which is used to identify channel interference of the same radio carrier frequency.

Symbol -A modulation symbol is the interval of time during which the modulated parameter of the radio waveform is held (approximately) constant. For a system with one symbol per bit, such as GSM and related systems, the symbol duration is the same as one bit duration. In GSM and relatives, the frequency is held approximately constant for one bit interval. In IS-54/IS-136, which use a differentially coded 4 phase-shift modulation symbol method, the symbol interval corresponds to two bits. The phase angle of the radio waveform is held approximately constant for a symbol interval corresponding to two bits. .

Synthesizer - A frequency synthesizer is a digitally controlled radio oscillator that can provide any one of the stable RF carrier frequencies required, upon direction from the microprocessor in the logic section.

System Identification (SID) [Field] - The bit field in the forward control channel which holds the system identification number.

Tandem Office Switches (TO) - Telephone switching systems which are used to interconnect end offices (EO) when direct trunk groups are not economically justified. Tandem office switches can be connected to other tandem office switches. Also called a Transit switch.

Task - A set of steps or processes a transceiver must take to accomplish a function.

Teleservices - Telecommunications services that provide facilities for the transport and processing of user information.

Thermal Noise - A theoretical power level of noise due to temperature. kTB, where k is Boltzmann's constant, T is degrees kelvin, B is bandwidth in Hertz (Boltzmann's constant named for Austrian physicist Ludwig Boltzmann, degrees kelvin named for British physicist Lord Kelvin).

Time Division Multiple Access (TDMA) - A process of sharing a cellular channel by sharing time solts or intervals between users. Each user is assigned a specific time interval .

Time Slot - A particular time period or interval assigned in the digital channel. There are 6 time slots allocated per frame on the IS-54/IS-136 digital traffic channel, and 8 for GSM and related systems.

Total Access Communications System (TACS) - The cellular system in use in England today. It is an enhanced version of AMPS and uses a different set of carrier frequencies which are separated by 25 kHz rather than the 30 kHz used for AMPS. TACS was originally a Motorola trade name.

Traffic Channel (TCH) - The combination of voice and data signals existing in a communication channel. Sometimes called traffic and associated channel(s) - TACH. The name TCH is used for GSM and relatives, and also for IS-54/IS-36. The name TACH is used for GSM and related systems.

Trunking - A process which allows a mobile to be connected to any unused channel in a group of channels for an incoming or outgoing call. Name is derived from the main trunk lines of railroad track between cities, which in turn is derived from the trunk of a tree, where all the branches come together.

Trunking Efficiency When a mobile has trunking access to a larger number of channels, the probability of blocking a call setup attempt due to all channels busy decreases. Curiously enough, if the number of working channels is doubled and the amount of offered traffic is also doubled, the probability of blocking does not stay the same, but it goes down! This "more than proportional" improvement in system performance as the number of trunked channels increases is called "Trunking efficiency.".

Um Interface - The RF interface between the base station and the mobile telephone. The name comes from the analogous U interface designation in ISDN for the wire between the customer and the PSTN. The small m at the end implies "mobile." The name U interface in ISDN is part of a sequence of interfaces described by some arbitrarily chosen letters of the alphabet (R,S,T,U...etc.). However, it has become customary to use mnemonic names for some of these interfaces, and the U interface in ISDN is often humorously called the "undefined" interface because it was one of the last items completed by the standards committee. The Um interface is defined in great and minute detail in all systems!.

Unequal Cell Loading - A process where cell sites in a general area share a provide a different amount voice channels. This can be dynamic where channels are redefined on command of by MTSO or by the established frequency plan.

Update Overhead Information {Task} - A procedure where the mobile tunes to the strongest dedicated control channel and gathers system overhead information. It uses the information gathered here to determine the paging channels, system ID, and other system parameters.

Update Protocol Capability Indicator {Task} - This is the same task as Update Overhead Information except it has an additional function of determining if the cell site is digital capable for dual mode North American IS-54 800 MHz cellular systems.

User Channel (UCH) - The raw data portion of a channel that is available to the user.

Viterbi (adaptive) equalizer - A Viterbi decoder designed to correct for multipath inter-symbol interference, mainly used in GSM and related systems.

Viterbi decoder - A decoder designed to produce the best digital output signal in the presence of noise and/or signal interference by choosing a sequence of decoded bits which have the minimum overall proportional errors in voltage, frequency, phase or whatever parameter is used to encode the digital information. Named for celebrated engineer and inventor Andrew Viterbi.

Voltage Controlled Oscillator (VCO) - An oscillator circuit which has an output frequency that changes proportionally with a input voltage.

Wake - See sleep.

Wavelength - The distance covered for one complete cycle of a propagated signal (electro-magnetic, sound, etc.). This can be calculated by dividing the propagation velocity c (speed of light - 300,000,000 meters/second) by the number of cycles in one second. For example, the wavelength at 840 MHz is 0.357 meters (approximately 14 inches).

Wireless Local Loop (WLL) -.A fixed wireless system using radio technology such as cellular or PCS. WiLL is a Motorola trade name for a WLL product.

Wireline Carriers - Cellular service providers that are also engaged in the business of landline telephone service in the same area as their cellular operations. Also called a B band operator. Some wireline carriers have been authorized to own a cellular license in band A, provided that this ownership is outside their area of landline telephone system operations (North American 800 MHz cellular systems).

Word Error Rate (WER) - The ratio of words received in error to the total number of words sent. A word, in this context, is any group of data bits for which the system can determine the accuracy of the received data, by means of error protection codes or otherwise. In different cellular and PCS system designs, the bit length of a "word" may be 32 bits, 36 bits, 40 bits, 48 bits, or more.

---

1. The oscillator may be replaced by a sinusoidal signal that may have been created in a variety of ways including by a demodulation circuit.

APPENDIX II - ACRONYMS

A/D - Analog to Digital

AC - Authentication Center (also AuC)

ACCOLC - Access Overload Class

ACK - Acknowledge

ADPCM - Adaptive Differential Pulse Code Modulation

A-Law - type of logarithmic compression codec used in PSTN outside of North America and Japan (see Mu-Law)

AM - 1) Amplitude modulation; 2) ante meridian, before noon.

AMA - Automatic Message Accounting

AMPS - Advanced Mobile Phone Service

ANI - Automatic Number Identifier

APC - Adaptive Predictive Correction speech codec

ARTS - Advanced Radio Technology Subcommittee

AuC (also AUC) - Authentication Center (also AC)

B/I - Busy Idle Bit

B-CDMA - Broadband CDMA

BCH - Broadcast channel (GSM and related systems)

BER - Bit Error Rate

BIS - Busy Idle Status

BS - Base Station

BTA - Basic Trading Area

C/I - Carrier to Interference Ratio

C/N - Carrier to Noise Ratio

C7 - Common Channel Signaling System No. 7

CAF - Cellular Anti-Fraud

CBCH - Cell Broadcast Channel (GSM and related systems)

CCIR - International Radio Consultative Committee (succeeded by ITU-R)

CCITT - International Telegraph and Telephone Consultative Committee (succeeded by ITU-T)

CCLIST - Control Channel List

CCS - 1) Common Channel Signaling System No. 7; 2) One hundred call seconds (traffic unit)

CDMA - Code Division Multiple Access

CDPD - Cellular Digital Packet Data

CDVCC - Coded Digital Voice Color Code

CEC - commission of the European Communities

CELP - Code Excited Linear Predictive

CEPT - European Conference of Posts and Telecommunications (standards activities succeeded by ETSI)

CMAC - Control Mobile Attenuation Code

CoDec (also codec, CODEC) - Coder/Decoder

COST - Committee on Science and Technology (part of ETSI)

CP - Cellular Provider

CPA - Combined Paging and Access

CPE - Cellular Provider Equipment

CQM - Channel Quality Measurement

CSMA - Carrier Sense Multiple Access

CSS - Cellular Subscriber Station

CT2 - Cordless Technology 2nd Generation

CT-2+ - Cordless telephone product trade name used by Nortel

CT-3 - Cordless telephone product trade name used by Ericsson Radio Systems

CTIA - Cellular Telecommunications Industry Association

CTRC - Canadian Television and Radio Commission (successor to DOC)

CUG - Closed User Group

D/R - Distance to Cell Radius Ratio

dB - Decibel, a logarithmic representation of the ratio of two power values. (named for Alexander Graham Bell)

dBm - Decibel level relative to a 1 milliwatt reference level.

DCC - Digital Color Code

DCS-1800 - Digital Communications System on 1800 MHz band, UK up-banded and lower power version of GSM.

DCT - Digital Cordless Telephone

DECT -Digital Enhanced or European Cordless Telephone

DID - Direct Inward Dialing

DOC - (Canadian) Department of Communications (succeeded by CTRC)

DQPSK - Differential quadrature (four angle) phase shift keying (type of modulation)

DQPSK - Differential Quadrature Phase Shift Keying

DRX (or DRx) - Discontinuous receive

DSI - Digital Speech Interpolation

DSS - Direct spread spectrum

DTC - Digital Traffic Channel

DTI - Department of Trade and Industry (United Kingdom government radio regulation organization)

DTMF - Dual Tone Multiple Frequency

DTX (also DTx) - Discontinuous Transmission

DVCC - Digital Voice Color Code

E^{2}PROM (also EEPROM) - Electrically Eraseable Programmable Read Only Memory

ECB - Enhanced Cordless Base (station)

EIA - Electronics Industries Association

EMI - Electromagnetic Interference (also Electronic and Musical Industries, a British manufacturer)

EP - Extended Protocol Indicator

EPROM - Eraseable Programmable Read Only Memory

ERP - Effective Radiated Power

ESMR - Enhanced Specialized Mobile Radio

ESN - Electronic Serial Number

ETACS - Enhanced (or European) Total Access Communications System

ETDMA - Extended Time Division Multiple Access

ETSI - European Telecommunications Standards Institute (successor to standards activities of CEPT)

FACCH - Fast Associated Control Channel

FCC - Federal Communications Commission (US government radio regulation organization)
FCCH - Frequency Correction Channel (GSM and related systems)
FDD - Frequency division duplex
FDM - Frequency Division Multiplexing
FDMA - Frequency Division Multiple Access
FDTC - Forward Digital Traffic Channel
FEC - Forward Error Correction
FH - Frequency hopping
FIRSTCHA - First Access Channel
FIRSTCHP - First Paging Channel
FM - Frequency modulation
FOCC - Forward Analog Control Channel
FPLMTS - Future Public Land Mobile Telephone System
FSK - Frequency Shift Keying
FVC - Forward Analog Voice Channel
GHz - Gigahertz. One billion (US usage) or 1,000,000,000 cycles per second.Official pronunciation is similar to "gigantic" but unofficial pronunciation similar to "giggle" is also widely used.
GMSK - Gaussian minimum shift keying (type of digital FM used in GSM and related systems, CT2, DECT, etc).
GSA - Geographical Service Area
GSM - Global System for Mobile communication (formerly Groupe Spècial Mobile)

Hz - frequency unit hertz (named for radio inventor Heinrich Hertz). One cycle per second
IBCN - Integrated Broadband Communications Network
IEC - Inter Exchange Carrier
IF - Intermediate frequency
IMBE - Improved Multi-Band Excitation speech codec
IMEI - International Mobile Equipment Identity (GSM and related systems)
IMSI - International Mobile Subscriber Identity
IMTS - Improved Mobile Telephone Service
IPR - Intellectual Property Rights
ISDN - Integrated Services Digital Network
ISI - Inter-Symbol Interference
ISM - Industrial, Scientific and Medical (portion of the 900 MHz band in North America)
ITU-R - International Telecommunications Union-Radio sector (successor to CCIR)
ITU-T - International Telecommunications Union-Telecom sector (successor to CCITT)
IVCD - Initial Voice Channel Designation
JEG - Joint Experts' Group
JEM - Joint Experts' Meeting
JTACS - Japanese Total Access Communications Systems
JTC - Joint Technical Committee
KSU - Key service unit (small business telephone system)
LAN - Local Area Network
LAPD - Link Access Protocol, version D, used in ISDN

LAPDm - Link Access Protocol, version D, final "m" indicates mobile version as used in GSM and related system

LAPM - Link Access Procedure for Modems

LASTCHA - Last Access Channel

LASTCHP - Last Paging Channel

LATA - Local Access and Transport Area

LEC - Local Exchange Carrier

LSB - Least Significant Bit

MAHO - Mobile Assisted Hand Off

MHz - Megahertz. One million or 1,000,000 cycles per second.

MIN - Mobile Identification Number

MIPS - Million Instructions Per Second

MNC - Mobile network code (GSM and related systems)

MS - Mobile Station

MSA - Metropolitan Statistical Area

MSB - Most Significant Bit

MSC - Mobile-service Switching Center

MSK - Minimum shift keying (type of digital FM)

MTA - Major Trading Area

MTSO - Mobile Telephone Switching Office

MTX - Mobile Telephone Exchange

Mu-Law - type of logarithmic compression coded used in PSTN in North America and Japan (see A-Law)

NAM - Number Assignment Module

NAMPS - Narrowband Advanced Mobile Phone Service

NAWC - Number of Additional Words Coming

NMT - Nordic Mobile Telephone

OA&M - Operations, Administration and Maintenance

OAM&P - Operations, Administration, Maintenance & Provisioning

OLC - Overload Class

ORDQ - Order Qualifier

OSI - Open System Interconnection

PA - 1) Public Address (system), 2) Power Amplifier

PABX - Private Automatic Branch Exchange

PACS - Personal Access Communications Systems

PAGCH - Paging and Access Grant Channel(s) (GSM related systems)

PBX - Private Branch Exchange

PCH - Paging Channel (GSM related systems)

PCI - Protocol Capability Indicator

PCM - Pulse Cod(ed) Modulation

PCN - Personal Communications Network

PCS - Personal Communications Services (or System)

PCS-1900 - Personal Communications System on 1900 MHz band, North American up-banded and lower power version of GSM.

PDC - Pacific Digital Cellular
PHP - Personal HandyPhone
PHS - Personal Handyphone System
PIN - Personal Identification Number
PLL - Phase Locked Loop
PLMN - Public Land Mobile Network
POP - Point of Presence
POTS - Plain Old Telephone Service
PSCC - Present SAT Color Code
PSK - Phase Shift Keying
PSTN - Public Switched Telephone Network
PTT - Postal Telephone and Telegraph
PUC - Public Utilities Commission
QPSK - Quadrature (four angle) phase shift keying (type of modulation)
RACE - Research and Development of Advanced Communication Technologies in Europe
RACH - Random Access Channel
RAM - Random Access Memory
RCF - Read Control Filler
RDTC - Reverse Digital Traffic Channel
RECC - Reverse Analog Control Channel
RELP - Residual Excited Linear Predictive speech codec, (or Regular Pulse Excited Linear Predictive)
RF - Radio frequency
ROM - Read Only Memory
RP - Radio Port
RPCU - Radio Port Control Unit
RSA - Rural Statistical Area
RSSI - Received Signal Strength Indicator (or Indication)
RTC - Reverse Traffic Channel Digital
RVC - Reverse Analog Voice Channel
SACCH - Slow Associated Control Channel
SAT - supervisory audio tone (analog cellular systems)
SBI - Shortened Burst Indicator
SCC - Sat Color Code
SCH - Synchronization Channel (GSM and related systems)
SCM - Station Class Mark
SCP - Service Control Point
SCT - Secretaria de Comunicaciones y Transportes (México government radio regulation organization)
SDCC(1,2) - Supplementary Digital Color Codes
SDCCH - Standalone Dedicated Control Channel (GSM and related systems)
SDMA - Spatial Division Multiple Access

SID - System Identification
SIM - Subscriber Identity Module (GSM and related systems)
SMR - Specialized Mobile Radio
SMS - Short Message Service
SOR - Standard Offset Reference
SS7 - Signaling System #7
ST - Signaling Tone
STP - Signaling Transfer Point
SWR - Standing Wave Ratio
TA - Time Alignment
TACH - Traffic and associated channel(s) (GSM and related systems)
TACS - Total Access Communications System
TCH - Traffic channel (IS-54, IS-136 and GSM and related systems)
TCM - Time compression multiplexing
TDD - Time Division Duplex
TDM - Time Division Multiplexing
TDMA - Time Division Multiple Access
TIA - Telecommunications Industries Association
TMSI - Temporary Mobile Service Identity (GSM and related systems)
TSI - Time Slot Interchange
uCell - Micro Cell (u is used here as a substitute for Greek letter Mu (written μ)
UCH - User Channel
UHF - Ultra-high frequency. Specifically frequencies between 300 MHz and 3 GHz.
uP - Microprocessor (u is used here as a substitute for Greek letter Mu (written μ)
UPR - User Performance Requirements
VCO - Voltage Controlled Oscillator
VCS - Voice Controlled Switch(ing)
VMAC - Voice Mobile Attenuation Code
VSELP - Vector-Sum Excited Linear Predictive Coding
VSWR - Voltage Standing Wave Ratio
WACS - Wireless Access Communications System
WER - Word Error Rate
WFOM - Wait For Overhead Message
WLL - Wireless Local Loop
WOTS - Wireless office telephone system
WPBX - Wireless PBX

INDEX

Access channel, 71, 111-112, 168, 176-177

Access grant channel, 176, 178

Activation Subsidy, 368

Adaptive equalization, 219

Adaptive Predictive Correction' adaptive predictive correction | APC, 273

ADaptive Pulse Coded Modulation (ADPCM), 96, 201, 215, 218, 256, 261, 271, 273, 335, 337

ADPCM. See Adaptive Pulse Coded Modulation

ADC. See American Digital Cellular

Adjacent channel, 33-35

Adaptive Equalizer, 72, 176, 215

Adjacent channel interference, 34-35

Advanced Mobile Phone System
--features, 50, 54, 59, 62, 103, 105, 108
--frequencies, 51-53, 59, 105, 108, 113, 332, 379, 386
--history of, 23, 103
--network, 23, 104, 296, 379
--base station, 49, 104, 120, 124, 302
--signaling, 50, 52, 54, 114, 227, 269, 276
--specifications, 49-50, 52, 54, 59, 62, 103-104, 275, 332

AGC. See Automatic Gain Control

AGCH. See Access Grant CHannel

AM. See Amplitude modulation

American Digital Cellular. See TDMA

Amplitude Modulation (AM), 234, 236, 253, 271, 376

AMPS. See Advanced mobile phone system

Antenna
--base station, 27, 170, 190, 221, 298, 302-305, 310, 328
--beamwidth, 285
--combiner, 27, 190, 307, 338
--microwave, 356
--portable, 285, 299, 385

APC. See Adaptive Predictive Correction

Application specific integrated circuits, 286, 291, 346, 370

AR. See Authentication Register

Associated control channel, 114-116, 161-162

AuC. See Authentication Center

AuR. See Authentication Register

Authentication Center (AuC), 314, 318, 324-325

Authentication Key (AKEY), 86-88

Authorization and Call Routing Equipment (ACRE), 275

Automatic Gain Control (AGC), 209, 284

Bandwidth, 30, 34-35, 54, 56, 58-60, 65, 68-69, 94, 100, 105, 108, 120, 152, 156, 163, 171, 173, 197, 211, 213-214, 220, 227, 234, 247-248, 253, 255, 263, 273, 276, 330, 380

Base Station (BS)
--AMPS, 49, 104, 120, 124, 302
--CDMA IS-95, 71, 302, 336
--GSM, 88, 152, 158-159, 171, 174, 180, 351
--IS-91A, 222, 274
--IS-94, 209, 225, 274
--NAMPS IS-88, 222
--PACS, 271-273